AF323728

Essentials of
CRYSTALLOGRAPHY

Essentials of
CRYSTALLOGRAPHY

M. A. Wahab

Alpha Science International Ltd.
Oxford, U.K.

Essentials of
CRYSTALLOGRAPHY
332 pgs. | 228 figs. | 48 tbls.

M.A. Wahab
Department of Physics
Jamia Millia Islamia (Central University)
New Delhi, India

Copyright © 2009
Reprint 2011 (Revised)

ALPHA SCIENCE INTERNATIONAL LTD.
7200 The Quorum, Oxford Business Park North
Garsington Road, Oxford OX4 2JZ, U.K.

www.alphasci.com

All rights reserved. No part of this publication may be reproduced, stored in a retrieval system, or transmitted in any form or by any means, electronic, mechanical, photocopying, recording or otherwise, without prior written permission of the publisher.

ISBN 978-1-84265-392-0

Printed in India

PREFACE

As the name indicates this book contains the essential topics needed to understand the crystals from macroscopic and microscopic point of view and the mode of their growth, structure determination and structural refinements from experimental and theoretical point of view. The aim of the present book is to provide a comprehensive introduction to the subject of crystallography in a simplified manner to the students of M.Sc, M.Phil, and Ph.D of Physical, Chemical, Biological, Materials, Engineering Sciences and Faculties in this field. The subject of crystallography is conceptual and imaginative and the students generally find it relatively difficult. However, the author with all his experience in the field and hard work has successfully made it easy for them to understand.

This book contains a total of fourteen chapters, out of which initial seven chapters deal with the basic aspects of crystallography. They include the development of space lattices, description of macroscopic and microscopic symmetries exhibited by crystalline solids and simplified approach to understand the derivation of point groups and space groups. The chapter number eight particularly deals with the basics of theory and experiments of crystal growth while the chapter number nine deals with the imperfections normally encountered in crystalline solids. The chapter number ten deals with the diffraction methods based on both film and counter techniques and for data collection. The last four chapters deal with topics related to crystal structure analyses, they are: factors affecting X-ray intensities, structure factors and Fourier synthesis,(where the concepts of structure factor have been developed using optical analogue), and important methods of crystal structure analysis and refinements.

The main features of the book are:

1. Solved examples for a better understanding of the text.
2. Matrix representation of symmetry elements and derivation of point group by simple matrix multiplication.
3. Summary and definitions of important terms at the end of each chapter for quick revision.
4. Problems and exercises.

Although proper care has been taken during the preparation of the manuscript, still some errors are expected to creep in. Any omissions, errors, suggestions brought to the knowledge of the author will be gratefully acknowledged.

First of all I sincerely thank our Vice Chancellor Prof. Musheerul Hasan for providing me the timely needed sabbatical leave. This indeed helped me in completing the MS in time. I sincerely express my sense of gratitude to my Research guide, Prof. G.C. Trigunayat, former Head, Deptt. of

Physics, Delhi University. I also thank colleagues of J.M.I from different departments for their encouragement.

I am indeed grateful to all the authors and publishers of Books and Journals mentioned in the bibliography for freely consulting them and even borrowing some ideas during the preparation of the manuscript. I am also grateful to M/s Narosa Publishing House, New Delhi for timely publication of this book.

My special thanks to the members of my family for their continuous support and encouragement during the preparation of the manuscript. I particularly thank my wife Mrs. Raees Begum for making a fair draft of the entire manuscript.

M.A. Wahab

CONTENTS

CHAPTER 1

BRAVAIS LATTICE IN TWO DIMENSIONS

1.1 INTRODUCTION

From our everyday experience, we are familiar with a one dimensional iron chain and two dimensional periodic patterns as in wall papers (Fig. 1.1) and fabrics etc. We find a large variety in design by changing both the forms of the basic motif (or unit of pattern) and the spacing of its periodic repetition. To analyse the nature of the given pattern, a single motif is selected such that the pattern could be obtained by two periodic transformations in two non collinear directions. Thus the essential features of the given pattern are specified by two non parallel translation vectors. Replacing each motif in the pattern with a point, a periodic pattern of points defined by the same translation

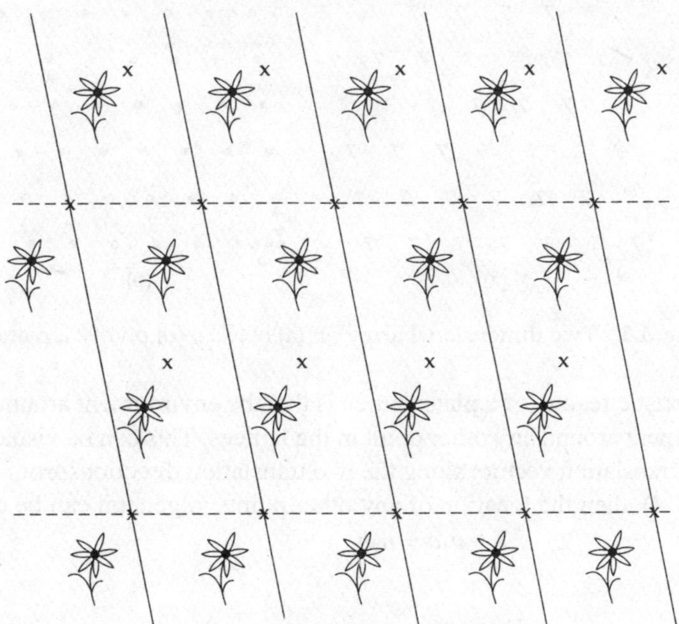

Fig. 1.1 A two dimensional periodic pattern

vectors with identical environment around every point can be obtained. This is known as a two dimensional lattice or a plane lattice. The nature of the lattice type is a fundamental feature of any pattern.

1.2   DEVELOPMENT OF ONE AND TWO DIMENSIONAL LATTICES

In order to develop a two dimensional (or a plane) lattice, let us start with a one dimensional lattice first, let us suppose that an object (say the number 7) is translated to a distance 'a'. It is further translated repeatedly to obtain a collection of one dimensional periodic array of 7's, as shown in Fig. 1.2a. If we replace each member in the array with a point, a collection of one dimensional points is obtained as shown in Fig. 1.2b. This is known as a linear lattice. It is to be kept in mind that a point has no (or zero) dimension, therefore the lattice of points is an imaginary concept. On the other hand, the array of the number '7' in Fig. 1.2a is real. Thus it is not a lattice of 7's, instead it is correctly called a lattice array of 7's.

Fig. 1.2   one dimensional array of: (a) objects, (b) points: a linear lattice

Now if we add another non-collinear translation '*b*' to the entire lattice array due to the translation 'a' then a two dimensional array of object is obtained (Fig. 1.3a). The corresponding collecton of two dimensional points shown in Fig. 1.3b is called a plane lattice.

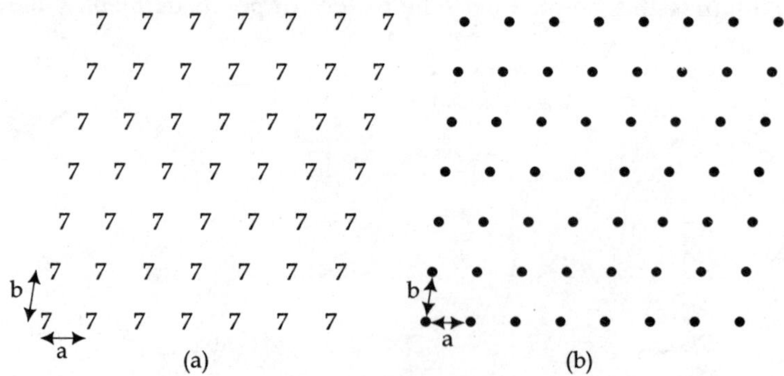

Fig. 1.3   Two dimensional array of: (a) objects, (b) points; a plane lattice

The characteristic feature of a plane lattice is that the environment around one point is identical with the environment around any other point in the lattices. This can be visualized by considering *a* and *b* as the unit translation vectors along the two translation directions from a lattice point taken as the origin (Fig. 1.4), then the location of any other points in general can be defined as

$$T = n_1 a + n_2 b$$

(1)

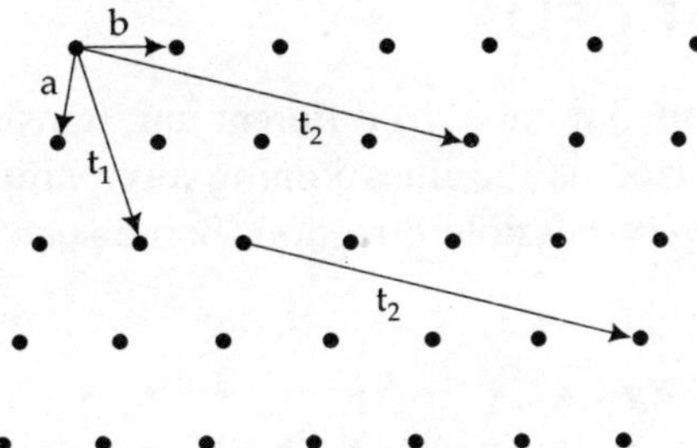

Fig. 1.4 Translation vectors, t_1 and t_2

where n_1 and n_2 are arbitrary integers. For example, Fig. 1.4 shows two such translations in the plane lattice, i.e.

$$t_1 = 2a + 1b$$

and

$$t_2 = 1a + 4b \tag{2}$$

1.3 BASIS AND THE CRYSTAL STRUCTURE

As pointed out in the preceeding sections that a lattice of pattern is an imaginary concept and is required to be distinguished it from a real crystal. A crystal structure is formed only when an atom or a group of atoms (or molecules) is attached to each and every lattice points. Such an atom or the group of atoms (or molecules) is called the basis. It is identical in composition, arrangement and orientation whose periodic repetition yields the crystal structure. The generation of crystal structure from a two dimensional lattice and a basis is illustrated in Fig. 1.5. A logical relation of the same is expressed as

Lattice + basis → crystal structure

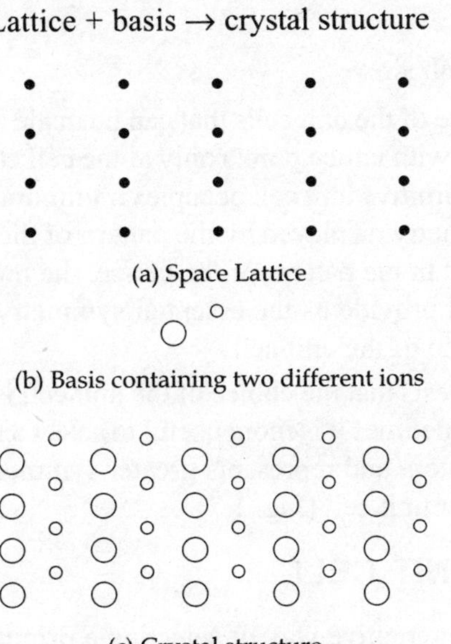

Fig. 1.5 Showing the space lattice, the basis and the crystal structure

1.4 CHOICE OF A UNIT CELL

In general, there exists an unlimited number of different unit translations in a given lattice. These unit translations are best visualized as the lines joining any particular lattice point taken as the origin to the nearest point in every possible direction. Some such unit translations are shown in Fig. 1.6.

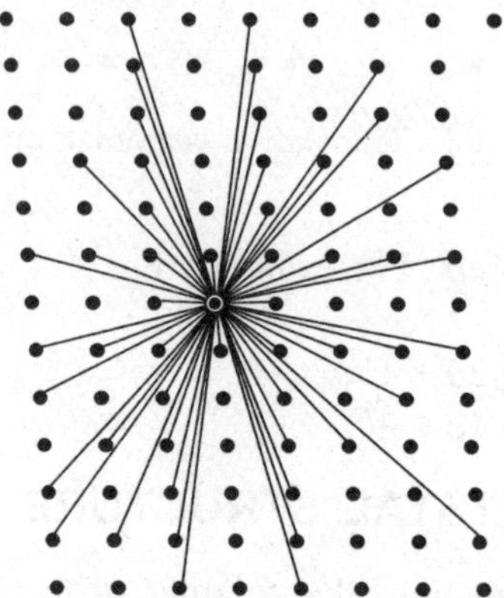

Fig. 1.6   Part of a plane lattice showing a few of the unit translations

In general, any non-collinear pair of unit translations could be selected to form a unit cell and describe the lattice. Since the number of such unit translations in a given lattice are infinite, therefore the number of possible unit cells are also infinite. The shapes and sizes of such unit cells are determined by the magnitudes of the unit translations '*a*' and '*b*' the interaxial angle γ. The area of such a unit cell is given by

$$ab \sin \gamma \tag{3}$$

Some of the possible choice of the unit cells that can be made are shown in Fig. 1.7. The chosen unit cells are usually primitive with lattice points only at the cell corners and therefore has only one lattice points per unit cell. A primitive unit cell occupies minimum area for a given plane lattice and is in accordance with the symmetry displayed by the pattern of the lattice. That is, once a particular symmetry element is displayed in the pattern of the lattice, the most suitable shape of the unit cell could be determined. This will provide us the essential symmetry element which is inherently related to the corresponding shape of the unit cell.

The above discussion suggests that the choice of the unit cell is not unique. In fact, according to the requirement of the case, sometimes it is more useful to select a non-primitive (centered) unit cell if it has simpler coordinate system and represents greater symmetry. A centered unit cell contains more than one lattice points per unit cell (Fig. 1.7).

1.5 WIGNER-SEITZ UNIT CELL

A wigner-seitz unit cell is an alternative way of selecting a primitive unit cell of area equal to the other conventional primitive cells (as in eq. 3) in a given lattice. This unit cell is constructed around a lattice point according to the following procedure.

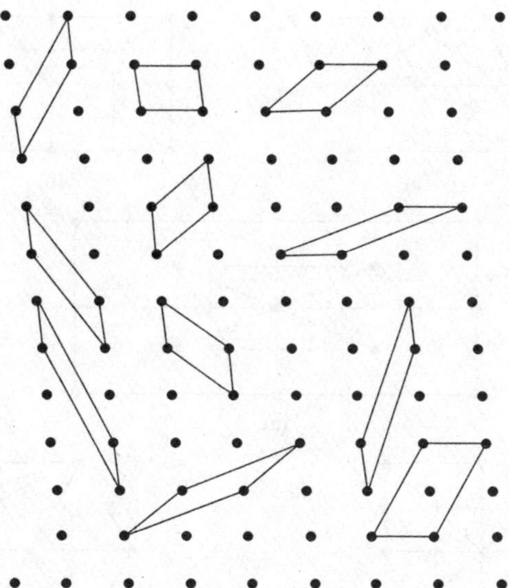

Fig. 1.7   Some unit cells of a plane lattice obtained from different combinations of two unit translation vectors

(i)   In a given lattice, select a reference points and draw lines to connect this points with all other neighbouring lattice points.

(ii)   Draw normal at the mid points of the lines connecting the lattice points. Smallest area enclosed around the central lattice point gives the required Wigner-Seitz primitive unit cell.

For a plane lattice, a Wigner-Seitz unit cell is found to have an irregular hexagonal shape as shown in Fig. 1.8. Hence, the choice of a convensional unit cell of the lattice is a matter of convenience. However, a Wigner-Seitz unit cell has its own importance in solid-state physics.

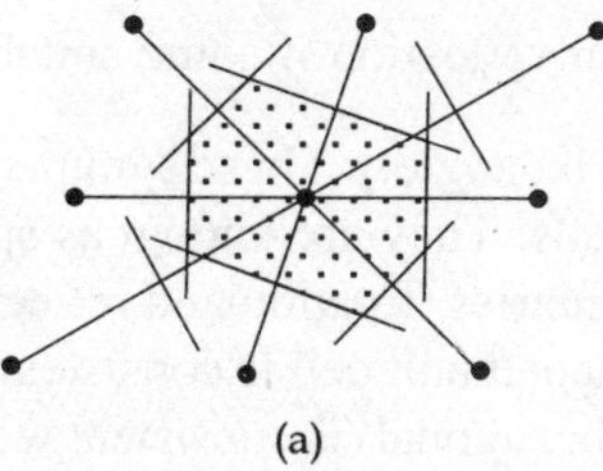

(a)

Fig. 1.8   A Wigner-Seitz unit cell in two dimension

Example 1:   Draw Wigner-Seitz unit cell for each of the five plane lattices.

Solution: Following the above mentioned procedure, we can draw the Wigner-Seitz cells of five plane lattices shown in Fig. 1.9.

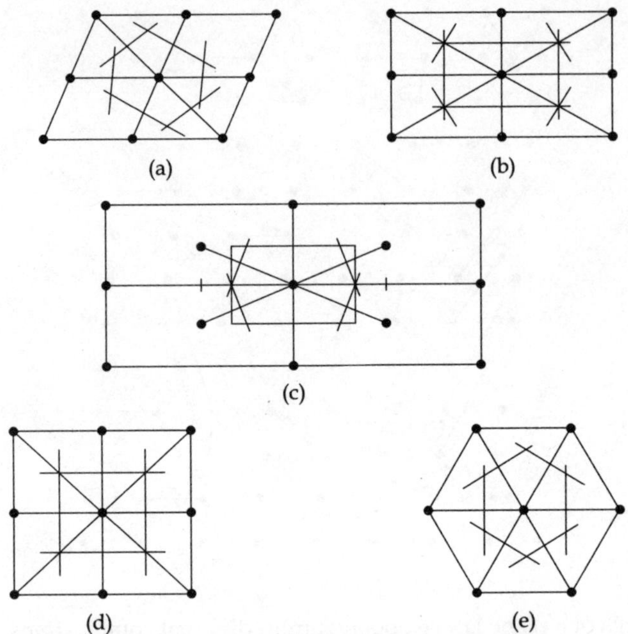

Fig 1.9   Wigner-Seitz unit cells of plane lattices: (a) oblique (b) rectangular (c) centered rectangular (d) square (e) hexagonal

1.6   PRIMITIVE LATTICE TYPES AND CRYSTAL SYSTEMS

The most general shape of the primitive unit cell of a plane lattice is a parallelogram (Fig. 1.7). This is also known as oblique lattice. This is invariant only under the rotation $\dfrac{2\pi}{n}$ (with $n = 1$ and 2) about any lattice point normal to the cell. However, this can also be made invariant under the rotation $\dfrac{2\pi}{n}$ with $n = 3, 4, 6$ or mirror reflection if some suitable restrictions are imposed on the lattice translation a, b and the inter axial angle γ. These symmetry elements in turn put restrictions on the shape of the respective unit cells. They are known as special lattices. In fact, the unit cell shapes consistent with the plane symmetries (rotation and reflection) are found to be small in number, i.e. four only. A parallelogram shaped unit cell is consistent with 1- or 2- fold axial symmetry, a 120° - rhombus shaped (also called hexagonal) is consistent with 3- or 6- fold axial symmetry and a square unit cell is consistent with 4-fold axial symmetry, respectively. On the other hand, a rectangular shaped unit cell is consistent with a mirror symmetry or a 2- fold axial symmetry. These unit cells are shown in Fig. 1.10. Thus in two dimensions there exist only four distinct possible shapes of unit cells in which (a) shows an oblique cell with $a \neq b$, γ arbitrary, (b) shows a rectangular cell with $a \neq b$, $\gamma = 90°$, (c) shows the square unit cell with $a = b$, $\gamma = 90°$ while (d) shows a hexagonal unit cell (rhombus shaped) with $a = b$, $\gamma = 120°$. Accordingly, these distinct primitive unit cells give rise to four different crystal system in two dimensions.

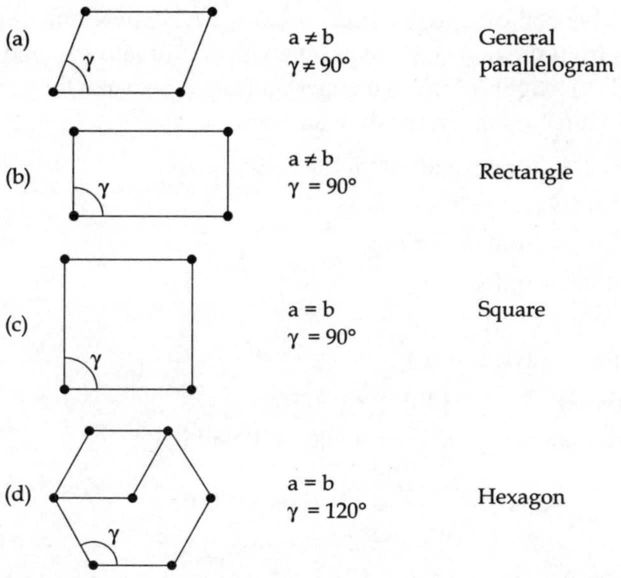

Fig 1.10 Two dimensional primitive unit cells

Example 1: Determine the area of a primitive rectangle whose sides are 4 Å and 3 Å respectively. Construct another primitive cell of equal area whose sides are 4 Å and 5 Å. Determine the angle between then.

Solution: Given: $a = 4$ Å, $b = 3$ Å. For other unit cell $a' = a$, $b' =$ diagonal of rectangle. Also for a rectangular lattice, $a \neq b$, $\gamma = 90°$. Construct two rectangular unit cells side by side (Fig. 1.11). Area of the rectangle ABCD $= ab \sin 90° = 4 \times 3 = 12$ Å^2 .

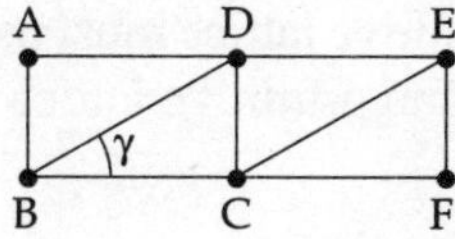

Fig 1.11 Rectangular unit cells

Now to construct another unit cell, joint BD and CE. The required unit cell is DBCE. Consider the triangle BCD, where the $\angle BCD = 90°$. Threrfore,

$$BD = (4^2 + 3^2)^{1/2} = (25)^{1/2} = 5 \text{ Å}$$

According to the question, the areas of the two unit cells are equal. Therefore,

$$12 = 5 \times 4 \sin \gamma$$

or
$$\sin \gamma = \frac{12}{5 \times 4} = \frac{3}{5} \text{ or } \gamma = \sin^{-1}\left(\frac{3}{5}\right) = 36.87°$$

Example 2: A primitive cell of square lattice has $a = 2$ Å. A new unit cell is chosen with edges defined by the vactors from the origin to the points with coordinate 1, 0 and 0,3. Calculate (a) area of the original unit cell (b) lengths of the two edges and angle between them, (c) area of the new unit cell (d) the number of lattice points in the new unit cell.

Solution: Given: $a = 2$ Å, for a square lattice $\gamma = 90°$.

(a) Area of the original unit cell, $a^2 = 4$ Å^2

(b) Length of the edges from the origin

    (i) for coordinates 0, 0 and 1,0

$$a = [(0-1)^2 \times 2^2 + 0]^{1/2} = (4)^{1/2} = 2\text{Å}.$$

    (ii) for coordinates 0,0 and 0,3

$$b = [0 + (0-3)^2\, 2^2]^{1/2} = (36)^{1/2} = 6\text{Å}.$$

Angle between the lines with end coordinates 1, 0 and 0,3

$$\cos \gamma = \frac{0}{\sqrt{1^2}\ \sqrt{3^2}} = 0 \Rightarrow \gamma = 90°.$$

(c) Area of new unit cell, $ab = 2 \times 6 = 12\text{Å}^2$.

(d) Ratio of the two areas, $\dfrac{12}{4} = 3$, therefore the number of lattice points in the unit cell is 3.

1.7 CENTERING OF PLANE LATTICES

In order to verify the existence of non-primitive (or centered) lattice in two dimensional system, let us systematically expose the possibility by adding lattice points at appropriate positions without-disturbing the essential symmetry of the original lattice (unit cell) and then examining whether a smaller primitive cell may be chosen with the same general shape or not. If the new primitive cell has the same general shape then it represents the same original lattice. However, if no new primitive cell is found, then the resulting non-primitive lattice must be recognised as a new lattice type. Let us discuss the possible centering in the plane primitive lattice one by one.

1. Oblique Lattice

We know that the oblique lattice is defined as $a \neq b$ and γ is arbitrary. The following four possibilities of centering may arise:

(a) An extra lattice point is added at the center of an edge at A to give a doubly primitive cell as shown in Fig. 1.12a. This is not a new cell because another oblique primitive cell with the same general shape may be defined by a, $\dfrac{b}{2}$ and γ. A similar argument will apply if the extra lattice point is added at the mid point of the other cell edge.

(b) Two extra lattice points are added at the centers of both the cell edges A and B, to give a triply primitive cell as shown in Fig. 1.12b. However, this does not constitute a lattice, because the environment of A and B are now different and hence does not fulfill the criterion of a lattice.

(c) Three extra lattice points are placed one each at A, B and C to produce a quadruply primitive cell as shown in Fig. 1.12c, where C is the center of the unit cell. However, this does not produce a new lattice because another new primitive cell with same general shape may be defined by $\dfrac{a}{2}, \dfrac{b}{2}$ and γ.

(d) An extra lattice point is added at C, the centre of the unit cell to produce a doubly primitive unit cell as shown in Fig. 1.12d. At first sight it appears that this represents a new non-primitive lattice type. However, a careful look suggests that a smaller primitive oblique cell of new dimensions a', b' and γ' (Fig. 1.12d) can be selected to describe this arrangement of lattice points. So, again no new non-primitive lattice is formed. Thus there is only one distinct oblique lattice exits in two dimensions.

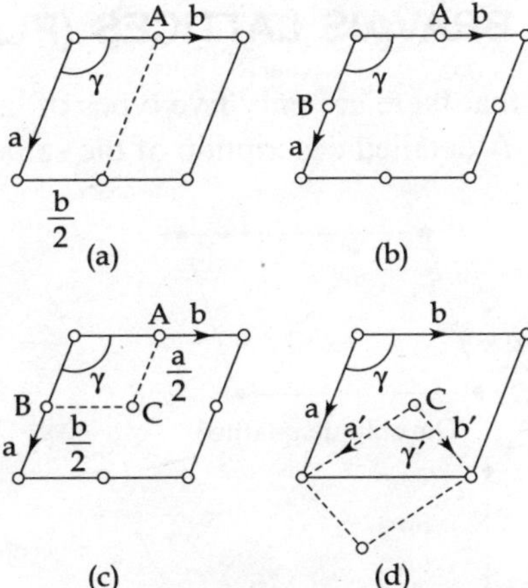

Fig. 1.12 The addition of extra lattice points to a primitive oblique cell. Alternative primitive cells are shown with broken lines

2. Rectangular Lattice

Examination of similar four possibilities of centering in rectangular lattice (as considered in oblique lattice), we observe a similar result in cases (a) to (c). However, the result in case of (d) is different. In the rectangular lattice the addition of extra lattice point at the centre of the unit cell does produce a new non-primitive lattice (or centered) lattice. However, the centered rectangular lattice may also be described as a primitive rhombic (or primitive diamond) lattice as shown in Fig. 1.13. These are just two alternative descriptions of the same arrangement of lattice points. In fact, the choice of the unit cell is arbitary, but a more useful centered lattice is preferred because of the simple geometry and greater symmetry.

3. Square and Hexagonal Lattice

Proper examination reveals that the application of above mentioned centerings do not produce new lattice in these cases.

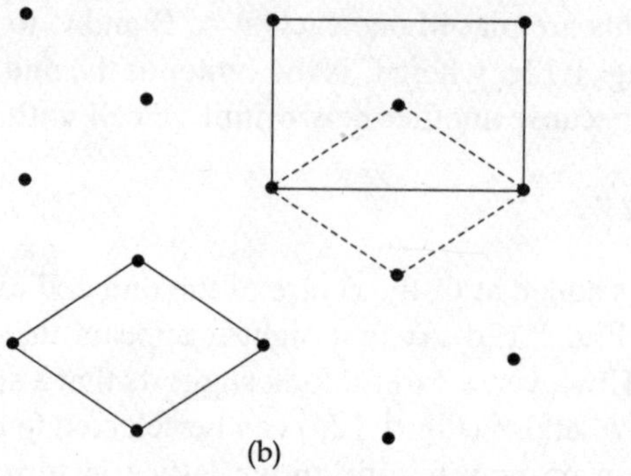

Fig. 1.13 Alternative way of description of centered rectangular lattice

1.8   TWO DIMENSIONAL BRAVAIS LATTICES (PLANE LATTICES)

From above discussions, it is clear that there are only five types of lattices possible in two dimensions. They are shown in Fig. 1.14. A detailed description of the same is given in Table 1.1.

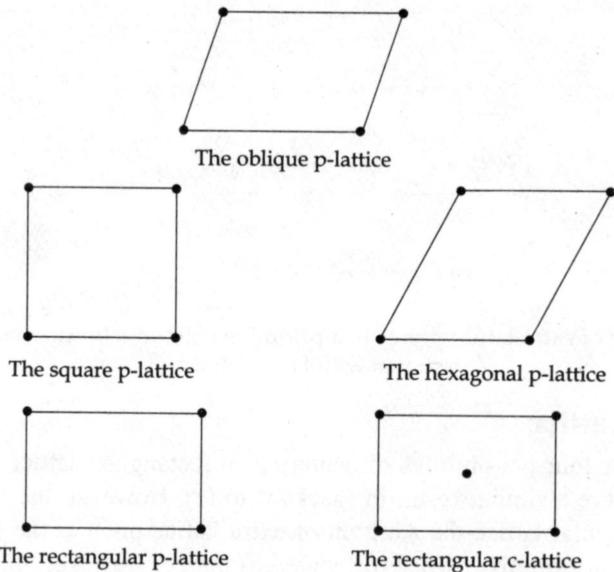

Fig. 1.14   Bravais lattices in two dimensions

1.9   UNIT CELL CALCULATIONS

Distance Between Two Lattice Points (Oblique System)

To determine the distance between two lattice points with known coordinates in an orthogonal system of axes is easy. However, it is somewhat difficult in an oblique system of axes. Let us consider an oblique axial system OX and OY where the angle between $XOY = \gamma$ as shown in

Table 1.1   Two-dimensional Bravais Lattices

Lattice	Unit cell	Lattice parameters	Point group
Oblique		$a \neq b, \gamma \neq 90°$	1, 2
Square		$a = b, \gamma = 90°$	4, 4 mm
Hexagonal		$a = b, \gamma = 120°$	3, 3 m, 6 mm, 6
Rectangular Primitive		$a \neq b, \gamma = 90°$	1 m, 2 mm
Centered Rectangular		$a \neq b, \gamma = 90°$	1 m, 2 mm

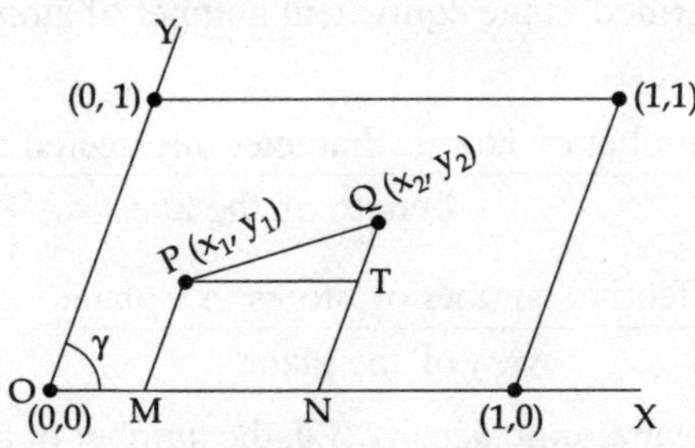

Fig. 1.15   Distance between two lattice points

Fig. 1.15. Let us consider two points $P(x_1, y_1)$ and $Q(x_2, y_2)$ within the oblique unit cell whose distance is to be determined. Draw the lines PM and QN parallel to OY axis and PT parallel to OX axis, respectively. From this construction, we have,

$$OM = ax_1, \; PM = by_1, \; ON = ax_2, \; NQ = by_2$$

$$PT = MN = a(x_2 - x_1) \text{ and } QT = QN - TN = QN - PM = b(y_2 - y_1)$$

Now, in the triangle PQT, the angle $\angle PTQ = 180 - \gamma$

Therefore we can write,

$$(PQ)^2 = (PT)^2 + (QT)^2 + 2\, PT \cdot QT \cos \gamma$$

$$= (x_2 - x_1)^2 \, a^2 + (y_2 - y_1)^2 \, b^2 + 2\, (x_2 - x_1)(y_2 - y_1)ab \cos \gamma$$

and
$$PQ = [(x_2 - x_1)^2 \, a^2 + (y_2 - y_1)^2 \, b^2 + 2\, (x_2 - x_1)(y_2 - y_1)ab \cos \gamma]^{1/2} \qquad (4)$$

This is a general equation expressing the distance between two lattice points in an oblique lattice. Equation for other lattices can be obtained by substituting the respective axial parameters (axes and angle between than) in eq. 4.

Example: A unit has dimensions $a = 3$Å, $b = 4$Å and $\gamma = 115°$. Calculate the distance between the points (i) 0.210, 0.150 and 0.410, 0.050, (ii) 0.210, $-$ 0.150 and 0.410, 0.050.

Solution: Given: $a = 3$Å, $b = 4$Å, $\gamma = 115°$. Fractional coordinates (i) 0.210, 0.150 and 0.410, 0.050 (ii) 0.210, $-$ 0.150 and 0.410, 0. 050. Making use of eq. 4, we can obtain the distance between first set of fractional coordinates as

$$l_1 = [(0.410 - 0.210)^2 \times 3^2 + (0.050 - 0.150)^2 \times 4^2 + 2(0.410 - 0.210)(0.050 - 0.150) \times \cos 115°]^{1/2}$$
$$= [(0.2)^2 \times 3^2 + (0.1)^2 \times 4^2 + 2(0.2)(-0.1) \cos 115°]^{1/2}$$
$$= [\,0.04 \times 9 + 0.01 \times 16 + 2(-0.02) \times (-0.432)]^{1/2}$$
$$= 0.733 \text{ Å}$$

Similarly, for second set of fractional coordinates

$$l_2 = [(0.410 - 0.210)^2 \times 3^2 + (0.050 + 0.150)^2 \times 4^2 + (0.410 - 0.210)(0.050 + 0.150)^2 \cos 115°]^{1/2}$$
$$= [(0.2)^2 \times 3^2 + (0.2)^2 \times 4^2 + 2(0.2)(0.2)(-0.423)]^{1/2}$$
$$= [0.04 \times 9 + 0.04 \times 16 - 2(0.04)(0.423]^{1/2}$$
$$= 0.983 \text{ Å}$$

Linear and Planar Atomic Density

Sometimes it is important to know the atomic density in a particular direction of the plane lattice or in the plane of the given lattice itself therefore, we define here the linear atomic density ρ_l as the number of atomic diameter intersected by the selected line in the given direction. Similarly, the planar atomic density ρ_p is defined as the equivalent number of atoms whose centers are intersected by the selected area. Accordingly,

$$\rho_l = \frac{\text{Number of atomic diameter intersected by a line}}{\text{Length of the line}} \tag{5}$$

and
$$\rho_p = \frac{\text{Effective number of atoms in a plane}}{\text{Area of the plane}} \tag{6}$$

Example: Calculate the linear atomic density (i.e. the number of atoms/m) along [11] direction of a centred rectangular lattice with $a = 3$ Å and 4 Å.

Solution: Given: A centred rectangular lattice with $a = 3$ Å, 4 Å.

$$\text{Length of the diagonal} = \sqrt{3^2 + 4^2} = \sqrt{9 + 16} = 5\text{Å}$$

$$\text{Effective atomic diameters interested by the diagonal} = \frac{1}{2} + 1 + \frac{1}{2} = 2 \text{ atoms. Therefore, using}$$

eq. 5, we can obtain

$$\rho_l = \frac{2 \text{ atoms}}{5\text{Å}} = \frac{2}{5 \times 10^{-10}\,\text{m}} = \frac{2}{5} \times 10^{10} \text{ atoms/m}$$
$$= 4 \times 10^9 \text{ atoms/m}$$

Example: Calculate the atomic density in *ab*-plane of the centered rectangular lattice with $a = 3$ Å, $b = 4$ Å.

Solution: Given: A centered rectangular lattice with $a = 3$ Å, $b = 4$ Å

Area of the plane, $ab = 3 \times 4 = 12$ Å^2

Effective number of atoms intersected by the area = 1 atom at the center $+ 4 \times \dfrac{1}{4}$ atoms at four corners of ab-plane = 2 atoms.

Now, making use of eq.6 the planar density is obtained as

$$\rho_p = \frac{2 \text{ atoms}}{12\text{Å}^2} = \frac{2}{12 \times (10^{-10})^2} = \frac{2}{12} \times 10^{20}$$
$$= 1.66 \times 10^{19} \text{ atoms/m}^2.$$

Planar Packing Efficiency

As a part of the unit cell calculation, sometimes it is needed to know the efficiency with which the available space in the unit cell is filled. In other words, it is to know the relative packing density (also known as packing factor or filling factor) of a given plane crystal lattice. This is a dimension-less quantity and is defined as the ratio of the area occupied by the plane projected atom in the unit cell to the area of the unit cell of the plane crystal lattice, i.e.

$$\text{Efficiency} = \frac{\text{Area occupied by plane projected atom (in the unit cell)}}{\text{Area of the unit cell}}$$

In order to find the efficiency, first of all check the coordination number of the given lattice and then select a unit cell to know the effective number of atoms in it. Also, side of the unit cell will provide the area (volume in there dimensions) of the unit cell.

Based on the above discussion, let us analyes two different packing arrangements of equal spheres of radius R on square and hexagonal lattices (Fig. 1.16) whose coordination numbers are 4 and 6 respectively. Further, knowing the side of the unit cell and its relationship with R, areas of the unit cells can be easily obtained. With these data, the efficiency of the above mentioned two lattices are found as

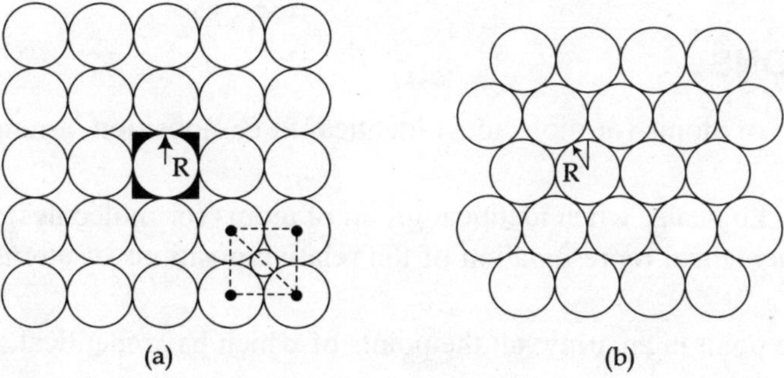

Fig. 1.16 Two-dimensional arrangements of equal spheres: (a) on a square lattice, (b) on a triangular (hexagonal) lattice

$$(\text{Efficiency})_s = \frac{\text{Area of the circle}}{\text{Area of the square}} = \frac{\pi R^2}{4R^2} = \frac{\pi}{4} = 78.5\%$$

and

$$(\text{Efficiency})_h = \frac{\text{Area of the circle}}{\text{Area of the hexagon}} = \frac{\pi R^2}{2\sqrt{3}R^2} = \frac{\pi}{2\sqrt{3}} = 90.7\%$$

where the suffixes s and h stand for square and hexagonal lattice, respectively. The above simple calculation makes it clear to us that the hexagonal packing is the most efficient way of packing in two dimensions.

1.10 SUMMARY

1. A plane lattice is an array of points related by the translations $T = n_1 a + n_2 b$, where n_1 and n_2 are arbitrary integers and 'a' 'b' are primitive translations along x and y axes.

2. A crystal structure is formed when a group of atoms (or molecules) called basis is associated with each and every lattice points. The basis is identical in composition, arrangements and orientation.

3. There is a definite relationship between the shape of a unit cell and associated symmetries.

4. The choice of a unit cell is not unique. It may be either primitive (containing one lattice point per unit cell) or non-primitive (centered) containing more than one lattice per unit cell.

5. A Wigner-Seitz unit cells is an alternative way of representing a primitive unit cell of area equal to other primitive cells in a given lattice.

6. In two dimensions, crystal systems are limited to four and plane lattices are limited to five only.

7. Two dimensional lattices are invariant under rotation of $\dfrac{2\pi}{n}$ (with $n = 1, 2, 3, 4$ and 6) and mirror reflection.

1.11 DEFINITIONS

Basis: It is a group of atoms (or molecules) identical in composition, arrangement and orientation.

Crystal structure: Logically when identical group of atoms (or molecules) are attached to each lattice point. A mathematical representation of the relative positions of atoms or ions in an ideal crystal.

Lattice point: One point in an array; all the points of which have identical surroundings.

Lattice translation: A vector connecting any two-lattice points in the same lattice.

Mirror line: It is a line, which divides a body into two halves that are mirror images of one another across a line.

Plane Lattice: A plane lattice is an infinite array of points in two dimensions such that every points of it has identical surroundings.

Unit cell: It is a convenient repeating parallelogram of a plane lattice having two non-collinear unit translations as its edges.

REVIEW QUESTIONS AND PROBLEMS

1. Discuss the procedure to develop a two dimensional lattice. What is the difference between a two dimensional array of objects and a two dimensional lattice? Point out the characteristic feature of a plane lattice.

2. Define the terms lattice, basis and crystal structure. How they are related to each other?

3. Differentiate between primitive and non-primitive cells. On what basis, a non primitive cells is selected?

4. Draw a plane lattice and indicate two kinds of double cells and one triple cell in the lattice.

5. Draw a pair of enantiomorphous objects about each lattice point in a plane lattice.

   (a) Show that any primitive cell in that lattice contains only one pair of objects.

   (b) Show that a double cell contains two pairs of objects.

6. What is the procedure to construct a Wigner-Seitz unit cell? Consruct Wigner-Seitz unit cell for five plane lattice.

7. Determine the area of primitive square unit cell whose side is 2 Å. Construct another unit cell of equal area whose one side is the diagonal of the square. Determine the length of this side.

   Ans. $4\ \text{Å}^2, 2\sqrt{2}\ \text{Å}.$

8. A primitive cell of rectangular lattice has $a = 2$ Å, $b = 3$ Å and $\gamma = 90°$. A new unit cell is chosen with edges defined by the vectors from the origin to the points with coordinates 2, 0 and 0, 3. Calculate (a) the area of the unit cell, (b) length of the two edges and angle between them (c) area of the new unit cell, (d) the number of lattice points in the new unit cell.

   Ans. (a) 6 Å, (b) 4 Å, 6 Å, 90°, (c) 24 Å, (d) 4

9. A unit cell has dimensions $a = 2$Å, $b = 3$Å and $\gamma = 90°$ calculate the distance between the points.

   (i) 0.100, 0.250 and 0.200, 0.050

   (ii) 0.100, – 0.250 and 0.200, 0.050.

   Ans. 0.632 Å, 0.922 Å

10. Calculate the linear atomic density (i.e. the number of atoms/m) along [11] of a centered square with $a = 4\sqrt{2}$ Å.

    Ans. 2.5×10^9 atoms/m

11. Calculate the planar atomic density of centered square lattice with $a = 4\sqrt{2}$ Å

    Ans. 6.25×10^{18} atoms/m^2

BRAVAIS LATTICES IN THREE DIMENSIONS

2.1 INTRODUCTION

The elements of two dimensional pattern theory given in the last chapter are relevant to any discussion of the three dimensional periodicity shown by crystalline state of matter. However, the addition of one extra dimension makes the analysis much more complex. In general, to analyes the nature of the given pattern in three dimensions, a single motif is selected such that the pattern could be obtained by three periodic translations in three non-coplanar directions. Replacing each motif in the pattern with a point, a periodic pattern of points defined by the some translation vectors are obtained. This is known as a three dimensional lattice or a space lattice.

2.2 DEVELOPMENT OF THREE-DIMENSIONAL LATTICES

In the last chapter, we learnt to develop a two dimensional lattice starting from a linear lattice by suitably adding another non-collinear translation to the entire linear array of points. Similarly, a three dimensional lattice can be obtained from a two dimensional lattice by suitably adding a third non-coplanar translation 'c' to the entire plane pattern due to translations 'a' and 'b' as shown in Fig. 2.1. This is also called a space lattice.

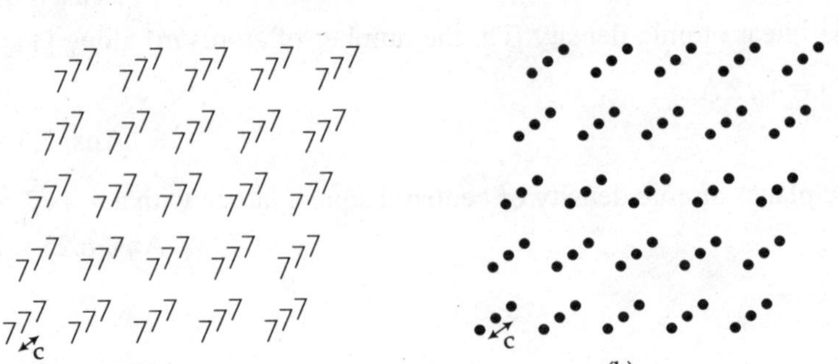

(a)   (b)

Fig. 2.1 Three dimensional array of (a) objects (b) points; a space lattice

Like a plane lattice, the characteristic feature of the lattice is that the environment around any one point is identical with the environment around any other point in the lattice. Thus, on a similar consideration, the location of any point in the lattice with respect to another is defined as

$$T = n_1\,a + n_2\,b + n_2\,c \tag{1}$$

where n_1, n_2 and n_3 are arbitrary integers and a, b and c are translation vectors along three crystallographic directions.

2.3 CHOICE OF AXES AND UNIT CELLS

In order to describe a space lattice two kinds of axial system are in use. They are right handed (where the senses of rotation from x to y and y to z are the same) and left handed axial systems (where the senses of rotation from x to y and y to z are opposite). They are shown in Fig. 2.2.

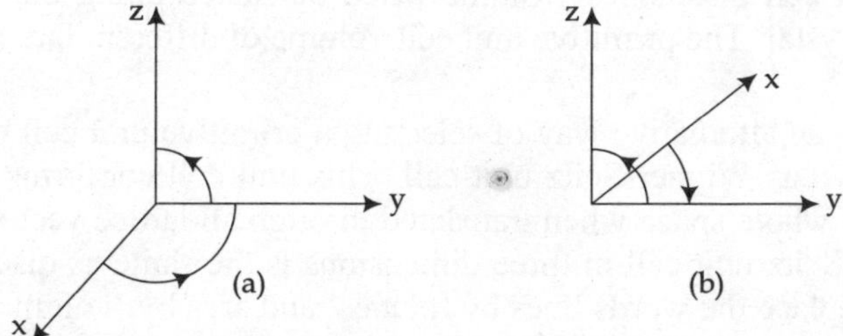

Fig. 2.2   Axial systems. (a) Right-handed axes: the senses of rotation from x to y and y to z are the same.
(b) Left-handed axes: the senses of rotation from x to y and y to z are opposite.

A set of three such non-coplanar axes are called the crystallographic axes. The crystallographers convensionally prefer the right handed axial system for the description of space lattices. However, the left handed axial system is also in use wherever needed.

In both the axial systems mentioned above any three non-coplanar translations (out of infinite such possibilities) can be selected as a coordinate system to describe a space lattice. However, in most of the cases, this would lead to a more complex and arbitrary descriptions than are necessary. Therefore, the choice of the coordinate axes to describe a space lattice has to be made judiciously. The simplest representation can be made when the natural axial systems are employed. In order to do this, we must chose the unit cell (and hence the kind of coordinate system) in harmony with the symmetry of the lattice. Thus any change in the symmetry of the lattice will change the shape of the unit cell also (and hence the kind of coordinate system).

Since in general the number of triplets (non-coplanar translations) in a given lattice are infinite. Therefore, the number of possible unit cells are also infinite. Some of the possible choices of the unit cell are shown in Fig. 2.3.

They represent the most general type of unit cells for a three dimensional lattice. They are in the form of a parallel-piped whose size and shape are defined by means of the lattice translations a, b, c as the shortest possible lengths along the three crystallographic axes x, y, z and the angles α, β, γ between them as shown in Fig. 2.4. This is a primitive unit cell and occupies minimum volume given by the equation,

$$V_c = |a \times b \cdot c| \tag{2}$$

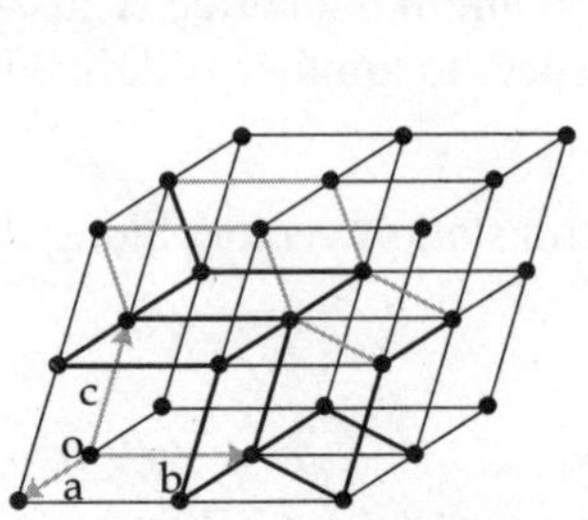

Fig. 2.3 Possibility of different choice of a primitive unit cell in a space lattice

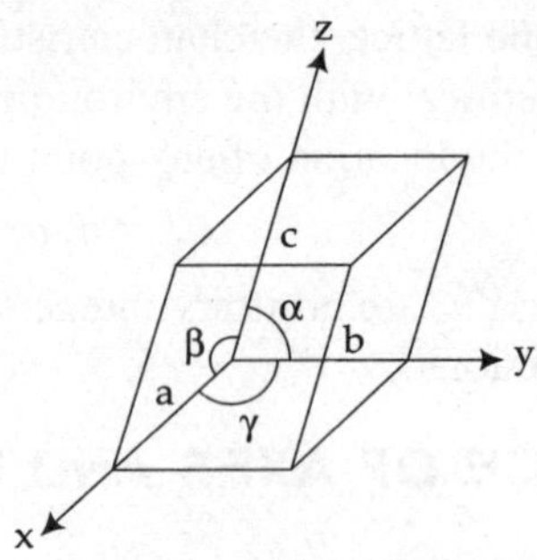

Fig. 2.4 The crystallographic axes and lattice parameters

The primitive unit cell of smallest volume when translated along the three crystallographic directions builds the crystal. The primitive unit cell volume of different lattice type are provided in Table 2.1.

Incidently, there is an alternative way of selecting a primitive unit cell of equal volume given by eq. 2. This is known as Wigner-Seitz unit cell. This unit cell conforms with the translational symmetry and fills the whole space when translated through all lattice vector. The method of construction of a Wigner-Seitz unit cell in three dimensions is the same as discussed for two dimensions except that we replace the words lines by 'planes' and area by 'volume'. Following the same procedure, the Wigner-Seitz unit cells for three-dimensional body centered cubic lattice and face centered cubic lattice can be constructed. They are truncated octahedron and rhombic dodecahedron, respectively. They are shown in Fig. 2.5.

Table 2.1   Unit cell volume of different lattice types

Lattice type	Volume
Cubic	a^3
Orthorhomic	abc
Tetragonal	a^2c
Hexagonal	$\dfrac{\sqrt{3}a^2c}{2}$
Rahombohedral	$a^3\sqrt{1-3\cos^2\alpha+2\cos^3\alpha}$
Monocline	$abc\sin\beta$
Triclinic	$abc\sqrt{1-\cos^2\alpha-\cos^2\beta-\cos^2\gamma+2\cos\alpha\cos\beta\cos\gamma}$

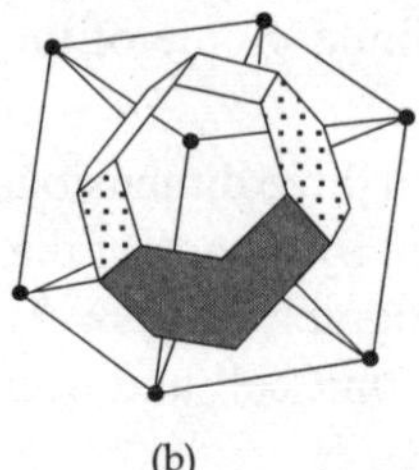

(b)

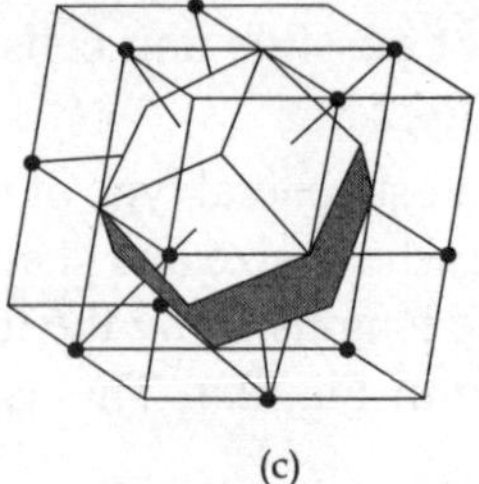

(c)

Fig. 2.5   Wigner-Seitz primitive cells for *bcc* and *fcc* lattices in three-dimensions

As we know that the choice of the unit cell is not unique. Sometimes it is more convenient to work with a unit cell of higher symmetry which may not always be a primitive. Therefore, depending on the requirement of the case, a non-primitive unit cell exhibiting higher symmetry can be selected. These cells contains more than one lattice points per unit cell. (Fig. 2.6).

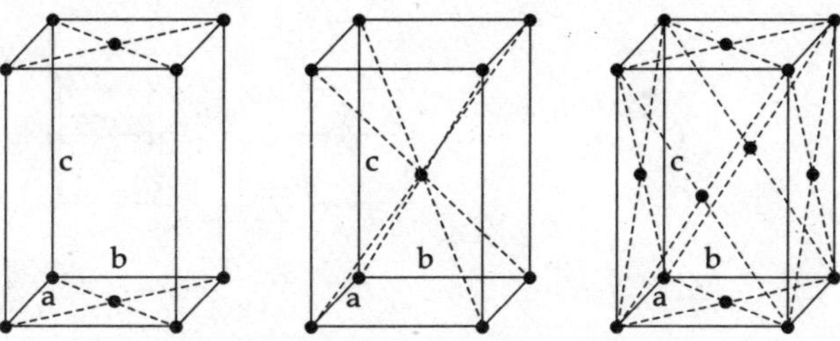

Fig. 2.6 Some non-primitive cells

2.4 DERIVATION OF SEVEN PRIMITIVE LATTICES/UNIT CELLS

We know that in general a space lattice can be obtained from a plane lattice simply by adding a third non-coplanar translation to it. Alternatively, we can think of two identical layers of a plane lattice placed on the top of the other. The height of the upper layer gives the c-dimension while its orientation with respect to the lower layer provides the angular relationships of the resulting (these dimensional) lattice. Making use of this principle, let us first consider each of the five plane lattices in its primitive form (primitive form of centered rectangular lattice is rhombic). Now, place identical plane over each of them to determine the corresponding three dimensional primitive lattices. Let us take them one by one.

(i) Oblique Lattice

For an oblique lattice, the axial and angular relationships are $a \neq$ b, $\gamma \neq 90°$. Now, placing an identical plane in an arbitrary manner, a triclinic lattice with $a \neq b \neq c$, $\alpha \neq \beta \neq \gamma \neq 90°$ is obtained (Fig. 2.7a). However, if the identical plane is placed such that the c-dimension is perpendicular to the lower plane, a monoclinic lattice with $a \neq b \neq c$, $\alpha \neq \beta = 90° \neq \gamma$ will be formed (Fig. 2.7b). This is generally known as the monoclinic lattice of I-setting.

(ii) Rectangular Lattice

For a rectangular lattice, the axial and angular relationships are $a \neq b$, $\gamma \neq 90°$. When an identical plane is placed such that the c-dimension is perpendicular to b-dimension only (not a-dimension of the rectangular lattice), a monoclinic lattice of II-setting results (Fig. 2.7c). However, when the identical plane is placed perpendicular to the lower plane (i.e. $\perp$ to both a and b), an orthorhombic lattice with $a \neq b \neq c$, $\alpha \neq \beta \neq \gamma = 90°$ is obtained (Fig. 2.7d). Further, there is a possibility of getting a tetragonal lattice from a rectangular lattice if the upper plane is placed such that the c-dimension is equal to either a or b plus a change of axes. However, we will see that tetragonal and cubic lattices can be obtained more conveniently from a square lattice.

(iii) Square Lattice

For a square lattice, the axial and angular relationships are $a = b$, $\gamma = 90°$. When an identical plane is placed perpendicular to the lower plane such that the c-dimension is different from a or b, a

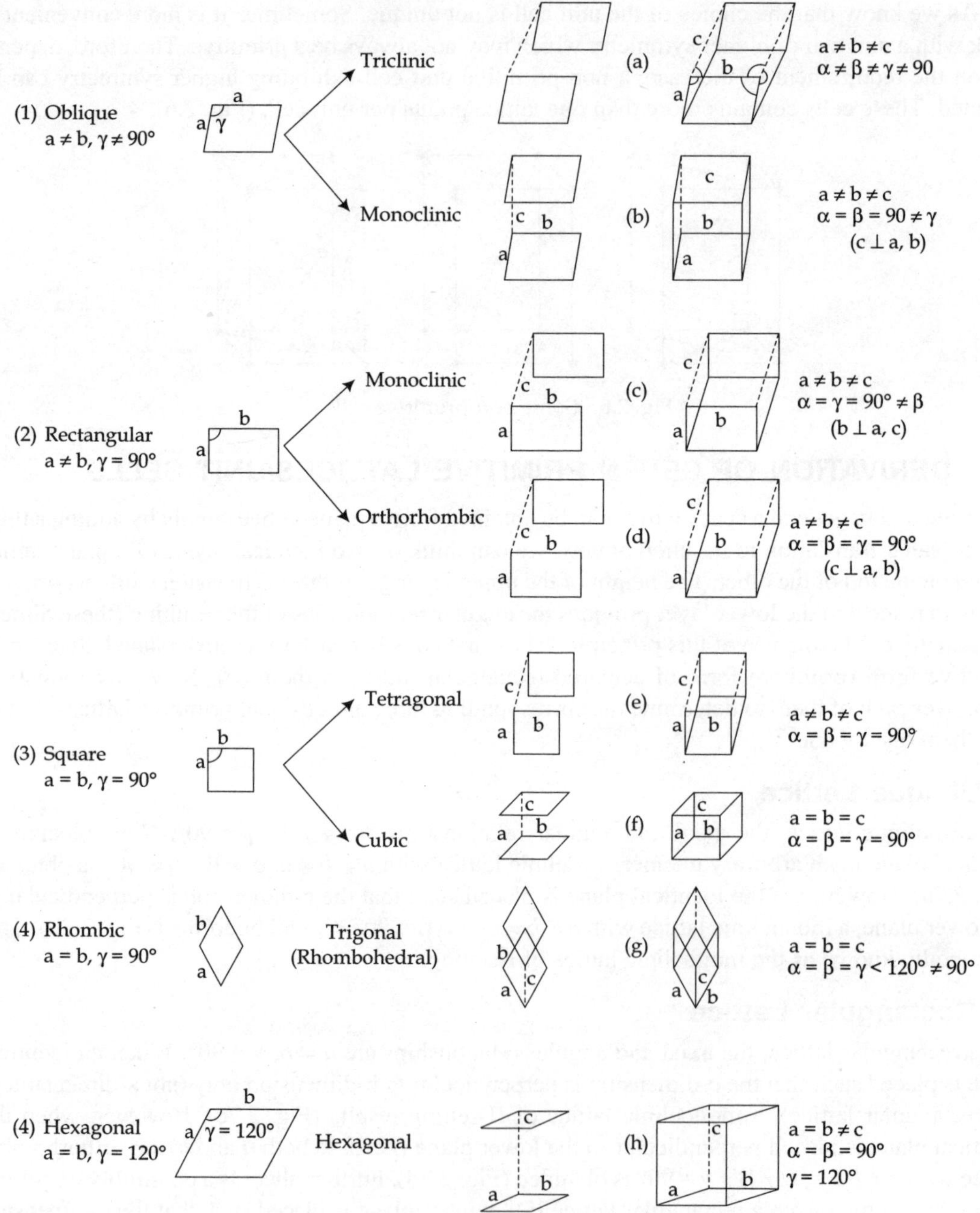

Fig. 2.7 Derivation of 3D-lattice from 2D-lattice

tetragonal lattice with $a = b \neq c$, $\alpha = \beta = \gamma = 90°$ is obtained (Fig. 2.7e). However, when the identical plane is placed perpendicular to the lower plane such that the c-dimension is equal to a and b both, a cubic with $a = b = c$, $\alpha = \beta = \gamma = 90°$ is formed (Fig. 2.7f).

(iv) Rhombic Lattice

For a rhombic lattice, the axial and angular relationships are $a = b$, $\gamma \neq 90°$. When an identical plane is placed such that the axial relationship $a = b = c$ and angular relationship $\alpha = \beta = \gamma < 120° \neq 90°$ is maintained a trigonal (rhombohedral) lattice is obtained (Fig. 2.7g).

(v) Hexagonal Lattice

For a hexagonal lattice the axial and angular relationship are $a = b$, $\gamma = 120°$. When an identical plane is placed perpendicular to the basal plane, a three dimensional simple hexagonal lattice is obtained (Fig. 2.7h).

A complete derivation scheme of all seven primitive space lattices from five plane lattices is diagrammatically illustrated in Fig. 2.7. The seven primitive lattices have distinct shapes and represent independent crystal system. Consequently, in three dimensional only seven crystal systems are possible.

2.5 TYPES OF LATTICE CENTERING

In the preceding section, we discussed about the possible primitive lattices/unit cells in three-dimension, where the lattice points lie only at their corners. However, when some more lattice points are added to these primitive cells without destroying their essential symmetries, non-primitive (or a centered) lattices/unit cells belonging to the same crystal system can be obtained. In these dimensional system of lattice, the following four types of centering are found to exist.

1. The body centering (I-centering)
2. The face centering (F-centering)
3. The base centering (A-centering, B-centering, C-centering)
4. The rhombohedral centering (R-centering)

The observations suggest that the process of centering the primitive unit cells of different crystal systems is not uniform in the sence that a particular crystal system may allow one, two or three types of centering while others may not allow any centering at all. Further, in some cases, it will be observed that one type of centering may well be represented by other centerings. In such cases, conventionally accepted lattice/unit cell occupying minimum volume and/or exhibiting greater symmetry will be preferred. Let us discuss each type of centering separately.

1. Body Centering (*I*)

In order to obtain the body centering in a primitive unit cell (Fig. 2.8a), an additional point must be placed at the body center (or the inner centre) I (strictly from the Germen Innenzentrierung) of the unit cell as shown in Fig. 2.8b. In other words, the additional lattice point must be placed at the end of the vector $\left(\dfrac{a}{2} + \dfrac{b}{2} + \dfrac{c}{2} \right)$. The resulting unit cell will contain two lattice points, one at the origin of the unit cell $(0, 0, 0,)$ and another at the body center $\left(\dfrac{1}{2} + \dfrac{1}{2} + \dfrac{1}{2} \right)$. Truly speaking, there is no intrinsic difference between the lattice points at the origin or at the body centre because they all

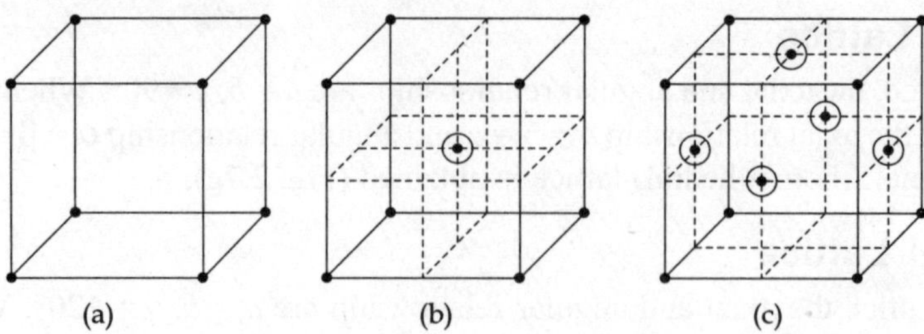

(a)   (b)   (c)

Fig. 2.8 (a) Simple cubic structure (sc), (b) Body centered cubic sturcture (bcc) and (c) Face centered cubic sturcture (fcc)

have the same environment in the lattice. However, once the origin is selected, other points in the lattice are accordingly fixed.

2. Face Centering (*F*)

In this case, three new points are added to the centers of each of the given unit cell shown in

Fig. 2.8c. In other words, they are placed at the end of the vectors $\left(\dfrac{a}{2}+\dfrac{b}{2}\right)$, $\left(\dfrac{a}{2}+\dfrac{c}{2}\right)$ and $\left(\dfrac{b}{2}+\dfrac{c}{2}\right)$.

The resulting face centered (*F*) unit cell will contain four lattice points, one each at (0, 0, 0,),

$\left(\dfrac{1}{2},\dfrac{1}{2},0\right)$, $\left(\dfrac{1}{2},0,\dfrac{1}{2}\right)$, and $\left(0,\dfrac{1}{2},\dfrac{1}{2}\right)$, respectively.

3. Base Centering (*A*–, *B*–, *C*–)

In this case, one face (actually it is the pair of opposite faces) of the unit cell is centered at a time. Accordingly, *A*–, *B*– and *C*– centering means the centering of *bc*-plane, *ac*-plane and *ab*-plane,

respectively. Alternatively, the lattice points are said to be placed at the end of the vectors $\left(\dfrac{b}{2}+\dfrac{c}{2}\right)$,

$\left(\dfrac{a}{2}+\dfrac{c}{2}\right)$ and $\left(\dfrac{a}{2}+\dfrac{b}{2}\right)$, respectively and are shown in Fig. 2.9. The resulting unit cell will contain

two lattice points (in each centering). Their coordinates are

$$(0, 0, 0,) \text{ and } \left(0,\dfrac{1}{2},\dfrac{1}{2}\right) \text{ for } A\text{-centering}$$

$$(0, 0, 0,) \text{ and } \left(\dfrac{1}{2},0,\dfrac{1}{2}\right) \text{ for } B\text{-centering}$$

$$(0, 0, 0,) \text{ and } \left(\dfrac{1}{2},\dfrac{1}{2},0\right) \text{ for } C\text{-centering}$$

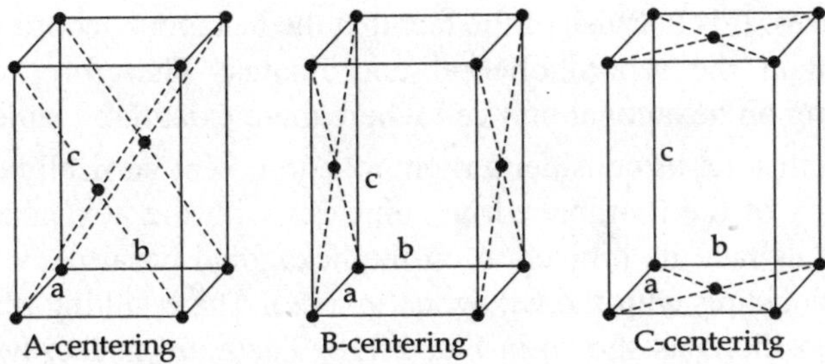

Fig. 2.9 *A–, B–, C–* centerings

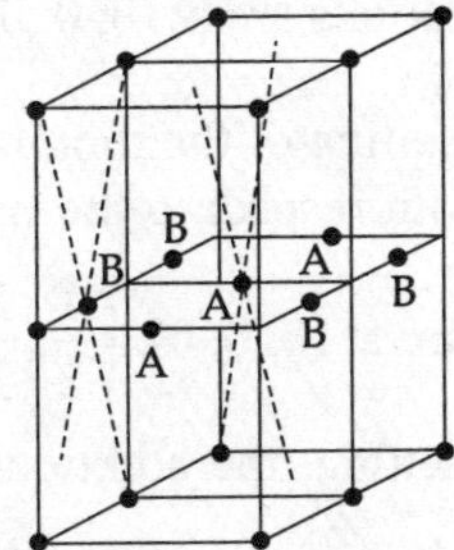

Fig. 2.10 An impossible way to center a lattice.

It is important to know that two-face centering together can never form a lattice, since the environment of all the points is not identical in this case, no matter how one choses the translation vectors. This is shown by the dashed line in the Fig. 2.10.

4. Rhombohedral Centering

From the literature, it appears that there exist some kind of complications and confusions while dealing with trigonal, rhombohedral and hexagonal systems. We know that in the hexagonal crystal system, there exists a unique 6-fold axis, while in the trigonal (rhombohedral) system, there is only a unique 3-fold axis but no 6-fold axis. In fact, in the trigonal system, both primitive hexagonal and primitive rhombohedral unit cells are possible. This is supported by the experimental observations, which suggest that some trigonal crystal structures are based on hexagonal lattice while others on rhombohedral lattice. It is also a fact that the crystal blonging to rhombohedral system could be described equally well on hexagonal axes as on rhombohedral axes. However, as per the requirement of the symmetry, the rhombohedral unit cell is consistent with the symmetry of the trigonal crystal system and not with that of hexagonal crystal system because of the absence of 6-fold axis in it. As a result, the rehombohedral system could be regarded as a special case of the trigonal system and therefore we shall consider trigonal and hexagonal system separately as it is considered in the International Tables for X-ray Crystallography.

As far as the hexagonal and rhombohedral systems are concerned, they are primitive on their own axis of representation. However, if the hexagonal system is described on rhombohedral (Miller notation, *a* 3-index system) axes or the rhombohedral system on hexagonal (Miller-Bravais notation, *a* 4-index system) axes, they no longer remain primitive. Their lattices/unit cells become non-primitive (or centered). However, it is conventionally preferred to represent the rehombohedral system on hexagonal axes. It is because of the fact that the hexagonal coordinates are considerably easier to deal with than the rehombohedral coordinates. Therefore, we shall discuss the rehombohedral centering on hexagonal unit cell where there exist three lattice points in it.

In order to explain this, let us consider a rhombohedron kept vertically on the hexagonal axes such that the 3-fold axis of the rhombohedron coincides with the hexagonal c-axis as shown in Fig. 2.11a. Further, let us take its projection on the hexagonal basal plane parallel to the 3-fold rhombohedral axis (coinciding with the hexagonal c-axis). The resulting plane projection clearly exhibits the centered positions as shown in Fig. 2.11b. There are in fact two equivalent ways of positioning the rhombohedron relative to the hexagonal axes. In the first setting the position of the rhombohedron is such that the centering points are at $(0, 0, 0)$, $\left(\dfrac{2}{3}, \dfrac{1}{3}, \dfrac{1}{3}\right)$ and $\left(\dfrac{1}{3}, \dfrac{2}{3}, \dfrac{2}{3}\right)$ as shown in Fig. 2.11b. This is called the obverse setting of the rhombohedron. In the second position of the rhombohedron is turned through $180°$ with respect to the first one (Fig. 2.12b). This is called the reverse setting and the centering points are at $(0, 0, 0)$, $\left(\dfrac{1}{3}, \dfrac{2}{3}, \dfrac{1}{3}\right)$ and $\left(\dfrac{2}{3}, \dfrac{1}{3}, \dfrac{2}{3}\right)$ as shown in Fig. 2.12b. This means that in the obverse setting, the additional points in a hexagonal unit cell are placed at the end of the vectors $\left(\dfrac{2a}{3} + \dfrac{b}{3} + \dfrac{c}{3}\right)$ and $\left(\dfrac{a}{3} + \dfrac{2b}{3} + \dfrac{2c}{3}\right)$. On the hand, in the reverse setting, the additional points are placed at the end of the vectors $\left(\dfrac{a}{3} + \dfrac{2b}{3} + \dfrac{c}{3}\right)$ and $\left(\dfrac{2a}{3} + \dfrac{b}{3} + \dfrac{2c}{3}\right)$.

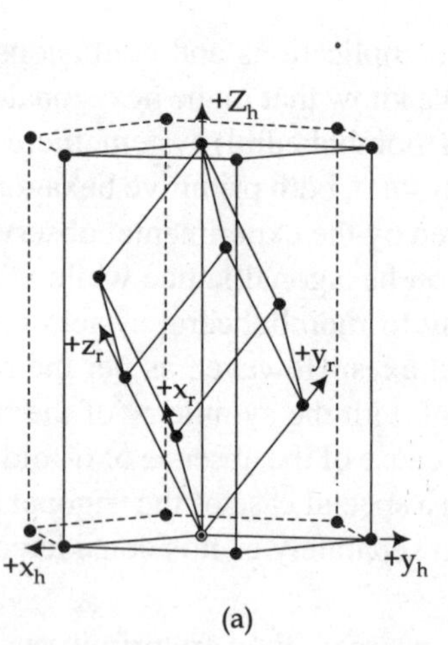
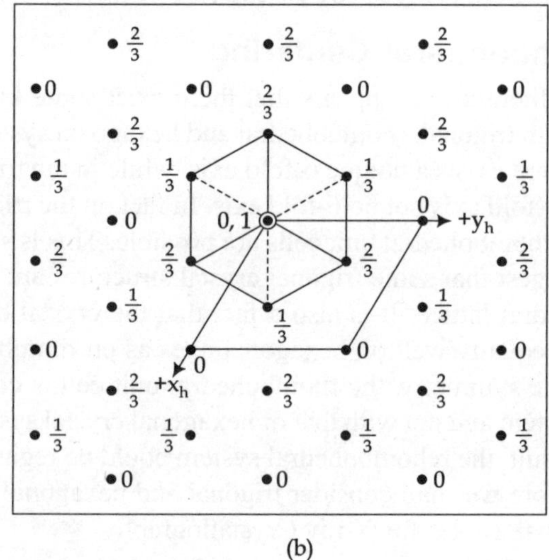

Fig. 2.11 Unit cells in the rhombohedral lattice. (a) Obverse setting of rhombohedron and corresponding hexagonal non-primitive unit cell. (b) Plan of obverse setting (- - - - - - - lower edges, _______ upper edges of rhombohedron)

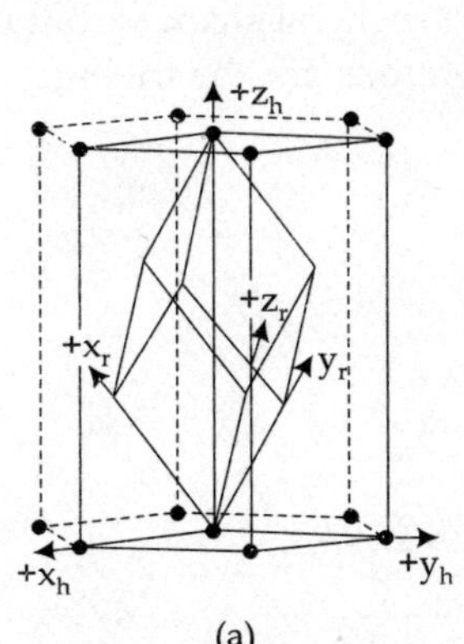

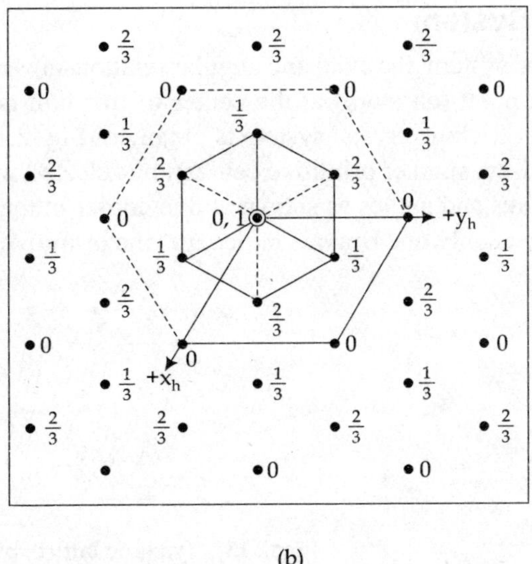

(a)   (b)

Fig. 2.12   (a) Reverse setting of rhombohedron and corresponding hexagonal non-primitive unit cell
(b) Plan of reverse setting (-------- lower edges, ________ upper edges of rhombohedron)

Table 2.2   Primitive and non primitive cells

Symbol	Name	Location of additional points	Total no of lattice points
P	primitive	–	1
I	Body centred	center of the cell	2
A	Base centred	center of (100) face	2
B		center of (010) face	2
C		center of (001) face	2
F	Face centred	center of all faces	4
R	Rhombohedral	Two points along the body diagonal of the cell	3

The kind of possible centerings, number of additional points and other details are provided in Table 2.2.

2.6   DERIVATION OF NON-PRIMITIVE (CENTERED) LATTICES

The basic principle to verify the existence of non-primitive (or centered) lattice in three-dimensional crystal system will remain the same as discussed in two dimensional cases. Further, according to the discussion in the last section that the possible positions for centering in crystal systems belonging to the orthogonal coordinate system are the body centering, face centering and base centering. Let us explore the possibility of existence of such centering in the primitive unit cells of crystal systems under orthogonal (or close to orthogonal) coordinate system. We shall start the process of examination with the lowest symmetry crystal system, i.e. triclinic crystal system and then progressively go on to the crystal systems of higher symmetry to obtain all possible non-primitive Bravais lattices and ultimately derive all the 14 Bravais lattices.

Triclinic System

In this crystal system, the axial and angular relationships are defined as $a \neq b \neq c$ and $\alpha \neq \beta \neq \gamma \neq 90°$. If we add a point (an atom) at the center of this unit cell, another unit cell compatible with the conditions of triclinic crystal system is obtained (Fig. 2.13). However, this does not form any new lattice because a smaller primitive cell can be selected with the same (general shape) arbitrariness of the cell axes and angles as shown at one corner in the figure. Therefore, for the triclinic crystal system there is only one Bravais lattice, i.e. the primitive lattice.

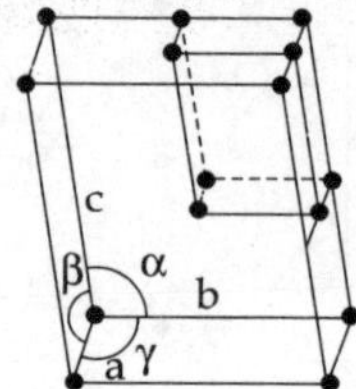

Fig. 2.13   Triclinic lattice, big and small

Monoclinic System

We know that in a monoclinic crystal system there exist a unique 2-fold axis, normally taken as the c-axis. Accordingly, the axes and angles are defined as $a \neq b \neq c$, $\alpha = \beta = 90° \neq \gamma$ (Fig. 2.14a). This is also called the monoclinic system of 1st setting. Let us first of all take the case of body centering (I-centering) as shown in Fig. 2.14b. This can equally well be described as a base centered (B-centering on ac-plane) lattice (Fig. 2.14d) by appropriate choice of the a and b axes. Thus, we can say $I \equiv B$. Next, considering the case of base ($A–$, $B–$ and $C–$ face) centering separately, we observe the C-face (ab-plane) centering does not provide anything new because the resulting lattice can still be described as a primitive lattice with a different value of the lattice parameter a and the angle γ (Fig. 2.14c) but the fundamental monoclinic conditions are retained. Thus, C-centering gives us the primitive monoclinic lattice only and hence we can say $C \equiv P$. However, if the B-face (ac-plane) is centered, a non primitive new lattice is formed as shown in Fig. 2.14d. A lattice centered in this way is called B-lattice. This retains the fundamental monoclinic conditions and is conventional centered monoclinic lattice. In a similar manner, A-face (bc-plane) can be centered to give rise A-lattice. This is found to be equivalent to a B-lattice by simply interchanging the axes $a – b$. Thus $A \equiv B$.

The above discussion also suggest that the F-centering (centering of $A–$, $B–$ and $C–$ faces together) will also be equivalent to a B-lattice. This finally gives us $B \equiv I \equiv A \equiv F$. It is only a matter of convention to use B-centering in preference to the others. Therefore, there are two unique monoclinc Bravais lattices i.e. a primitive lattice and a B-centered lattice.

It is to be noted that monocline system of 2nd setting is conventionally preferred by crystallographers, where the b-axis is the unique axis. The above discussion is still appropriate except for the interchange of axes. Thus, in this setting the primitive lattice and the C-lattice are the two monoclinic Bravais lattices. They are illustrated in Fig. 2.15.

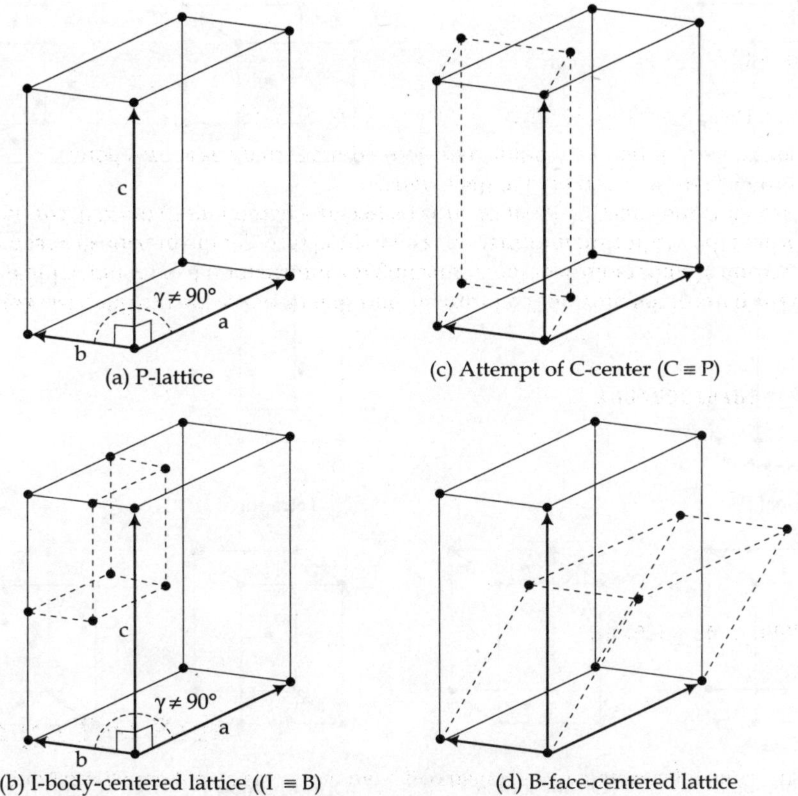

Fig. 2.14   Several aspects of centering a monoclinic lattice

Orthorhombic System

We know that the axes and angles in orthorhombic case is defined as $a \neq b \neq c$, $\alpha = \beta = \gamma = 90°$, implying that all the axes are unequal in length but perpendicular to each other. To make the problem easy let us compare this with the 1st setting of monoclinic system where the axes and angles are defined as $a \neq b \neq c$, $\alpha = \beta = 90° \neq \gamma$. Accordingly, the angle between a and b axes is not 90° and the corresponding C-centering gives rise to a primitive lattice. On the other hand, the faces $(b - c)$ and $(c - a)$ are perpendicular to each other and the corresponding $A-$ and $B-$ centerings give rise to a new centered lattice. Applying this criterion to the orthorhombic crystal system, all $A-$, $B-$, or $C-$ centering will give rise to a new centered lattice. However, they could be described in terms of any one simply by interchanging the orthogonal axes. Normally, the C-centering is conventionally preferred over others. In fact, on a similar ground in this system one can verify that the all face centering (F-centering) and body centering (I-centering) will give rise to distinct possible centered lattices. Thus for the orthorhombic crystal system there are four unique Bravais lattices, i.e. P, C, I and F. They are illustrated in Fig. 2.15.

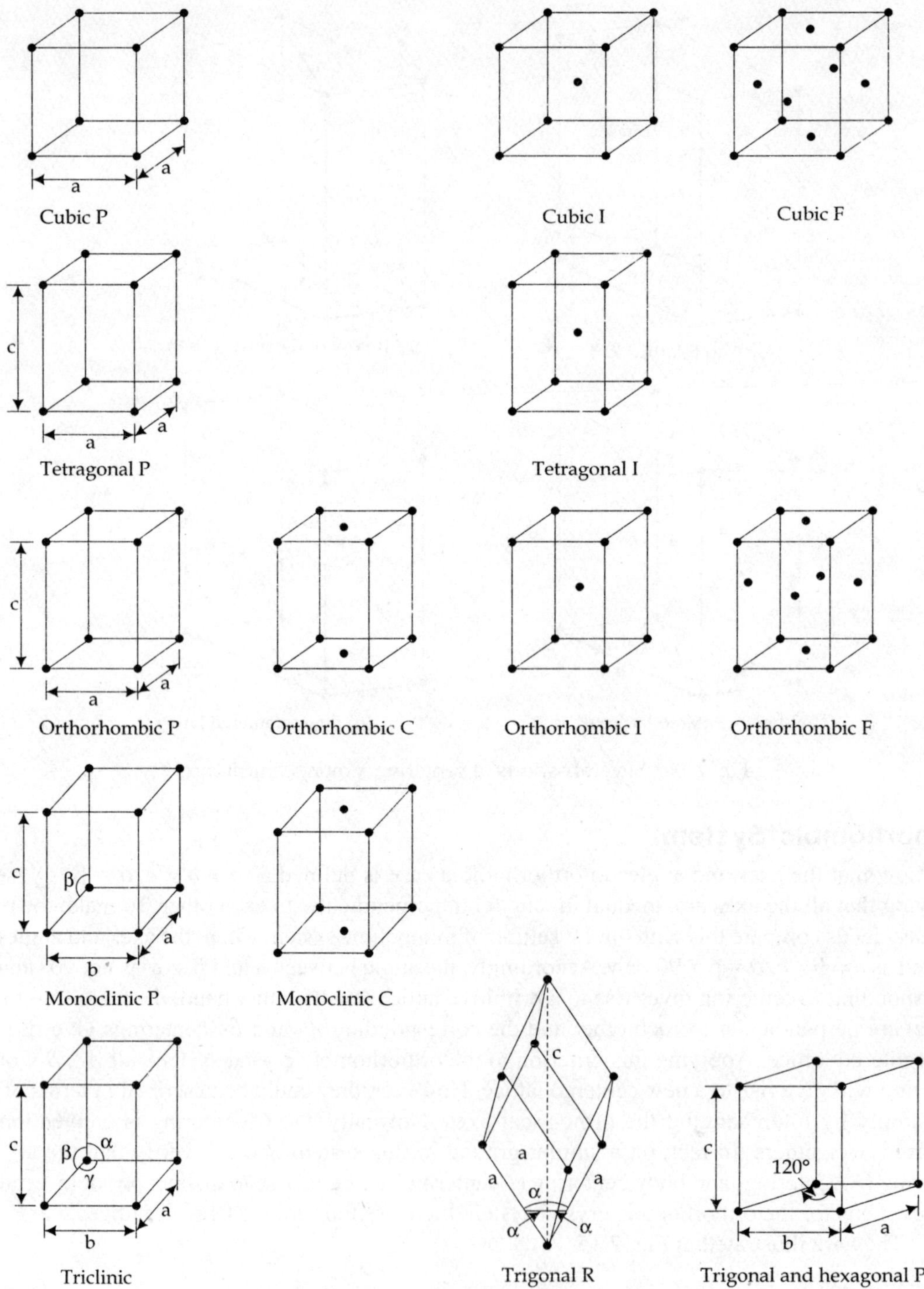

Fig. 2.15 Bravais lattices in three dimensions (angles indicated where $\gamma \neq 90$)

Tetragonal System

The axes and angles in tetragonal system are defined as $a = b \neq c$, $\alpha = \beta = \gamma = 90°$ and there exists one 4-fold symmetry axis. Considering the case of body centering, we observe that this gives a new lattice and also complying with the tetragonal condition. Therefore, the *I*-lattice is a new lattice in tetragonal system. Further, considering the case of *F*-centering, we observe that a body centered lattice could be constructed out of it in such a way that the new axes are rotated through an angle of 45° with the old axes as shown in Fig. 2.16a. Thus $F \equiv I$ in the tetragonal system. Since the body centered unit cell is smaller than the face centered, the *I*-cell is usually preferred.

Now, let us consider the *A*–, *B*– and *C*– base centering in tetragonal system. Since the axes $a = b$ and all the axes are perpendicular to each other, *A*– and *B*– centering together does not form any proper lattice, because the environment of all points in the lattice (see the points connected with the dashed lines) is not the same, no matter how we choose the translation vectors (Fig. 2.10). Further, if we consider the *A*– or *B*– centering separately, we observe that the tetragonal condition of 4-fold symmetry is not maintained in either case. Only the *C*-centering does represent a lattice as shown in Fig. 2.16b. However, this lattice can also be selected by a primitive lattice rotated though an angle of 45° with respect to *c*-axis. Thus, in the tetragonal system, $C = P$. Further, since the primitive cell is smaller of the two, the *P*-cell is usually chosen. Thus ultimately we have only two lattices in tetragonal system. Thay are primitive and the body centered lattices.

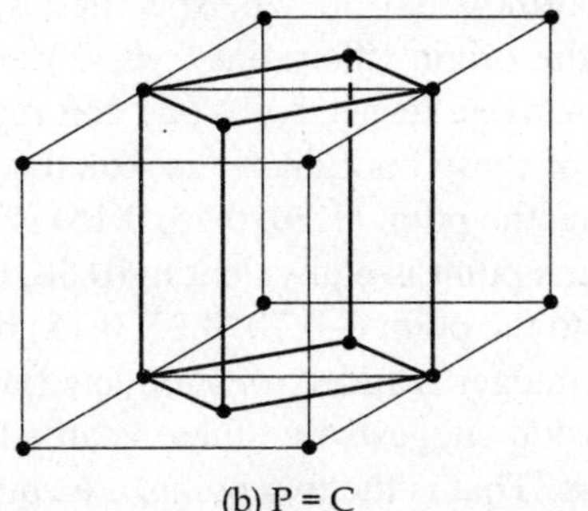

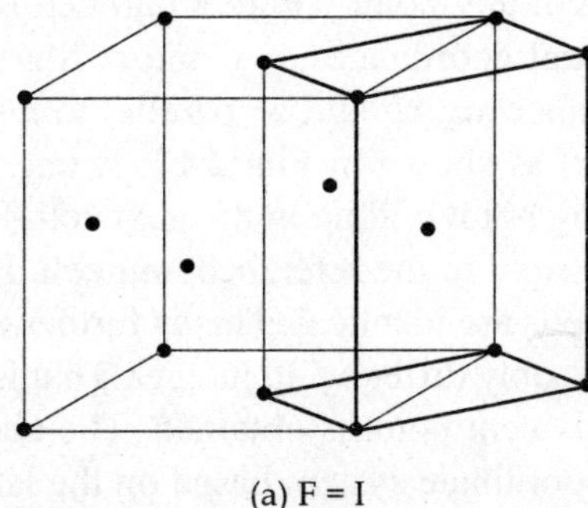

Fig. 2.16 Several aspects of centering a tetragonal lattice

Cubic System

Cubic crystal system is the most symmetric of all crystal systems, where all the axes are equal and perpendicular to each other. In this case, the body centering clearly forms a new lattice without disturbing the lattice condition. This is generally called the bcc (body centered cubic) lattice. In this case, each point is surrounded by eight first nearest neighbour points. Similarly, the *F*-centering is also found to produce a new lattice. In this case, each point has twelve first nearest neighbour points. The *F*-lattice is generally called fcc (face centered cubic) lattice. A close examination reveals that a base centering destroys the basic cubic condition of 3-fold symmetry and hence is not possible. Thus we conclude that in a cubic system, three possible Bravais lattices are the primitive (*P*) the body centered (*I*) and the face centered (*F*). They are illustrated in Fig. 2.8.

2.7 NUMBER OF LATTICE POINTS PER UNIT CELL

In the preceding section, we discussed various possible types of centering in three-dimensional system of lattice. We observed that there are different possible sites where the lattice points can exist in a unit cell. They are at the corners of the unit cell (called C-sites) which are shared by eight-neighbouring unit cells of the space lattice, at the center of unit cell faces (called F-sites) shared by two adjacent unit cells and at the body center (called B-site) completely contained within that unit cell. Therefore in general the effective total number of lattice points associated with a unit cell is given by

$$n = \frac{N_c}{8} + \frac{N_f}{2} + N_b \tag{3}$$

where N_c is the number of lattice points occupping corner (C-site) positions, N_f is the number of lattice points occupying face centered (F-site) positions and N_b is the number of lattice points occupying the body centered (B-site) position in the unit cell.

Using the relation, the effective total number of lattice points in a simple cubic, body centered cubic and face centered cubic are found to be 1, 2 and 4 respectively (Fig. 2.8).

2.8 FRACTIONAL COORDINATES (OBLIQUE SYSTEM)

The location of any point within a unit cell (oblique or orthogonal) may be specified by means of three fractional coordinates x, y and z. Starting from the origin (the point 0, 0, 0,) and moving through distance xa, yb and zc parallel to a–, b– and c– axes, respectively one can represent the point (x, y, z) as shown in Fig. 2.17. If one (or more) of these coordinates is exactly 1, then the corresponding point will lie in the next cell. For example, the point (1.30, 0.25, 0.15) is in the next unit cell just right to the reference unit cell. However, this point is equivalent to (0.30, 0.15) since all the unit cells are identical. This is further equivalent to the point (−0.70, 0.25, 0.15) because the x-coordinates only differ by an integer. That is, when an integer is added (or subtracted) to a coordinate, an equivalent point is obtained. The above discussion suggests that there is an advantage of having the coordinate system based on the lattice vectors. That is the two points are equivalent (or identical) if the fractional part of their coordinates are equal. Making use of the concept of equivalent points, crystal structures can be entirely specified with fractional coordinates.

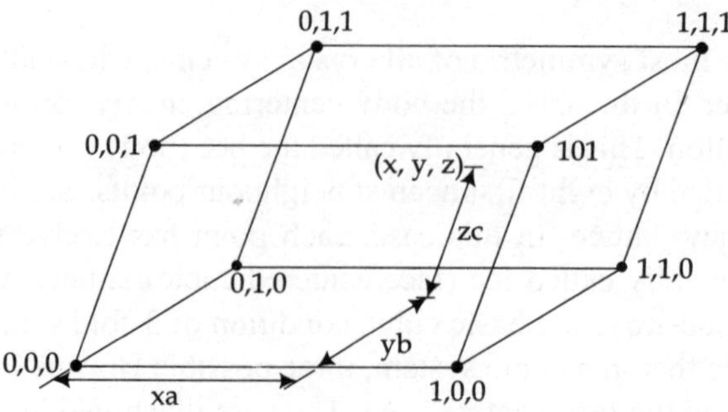

Fig. 2.17   Location of a point with coordinates x, y, z. Number indicate coordinates of unit cell corners

2.9 UNIT CELL CALCULATIONS

The unit cell calculations such as the volume or the distance between two lattice points with a unit cell are difficult in the oblique coordinate system as compared to the orthogonal coordinate system. To makes the calculations relatively simpler, we take the help of vector algebra and obtain their general expressions.

1. Volume of the Unit Cell

Making use of the vector algebra, let us express the lattice parameters a, b and c in eq. 2 in terms of any orthogonal set of unit vectors, i, j and k, so that

$$a = a_x i + a_y j + a_z k$$
$$b = b_x i + b_y j + b_z k$$

and
$$c = c_x i + c_y j + c_z k \tag{4}$$

From this the volume of the unit cell can be written as

$$V = \begin{vmatrix} a_x & a_y & a_z \\ b_x & b_y & b_z \\ c_x & c_y & c_z \end{vmatrix} \tag{5}$$

From the properties of determinant, we know that the value of the determinant remains unchanged when its rows and columns are interchanged. Consequently, it follows that

$$V^2 = \begin{vmatrix} a_x & a_y & a_z \\ b_x & b_y & b_z \\ c_x & c_y & c_z \end{vmatrix} \times \begin{vmatrix} a_x & b_x & c_x \\ a_y & b_y & c_y \\ a_z & b_z & c_z \end{vmatrix} \tag{6}$$

On multiplication the right hand side of eq. 6 becomes

$$V^2 = \begin{vmatrix} a_x a_x + a_y a_y + a_z a_z & a_x b_x + a_y b_y + a_z b_z & a_x c_x + a_y c_y + a_z c_z \\ b_x a_x + b_y a_y + b_z a_z & b_x b_x + b_y b_y + b_z b_z & b_x c_x + b_y c_y + b_z c_z \\ c_x a_x + c_y a_y + c_z a_z & c_x b_x + c_y b_y + c_z b_z & c_x c_x + c_y c_y + c_z c_z \end{vmatrix} \tag{7}$$

In reduced from, eq. 7 can be written as

$$V^2 = \begin{vmatrix} a.a & a.b & a.c \\ b.a & b.b & b.c \\ c.a & c.b & c.c \end{vmatrix} \tag{8}$$

where $a.a = a^2$, $b.b = b^2$, $c.c = c^2$

$$a.b = b.a = ab \cos \gamma$$
$$b.c = c.b = bc \cos \alpha$$

and
$$c.a = a.c = ca \cos \beta \tag{9}$$

Substituting these values in eq. 7 and simplifying the determinant, we get

$$V^2 = a^2 b^2 c^2 (1 - \cos^2\alpha - \cos^2\beta - \cos^2\gamma + 2\cos\alpha\cos\beta\cos\gamma)$$

or
$$V = abc(1 - \cos^2\alpha - \cos^2\beta - \cos^2\gamma + 2\cos\alpha\cos\beta\cos\gamma)^{1/2} \qquad (10)$$

This is a general equation expressing the volume of a triclinic unit cell. Equations for other unit cells can be obtained by substituting the values of respective axial parameters (axes and angles) in eq. 10. Unit-cell volume of different lattice types are provided in Table 2.1.

2. Distance Between Two Lattice Points (Oblique System)

In order to obtain the expression of distance between two lattice points, let us represent the point (x, y, z) within the unit cell (Fig. 2.17) in terms of a radius vectors $\vec{r}$ as

$$\vec{r} = i\,xa + j\,yb + k\,zc \qquad (11)$$

Then two similar points $(x_1, y_1, z_1,)$ and (x_2, y_2, z_2) can also be represented as

$$\vec{r_1} = i\,x_1 a + j\,y_1 b + k\,z_1 c$$

and
$$\vec{r_2} = i\,x_2 a + j\,y_2 b + k\,z_2 c \qquad (12)$$

Therefore, the distance between the two vectors, d_{12} is

$$d_{12} = \vec{r_1} - \vec{r_2} = i(x_1 - x_2)a + j(y_1 - y_2)b + k(z_1 - z_2)c \qquad (13)$$

In terms of dot product, we can write

$$(\vec{r_1} - \vec{r_2}).(\vec{r_1} - \vec{r_2}) = [i\,(x_1 - x_2)a + j(y_1 - y_2)b + k(z_1 - z_2)c].[i(x_1 - x_2)a + j(y_1 - y_2)b + k(z_1 - z_1)c]$$

On simplifying, this gives us

$$|\vec{r_1} - \vec{r_2}|^2 = (x_1 - x_2)^2 a^2 + (y_1 - y_2)^2 b^2 + (z_1 - z_2)^2 c^2 + 2abi.j$$

$$(x_1 - x_2)(y_1 - y_2) + 2bcj.k(y_1 - y_2)(z_2 - z_1) + 2cak.i(z_1 - z_2)(x_1 - x_2) \qquad (14)$$

where $i.j = \cos\gamma$, $j.k = \cos\alpha$ and $k.i = \cos\beta$. Substituting these values in eq. 14, we obtain

$$|r_1 - r_2| = [(x_1 - x_2)^2 a^2 + (y_1 - y_2)^2 b^2 + (z_1 - z_2)^2 c^2$$

$$+ 2(x_1 - x_2)(y_1 - y_2)\,ab\cos\gamma + 2(y_1 - y_2)(z_1 - z_2)\,bc\cos\alpha$$

$$+ 2(z_1 - z_2)(x_1 - x_2)\,ca\cos\beta]^{1/2} \qquad (15)$$

This is a general equation valid for triclinic unit cell. The equations for other unit cells can be obtained by substituting the respective axial parameters (axes and angles) in eq.15.

In a similar manner the angles can also be obtained by using the law of cosines. If θ is the angle subtended at atom 2 (Fig. 2.18) by the bonds of atoms $1 - 2$ and $3 - 2$, then

$$\cos\theta = \frac{d_{12}^2 + d_{32}^2 - d_{13}^2}{2d_{12}d_{32}} \qquad (16)$$

Example: A primitive unit cell has $a = 5$ Å, $b = 6$ Å and $c = 7$ Å; $\alpha = \beta = \gamma = 90°$. A new unit cell is chosen with edges defined by the vectors from the origin to the points with coordinates 3, 1, 0; 1, 2, 0 and 0, 0, 1. Calculate:

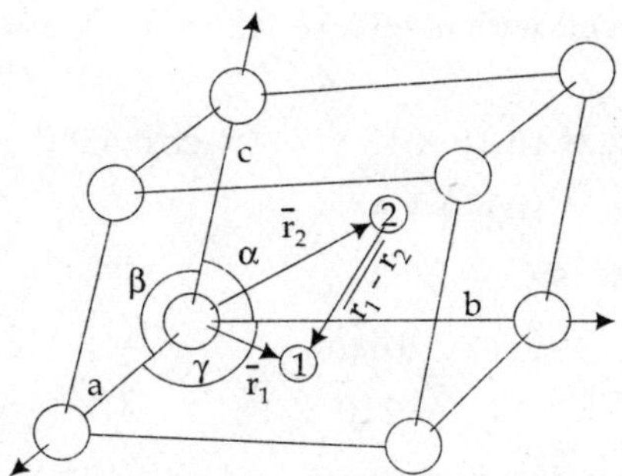

Fig. 2.18 A triclinic unit cell define by the lattice vectors a, b, c pointng out from the origin to the neighbouring cells. The distance between two atoms in the cell is given by the magnitude of vector $r_1 - r_2$

(i) the volume of the original unit cell.

(ii) the length of the three edges and the three angles of the new unit cell.

(iii) the volume of the new unit cell.

(iv) the ratio of new and original unit cell volumes. How many lattice points are there in the new unit cell?

Solution: Given: $a = 5$ Å, $b = 6$ Å, $c = 7$ Å, $\alpha = \beta = \gamma = 90°$. Coordinate of the new unit cell $-0, 0, 0; 1,2, 0$, and $0, 0, 1$. With the given data we can calculate

(i) Volume of the original unit cell, $V_o = abc = 5 \times 6 \times 7 \times = 210$ Å^3

(ii) Using eq. 15, the length of the vector a between the coordinates 0, 0, 0 and 3, 1, 0, is

$$a = [(0 - 3)^2\, 5^2 + (0 - 1)^2\, 6^2 + 0]^{1/2} = [9 \times 25 + 36]^{1/2} = (261)^{1/2} = 16.16 \text{ Å}$$

Similarly, the length of the vector b between the coordinates 0, 0, 0 and 1, 2, 0, is

$$b = [(0 - 1)^2\, 5^2 + (0 - 2)^2\, 6^2 + 0]^{1/2} = [25 + 4 \times 36]^{1/2} = (169)^{1/2} = 13 \text{ Å}$$

and the length of vector c between the coordinates 0, 0, 0 and 0, 0, 1 is

$$c = [\, 0 + 0 + (0 - 1)^2\, 7^2]^{1/2} = (7^2)^{1/2} = 7 \text{ Å}$$

Now the angle between the edges with the end coordinates 3, 1, 0 and 1, 2, 0 is given by

$$\cos \gamma = \frac{3 + 2}{(9 + 1 + + 0)^{1/2}\,(1 + 4 + 0)^{1/2}} = \frac{5}{(10)^{1/2}\,(5)^{1/2}} = \frac{1}{\sqrt{2}}$$

so that $\gamma = 45°$

Similarly, the angle between the edges with end coordinates 1, 2, 0 and 0, 0, 1 is given by

$$\cos \alpha = \frac{0}{(1 + 4 + 0)^{1/2}\,(0 + 0 + 1)^{1/2}} = 0$$

so that $\alpha = 90°$

and the angle between the edges with the end coordinates 3, 1, 0 and 0, 0, 1 is given by

$$\cos \beta = \frac{0}{(9 + 1 + 0)^{1/2}\,(0 + 0 + 1)^{1/2}} = 0$$

so that $\beta = 90°$

(iii) Volume of the new unit cell with $a = 16.16$ Å, $b = 13$ Å, $c = 7$ Å and $\alpha = \beta = 90°$, $\gamma = 45°$ is obtained as

$$V_n = 16.16 \times 13 \times 7 \, (1 - \cos^2 45°)^{1/2} = 1470.\,56 \sin 45°$$
$$= 1039.84 \text{ Å}^3$$

(iv) Ratio of the two volumes is

$$\frac{V_n}{V_o} = \frac{1040}{210} = 4.95 \approx 5$$

$\Rightarrow$ There are five lattice points in the new unit cell.

Example: A rhombohedral unit has $a_R = 5$ Å and $\alpha = 75°$. Calculate:
  (i) the volume of the cell.
  (ii) the dimensions of triply primitive hexagonal unit cell that may be chosen.
  (iii) the volume of the hexagonal cell.
  (iv) the ratio of hexagonal and rhombonherdral volumes. How many lattice points are there in the hexagonal unit cell?

Solution: Given: $a_R = 5$ Å, $\alpha = 75°$

  (i) The volume of a rhombondral unit cell is given by

$$V_R = a_R^3 \sqrt{1 - 3\cos^2 \alpha + 2\cos^3 \alpha} = 125\sqrt{1 - 3\cos^2 75 + 2\cos^3 75}$$

$$= 125 \, \sqrt{1 - 0.2 + 0.035} = 114.2 \text{ Å}^3$$

  (ii) Taking plane projection of the rhombonedral unit cell, we can choose a triply primitive hexagonal unit (Fig. 2.19) whose unit cell parameters are $a = b \neq c$, $\alpha = \beta = 90°$ and $\gamma = 120°$. In order to calculate a_H and c_H take projections of a_R on c-axis. Therefore

$$\frac{C_H}{3} = (a_R)_P = a_R \cos 45 = 5 \times \cos 45 = 3.535$$

or $\qquad\qquad C_H = 3 \times 3.\,535 = 10.61$ Å

Again taking the ab-plane projection of a_R, we have

$$(a_R)_P = 5 \cos 45 = 3.535 \text{ Å}$$

From Fig. 2.19, we can obtain

$$a_H = [(3.535)^2 + (3.535)^2 + 2\,(3.535)^2 \cos 60]^{1/2}$$

$$= 3.535 \times \sqrt{3} = 6.12 \text{Å}$$

  (iii) Volume of the hexagonal cell,
$$V_H = 0.866 \, a^2 c = 0.866 \times (6.12)^2 \, 10.61 = 344.14 \text{ Å}^3$$

  (iv) Ratio of the two volumes is

$$\frac{V_H}{V_R} = \frac{344.14}{114.2} \approx 3$$

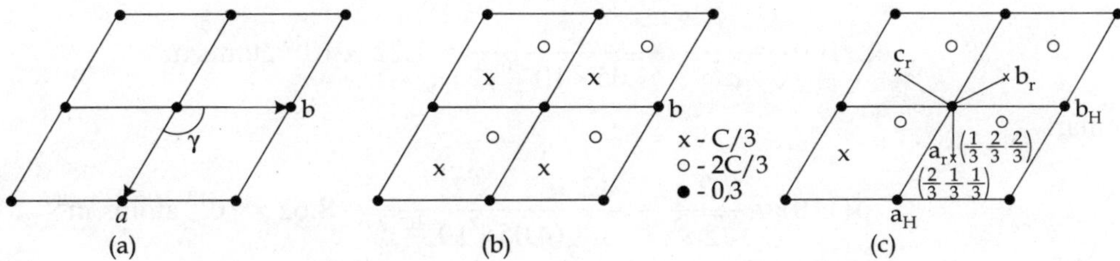

Fig. 2.19 Showing hexagonal unit cell and projected rhombohedral positions

3. Linear, Planar and Volume Atomic Density in Crystals

The equations for linear and planar densities remain the same as defined in the last chapter. However, here we shall use them in a three dimensional context (for crystal directions and planes the volume atomic density (ρ_v) of the crystal material is defined as

$$\rho_v = \frac{\text{Mass of the unit cell}}{\text{Volume of the unit cell}}$$

$$= \frac{Mn}{NV} = \frac{Mn}{Na^3} \tag{17}$$

where M is the atomic (molecular) weight, n is the number of atoms in the given unit cell, N is the Avogadro's number ($= 6.023 \times 10^{26}$ in SI system) and a is the side of the unit cell.

Example: Calculate the linear atomic density (atoms/meter) along [100], [110] and [111] directions in a *bcc* iron whose lattice constant $a = 2.87$ Å.

Solution: Given: Crystal is *bcc*, $a = 2.87$ Å $= 2.87 \times 10^{-10}$ m, $\rho_l = ?$ Draw a bcc unit cell and check the number of atomic diameters intersected by the given lines. Making use of eq. 5 (Ch.1), we can write

$$\rho[100] = \frac{1}{a} = \frac{1}{2.87 \times 10^{-10}} = 3.48 \times 10^9 \text{ atoms/m}$$

Similarly,

$$\rho[110] = \frac{1}{\sqrt{2}a} = \frac{1}{\sqrt{2} \times 2.87 \times 10^{-10}} = 2.46 \times 10^9 \text{ atoms/m}$$

and

$$\rho[111] = \frac{1}{\sqrt{3}a} = \frac{2}{\sqrt{3} \times 2.87 \times 10^{-10}} = 4.02 \times 10^9 \text{ atoms/m}$$

Example: Calculate the planar atomic density (atoms/m^2) in the planes (100), (110) and (111) of *fcc* aluminium lattice constant $a = 4.05$ Å.

Solution: Given: Crystal is *fcc*, $a = 4.05$ Å $= 4.05 \times 10^{-10}$ m, $\rho_p = ?$ Draw an *fcc* unit cell and check the effective number of atoms in the given planes. Making use of eq. 6 (Ch.1), we can write

$$\rho(100) = \frac{2}{a^2} = \frac{2}{(4.05 \times 10^{-10})^2} = 1.22 \times 10^{19} \text{ atoms/m}^2$$

Similarly,

$$\rho(110) = \frac{2}{\sqrt{2}a^2} = \frac{\sqrt{2}}{a^2} \frac{\sqrt{2}}{(4.05 \times 10^{-10})^2} = 8.62 \times 10^{18} \text{ atoms/m}^2$$

and

$$\rho(111) = \frac{4}{\sqrt{3}a^2} = \frac{4}{\sqrt{3}(4.05 \times 10^{-10})^2} = 1.41 \times 10^{19} \text{ atoms/m}^2$$

Example: Calculate the (volume atomic) density (atoms/m^3) of *fcc* nickel whose lattice constant $a = 3.52$ Å and atomic mass = 58.71 kg/kmol.

Solution: Given crystal is *fcc*, $n = 4$, $a = 3.52$ Å, atomic mass = 58.71 kg/k mol, $\rho_v = ?$

Substituting these values in eq. 15, we obtain

$$\rho_v = \frac{Mn}{NV} = \frac{58.71 \times 4}{6.23 \times 10^{26} \times (3.52 \times 10^{-10})^3}$$

$$= 8.94 \times 10^3 \text{ kg/m}^3$$

4. Volume Packing Efficiency

The procedure of determining the volume packing efficiency remains the same as described for planar efficiency in the last chapter except that we change the word 'area' by 'volume'. The packing efficiency for three dimensional crystal lattice can be define as

$$\text{Efficiency} = \frac{\text{Volume of the atoms (in the unit cell)}}{\text{Volume of the unit cell}}$$

Following the procedure discussed in section 1.9, let us calculate the efficiencies of *sc*, *bcc* and fcc crystal structures (Fig. 2.20). The required data of these structures can be easily obtained. For a simple cubic structure, the required data are:

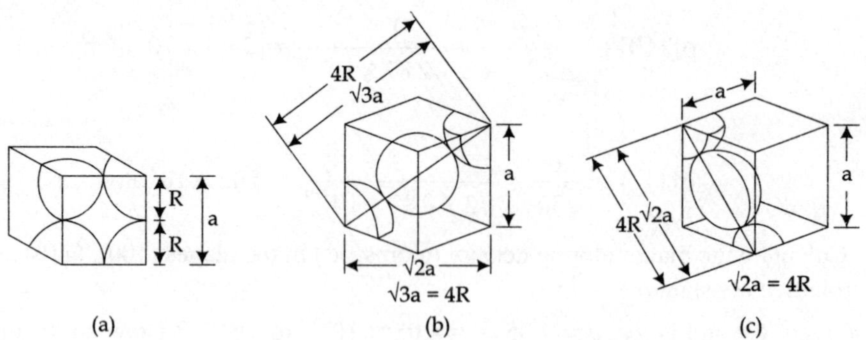

Fig. 2.20   Rationship between a and *R* in different cubic crystal system

Number of atoms per unit cell = 1

Volume of an atom = $\dfrac{4}{3}\pi R^3$

Side of the unit cell, $a = 2R$

Volume of the unit cell = a^3

Therefore, for *sc*

$$\text{Efficiency} = \frac{\dfrac{4}{3}\pi R^3}{a^3} = \frac{\dfrac{4}{3}\pi R^3}{(2R)^3} = \frac{\pi}{6} = 52\%$$

Similarly, for *bcc*

$$\text{Efficiency} = \frac{2 \times \dfrac{4}{3}\pi R^3}{\left(\dfrac{4R}{\sqrt{3}}\right)^3} = \frac{\sqrt{3}\pi}{8} = 68\%$$

and for *fcc*

$$\text{Efficiency} = \frac{4 \times \dfrac{4}{3}\pi R^3}{\left(\dfrac{4R}{\sqrt{2}}\right)^3} = \frac{\pi}{3\sqrt{\pi}} = 74\%$$

2.10 INTERPLANAR SPACING

Simple cartesian geometry is used to determine the interplanar spacing in the unit cells involving orthogonal coordinate system. However, the unit cells involving oblique system, simple cartesion geometry is not very much helpful. Therefore, let us use the reciprocal lattice concept to get the general expression for interlayer spacing in complex cases. Accordingly, let us write

$$\frac{1}{d_{hkl}^2} = \sigma_{hkl} \cdot \sigma_{hkl} = (ha^* + kb^* + lc^*)(ha^* + kb^* + lc^*)$$

$$\begin{aligned}
= \; & h^2 a^* \cdot a^* + hk a^* \cdot b^* + hl a^* \cdot c^* \\
& + kh b^* \cdot a^* + k^2 b^* \cdot b^* + kl b^* \cdot c^* \\
& + lh c^* \cdot a^* + lk c^* \cdot b^* + l^2 c^* \cdot c^*
\end{aligned} \tag{18}$$

where the values of a^*, b^*, c^* etc. are provided in Table 7.2. Also taking into account the identities

$$a^* \cdot a^* = a^{*2} \text{ etc. and } a^* \cdot b^* = b^* \cdot a^* = a^* \cdot b^* \cos \gamma^* \text{ etc.}$$

Then collecting the like terms, eq. 18 becomes

$$\frac{1}{d_{hkl}^2} = \sigma_{hlk}^2 = h^2 a^{*2} + k^2 b^{*2} + l^2 c^{*2} + 2hk a^* b^* \cos \gamma^* +$$

$$2kl b^* c^* \cos \alpha^* + 2lh c^* a^* \cos \beta^* \tag{19}$$

Expressing the reciprocal lattice vectors in terms of direct lattice parameters from Table 7.2, eq.19 becomes to

$$\frac{1}{d^2_{hkl}} = \frac{1}{V^2}\,[h^2b^2c^2\sin^2\alpha + k^2c^2a^2\sin^2\beta + l^2a^2b^2\sin^2\gamma + 2hk\,abc^2(\cos\alpha - \cos\beta)$$

$$+\, 2kla^2bc\,(\cos\beta\cos\gamma - \cos\alpha) + 2lhab^2c(\cos\gamma\cos\alpha - \cos\beta)] \qquad (20)$$

where $V^2 = a^2b^2c^2\,(1 - \cos^2\alpha - \cos^2\beta - \cos^2\gamma + 2\cos\alpha\cos\beta\cos\gamma)$

Formula for other crystal system can be obtained by substituting the respective axial parameters (axes and angles). They are provided in Table 2.3

Table 2.3   Reciprocals of the square of interplanar spacings ($1/d^2_{hkl}$) for each cell shape

Cell shape	$1/d^2_{hkl}$	System
Cube of side a	$\dfrac{h^2 + k^2 + l^2}{a^2}$	Cubic
Right prism with square base a height c	$\dfrac{h^2 + k^2}{a^2} + \dfrac{l^2}{c^2}$	Tetrogonal
Rectangular parallelipiped of sides a, b and c	$\dfrac{h^2}{a^2} + \dfrac{k^2}{b^2} + \dfrac{l^2}{c^2}$	Orthorhombic
Rhomobohedron with polar edges a inclined at angle α	$\dfrac{(h^2 + k^2 + l^2\sin^2\alpha + 2(hk + hl + kl)(\cos^2\alpha - \cos\alpha)}{a^2(1 + 2\cos^3\alpha - 3\cos^2\alpha)}$	Rhombohedral
Right prism with 60° rhombus base a and height c	$\dfrac{4}{3}\dfrac{(h^2 + hk + k^2)}{a^2} + \dfrac{h^2}{c^2}$	Hexagonal
Oblique parallelipiped with side b normal to a and c inclined at β	$\dfrac{1}{\sin^2\beta}\left(\dfrac{h^2}{a^2} + \dfrac{l^2}{c^2} - \dfrac{2hl}{ac}\cos\beta\right) + \dfrac{k^2}{b^2}$	Monoclinic
General parallelipiped with sides a b c angles α, β and γ	$\dfrac{\frac{h}{a}\begin{vmatrix}\frac{h}{a} & \cos\gamma & \cos\beta \\ \frac{k}{b} & 1 & \cos\alpha \\ \frac{l}{c} & \cos\alpha & 1\end{vmatrix} + \frac{k}{b}\begin{vmatrix}1 & \frac{h}{a} & \cos\beta \\ \cos\gamma & \frac{k}{b} & \cos\alpha \\ \cos\beta & \frac{l}{c} & 1\end{vmatrix} + \frac{l}{c}\begin{vmatrix}1 & \cos\gamma & \frac{h}{a} \\ \cos\gamma & 1 & \frac{k}{b} \\ \cos\beta & \cos\alpha & \frac{l}{c}\end{vmatrix}}{\begin{vmatrix}1 & \cos\gamma & \cos\beta \\ \cos\gamma & 1 & \cos\alpha \\ \cos\beta & \cos\alpha & 1\end{vmatrix}}$	Triclinic

Example: In a tetragonal, $a = b = 2.5$ Å and $c = 1.8$ Å. Determine the interplanar spacing between (111) planes

Solution: Given: $a = b = 2.5$ Å, $c = 1.8$ Å, plane $\equiv$ (111). We know that the interplanar spacing for a given set of parallel planes (111) on a tetragonal crystal system is given by

$$d = \left[\frac{h^2 + k^2}{a^2} + \frac{l^2}{c^2}\right]^{-1/2} = \left[\frac{2}{6.25} + \frac{1}{3.24}\right]^{-1/2}$$

$$= [\,0.32 + 0.308\,]^{-1/2} = (1.59)^{1/2}$$

$$= 1.26 \times 10^{-10}\,\text{m}.$$

2.11 SUMMARY

1. A space lattice is an array of points related by the translation $T = n_1 a + n_2 b + n_3 c$, where n_1, n_2, n_3, are integers and a, b, c are primitive translations along x, y, z axes.

2. The volume V_c of a primitive unit cell defined by primitive translations a, b, c is

$$V_c = |a \times b \cdot c|$$

3. Space lattice are limited to fourteen but crystal structures run into thousands.

4. There are four lattice centering in space lattices. They are body centering, face centering, base centering and reombohedral centering.

5. Fractional coordinates are extremely helpful in specifying the atomic positions in the unit cells.

2.12 DEFINITIONS

Axial system: To describe space lattices a set of three non-coplanar axes (called crystallographic axes) in two different system (right handed and left handed) are used.

Crystal: A crystal is a solid in which the component atoms are arranged in a three dimensional periodic manner.

Crystal system: Categorization of unit cells by axial and dimensional symmetry.

Interplanar spacing (d_{hkl}):   It is the perpendicular distance between two adjacent crystal planes with the same index.

Lattice constants: Edge dimensions of the unit cell.

Mirror plane: A plane which divides a body into two halves that are mirror images of one another across a plane.

Packing efficiency (packing factor):   It is the ratio of the volume occupied by the atoms present in the unit cell to the total volume of the unit cell.

Primitive cell: A unit cell which has lattice points only at its corners; equivalently a unit cell which contains only one lattice point.

Spaces lattice: A space lattice is an infinite array of points in three dimensions such that every point is absolutely indistinguishable form every other point; also called Bravais lattice.

Unit cell: It is a convenient repeating parallelepiped unit of a space lattice having three non-coplanar unit translations as its edges.

Volume density: It is the mass per unit volume. In SI unit it is expressed as kg/m^3.

REVIEW QUESTIONS AND PROBLEMS

1. Discuss the procedure to develop a three dimensional lattice. What is the difference between an array of objects and a lattice? What is the characteristic feature of a space lattice?

2. Construct two different sets of primitive translation vectors in a simple cubic lattice. Verify explicitly that the volume $|a \cdot b \times c|$ is the same for both.

3. What is the procedure to construct a Wigner-Seitz unit cell? Construct Winger-Seitz unit cell of *sc*, *bcc*, and *fcc* lattices.

4. What are Bravais lattices and crystal systems? Explain why the base centered tetragonal and the face-centered tetragonal have not been included in Bravais lattices.

5. By means of simple diagrams show that: (a) In an orthorhombic system it is possible to have a side-centered and a face-centered lattice but not a two side-centered lattice, (b) In tetragonal system, $C = P$ and $F = I$.

6. Why it is not possible to have an orthorhombic cell that is I, A, B, and C-centered simultaneously?

7. Why is A-centered tetragonal is not possible? How about A and B centering together?

8. How many atoms per unit cell are there in *sc*, *bcc*, and *fcc* crystal structures? Determine their coordination numbers also. **Ans.** 1,2,4; 6,8,12

9. How are atomic positions located in the unit cell of *sc*, *bcc*, and *fcc* crystal structures? List all of them.

10. For the bcc and fcc lattices, it is possible to choose a primitive rhombohedral unit cell where the primitive translation vectors are $a = b = c$ and the angle between any two of then is α. Draw them and find the value if α in each case. **Ans.** 60°, 109°27′

11. A unit cell has dimensions $a = 6$ Å, $b = 7$ Å, $c = 8$ Å, $\alpha = 90°$, $\beta = 115°$ and $\gamma = 90°$. Calculate the distance between the points: (a) 200, 0.150, 0.333 and 0.300, 0.050, 0.123, and (b) 0.200, 0.150, 0.333 and 0.300, 0.050, −0.123. **Ans.** (a) 2.13 Å and (b) 4 Å

12. A reombohedral unit cell has $a_R = 6$ Å and $\alpha = 60°$. Show that an fcc lattice may be chosen. Calculate the dimension of the cubic cell. *Ans.* $a_c = 8.48$ Å

13. Calculate the linear atomic density (atoms/m) along [100], [110] and [111] directions in an *fcc* copper whose lattice constant $a = 3.61$ Å

Ans. 2.77×10^9, 3.92×10^9, 1.60×10^9 atoms/m

14. Calculate the planar atomic density (atoms/m^2) in the planes (100), (110) and (111) of iron whose lattice constant $a = 2.87$ Å.

Ans. 1.21×10^{19}, 1.72×10^{19}, 7×10^{18} atoms/m^2

15. Calculate the planar atomic (atoms/m^2) in the (0001) plane of hcp zinc whose lattice constant $a = 2.66$ Å and $c = 4.95$ Å. **Ans.** 8.16×10^{18} atoms/m^2

16. Calculate the volume atomic density (atom/m^3) of *fcc* copper whose lattice constant $a = 3.61$ Å and atomic mass = 63.54 kg/kmol. **Ans.** 8.96×10^3 atoms/m^2

17. In a tetragonal crystal, the lattice parameters $a = b = 2.42$Å and c = 1.74 Å. Deduce the interplanar spacing between consecutive (101) planes **Ans.** 1.41Å

CHAPTER 3

SYMMETRY ELEMENTS IN TWO DIMENSIONS

3.1 INTRODUCTION

In the last two chapters, we observed that a given pattern (of lattice) exhibits certain symmetries and has a definite unit cell shape. Different pattern have been found to have different unit cell shapes and hence different set of symmetries. In other words, there seems to be an inherent relationship between the shape of a unit cell (or a regular object) and the symmetries associated with it or vice-versa. Thus any change in the shape (from one regular shape to another) changes the associated symmetries.

In this chapter, we shall study in detail the exact relationships between the cell shapes and the associated symmetries in two dimensional system. We shall also study about the symmetry elements specifically related to two dimensional system and analyse the results obtained from their compatible combinations. Ultimately, we shall obtain two dimensional point groups and space groups.

3.2 SYMMETRY ELEMENTS

A two dimensional object (or a regular body) is said to possess a symmetry element when an operation (such as translation, rotation, reflection or a compatible combination of these) is performed on the body to transform it to self coincidence. Some simple symmetry elements are illustrated in Fig. 3.1. The Characteristic feature of these symmetry elements is that the translation operation repeats an object infinite number of times while the rotation and the reflection operations repeat the same only finite number of times. Further, the translation or rotation operation leaves the character of the 'motif' unchanged. Such operations are called the operations of the first sort. On the other hand, the reflection operation changes the character of the motif from a right handed to a left handed one and vice-versa. Such an operation is called the operation of the second sort (see Table 4.9).

A two dimensional periodic pattern can have the following symmetry elements.

(i) Translation: $T = n_1 a + n_2 b$

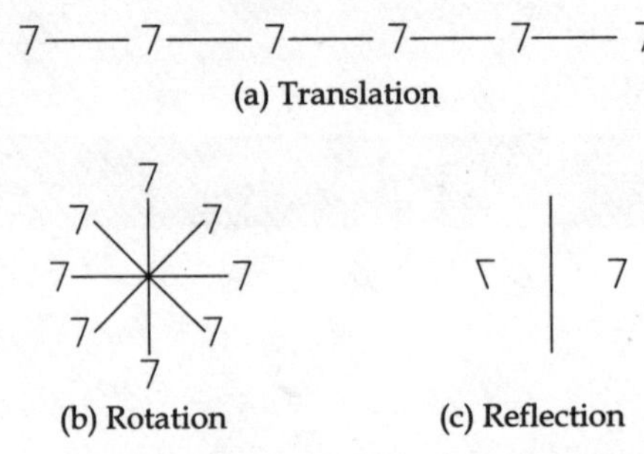

(a) Translation

(b) Rotation

(c) Reflection

Fig. 3.1 Simple symmetry operations

(ii) Proper rotation: $= \alpha = \dfrac{2\pi}{n}$, $n = 1, 2, 3, 4$, and 6

(iii) Reflection: m

1. Translation

A lattice is said to possess a translational symmetry if it is periodic along all its crystallographic directions. Transnational symmetry is an integral part of a lattice. The existence of all other symmetry elements must conform with the translational periodicities.

2. Proper Rotation

A plane lattice (or a two dimensional regular object) is said to possess the rotational symmetry if its rotation by an angle α about a point (the center) transforms the lattice (or object) to itself. Since n such rotations will bring back the lattice (object) to its original position (Fig. 3.2). This gives us

$$n\alpha = 2\pi$$

or $$\alpha = \frac{2\pi}{n} \qquad (1)$$

where n is an integer and the axis passing through the point of rotation is said to have n-fold rotational symmetry (Fig. 3.2).

For $n = 1$ in eq.1 means that the lattice (object) must be rotated through an angle of 2π (or $360°$) to achieve the congruency. Such an axis is called as an identity (or a monad) axis. Every lattice (regular object) possesses an infinite number of such axes. For $n = 2$, the lattice rotated through

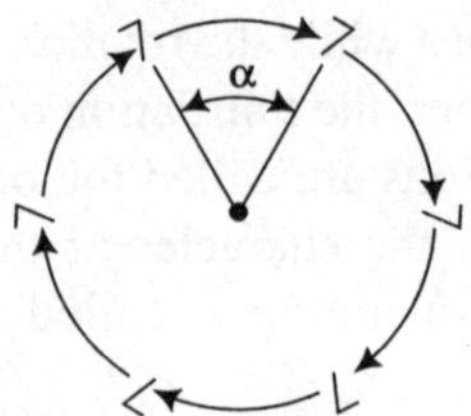

Fig. 3.2 Showing n-rotations

180° and the axis is said to have 2-fold symmetry. Such an axis is termed as a diad axis. In a similar way, for $n = 3$, 4 and 6, the corresponding angles of rotation are 120°, 90° and 60°, they are said to have 3–, 4– and 6-fold symmetry and the respective axes are termed as triad, tetrad and hexed. All five possible rotational symmetry elements of a plane lattice (object) are symbolically represented as shown in Fig. 3.3.

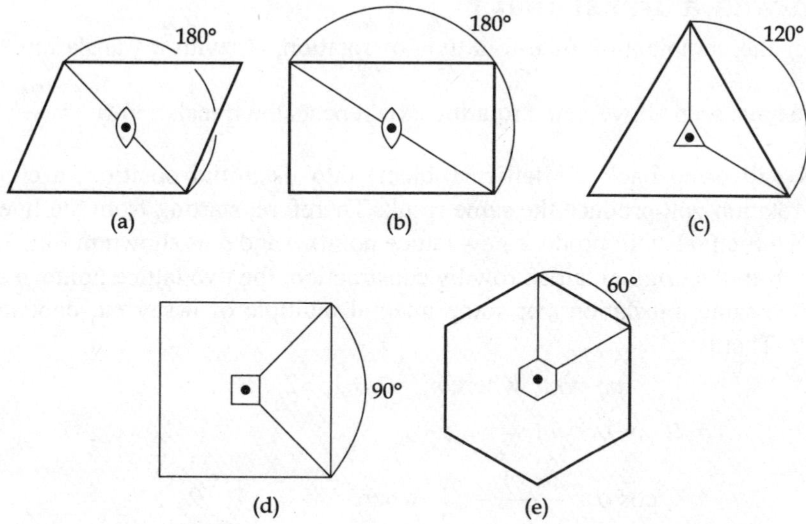

Fig. 3.3 Polygons having 2-fold, 3-fold, 4-fold, and 6-fold rotations

3. Reflection (Mirror Line)

A plane lattice is said to possess reflection symmetry if there exists a line dividing the lattice (or object) into two identical halves such that they are mirror images of each other (Fig. 3.1c) such a line is called the mirror line and is represented by the symbol *m*. It is to be noted that a single 7 is asymmetric, but the pattern in Fig. 3.1c is symmetric in the sense that each side of figure looks like a mirror image of the other. Mirror coincident figures of this type is Known as enantiomorphous pairs. The characteristic feature of a mirror symmetry is to change a right handed object into a left handed one and vice-versa.

3.3 CONSISTENT COMBINATIONS OF SYMMETRY OPERATIONS

In order to understand the plane lattice in terms of its symmetry elements, let us analyse the possible combinations of the above discussed three symmetry operations keeping in view the fundamental requirement that they all must conform with the translational periodicities. For a plane lattice, the following consistent combinations of macroscopic and microscopic symmetry operations are considered.

1. Macroscopic symmetry operations
    (i) Rotation with a translation *t*
    (ii) Rotation and reflection (two reflections)

2. Microscopic symmetry operations

Reflections with a parallel translation (glide line)

This is discussed in detail under a separate head.

3.4 COMBINATION OF MACROSCOPIC SYMMETRY OPERATIONS

(i) Rotation with a Translation *t*

Let us consider the combination of *n*-fold axis of rotation, A_n with a translation *t* as shown in Fig. 3.4. As we have seen above that a rotation axis repeats the translation $\alpha° \left(= \dfrac{2\pi}{n} \right)$ away, and *n* such rotations will bring back the lattice (object) into its initial position, a clockwise or an anticlockwise rotation will produce the same result. Therefore, starting from the linear array as in Fig. 3.4 two such rotations will produce new lattice points *p* and *q* as shown in Fig. 3.5. Since they are equidistant from the original lattice row by construction, the two lattice points *p* and *q* must be connected by the same translation *t* or some integral multiple of it say *mt*, depending upon the magnitude of α. That is

$$pq = mt, \text{ where } m = 0, \pm 1, \pm 2, \ldots$$

or
$$t + 2t \cos \alpha = mt$$

or
$$\cos \alpha = \frac{m-1}{2} = \frac{N}{2} \text{ where } N = 0, \pm 1, \pm 2, \ldots$$

Since $|\cos \alpha| \leq 1$, the allowed values of *N* are 2, 1, 0, –1, –2. The corresponding values of α and *n* can be easily determined. They are provided in Table 3.1. The results obtained from this study clearly reveals the non-existence of 5-fold symmetry axis and all higher than 6-fold axes. This provides us five points groups. They are: 1, 2, 3, 4 and 6. A detail discussion about the point groups is made in the next section.

(ii) Rotation and Reflection (Two Reflections)

Let us consider the combination of a mirror (line) placed parallel to a proper rotation axis. This combination produces another mirror line (also parallel to the rotation axis) such that the angle between the mirrors is equal to one-half the throw of the rotation axis. This can be best understood by reversing the order of operation. Consequently, let the two mirror lines m_1 and m_2 intersect at a point *A* and make an angle θ. Through reflection, the first mirror m_1 changes (say) a right handed

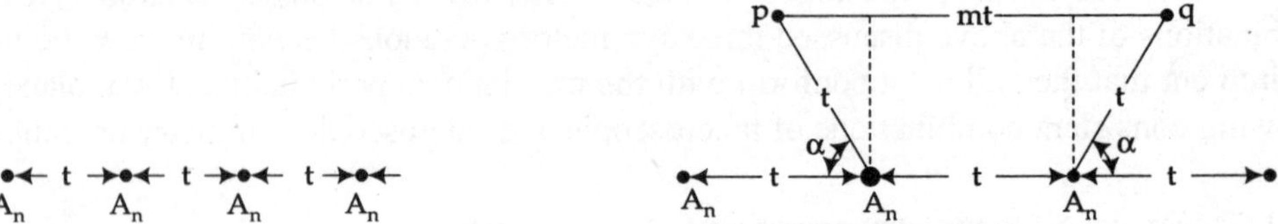

Fig. 3.4 *n*-fold axis of rotations with a translation *t*

Fig. 3.5 Showing how rotations axis repeats the translations

Table 3.1   Allowed rotational axes in a crystal lattice

N	$\cos \alpha$	α	*n allowed rotational axes*
-2	-1	180	2
-1	$-1/2$	120	3
0	0	90	4
$+1$	$+1/2$	60	6
$+2$	$+1$	360 or 0	1

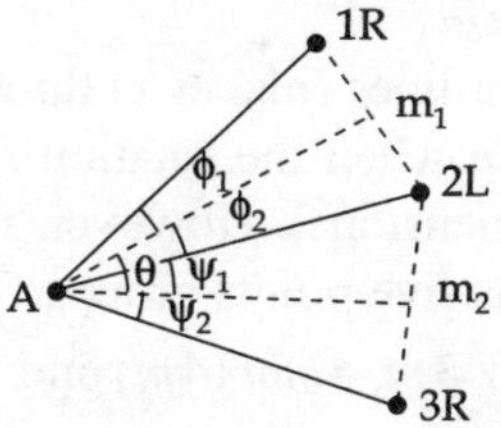

Fig. 3.6   Showing a combination of rotation and reflection

object $1R$ into a left handed object $2L$ and the second mirror m_2 changes $2L$ to $3R$ as shown in Fig. 3.6. For such reflections, the angles

$$\phi_1 = \phi_2 \quad \text{and} \quad \psi_1 = \psi_2$$

such that

$$\phi_2 + \psi_1 = \theta$$

The two reflections are equivalent to a rotation about A by the amount

$$\phi_1 + \phi_2 + \psi_1 + \psi_2 = 2\theta \tag{2}$$

That is the two mirrors at an angle θ together produce a rotation of 2θ. The combination of two reflections can be expressed as

$$m_1 \quad m_2 = A_{2\theta} \tag{3}$$

$$\theta$$

If $2\theta = \alpha$, then eq. 3 can be written as

$$m_1 \quad m_2 = A_{\alpha} \tag{4}$$

$$\alpha/2$$

or

$$m_1 \quad m_2 \cdot A_{-\alpha} = 1 \tag{5}$$

$$\alpha/2$$

This is a compact way of expressing the whole sequence of operations. That is m_1 followed by m_2, followed by A_α in reverse sense, is a sequence of operations which restores the original point to itself.

Let us check the three operations considered in eqs 3 or 4. They are connected in such a way that if any two are given the third follows. For example, if instead of combining the operations of m_1 and m_2, one combines the operations of A_α and m_2, then m_1 appears because A_α repeats $1R$ to $3R$ and m_2 repeats $3R$ to $2L$. Therefore, the combination $A_\alpha.m_2$ repeats $1R$ to $2L$ which is shown above as the result of m_1 (Fig. 3.6). Consequently, one can obtain an equally valid relation from eq. 4 as

$$A_\alpha.m_2 = (m_1)_{\alpha/2} \tag{6}$$

This explains the creation of mirror lines (planes in three dimensions) at an angular separation half the order of proper rotation axis n when the rotation axis of even order is combined with a parallel mirror line (plane in three dimensions). However, when n is odd, the two sets of mirrors coincide. This combination provides us five points groups. They are:

$$2mm \ (mm), \ 3m, \ 4mm \ (4m) \text{ and } 6mm \ (6m)$$

3.5 POINT GROUPS IN TWO DIMENSIONS

We discussed above the macroscopic symmetry elements/operations related to two dimensional system (object or lattices) and their consistent combinations. All the symmetry operations are found to operate around a fixed point and hence they are known as point groups. The word 'point' actually indicates that all the symmetry elements have one point in common, which remains fixed throughout the applications of all symmetry operations. This point is taken as the origin. Similarly, the word 'group' in general we mean a collection or a set of symmetry operations (such as rotation, reflection and their consistent combinations), the application of which leave one point unmoved. However, it must be remembered that the set of symmetry operations for every point group (crystallographic or non crystallographic) is a group in mathematical sense.

In general, unlimited number of point groups are possible for geometrical objects. However, we shall restrict our discussion only to those point groups which are consistent with the translational periodicity found in a two dimensional lattice (in general three dimensional crystalline solids). Surprisingly, we observe that the number of possible point groups in one two and three dimensional periodic systems are very much limited. For example, in a one dimensional system (i.e. a linear lattice) reflection across a point (a plane in three dimension becomes a point in one dimension) is the only point operation. Consequently, there are only two possible point groups in one dimension. They are: 1 and m.

In a two dimensional system (i.e. a plane lattice) the proper rotation about a point and the reflection across a line (a plane in three dimensions becomes a line in two dimensions) are the two possible point operations. Their consistent combinations considered above provide us 10 different possible point groups in two dimensions. The first combination provides us five point groups 1, 2, 3, 4 and 6, and the second combination also provides us five points groups m, $2mm$ (mm), $3m$, $4mm$ ($4m$) and $6mm$ ($6m$). They are illustrated in Fig. 3.7.

In the full symbol, the first position is the principle position and refers to the rotation about a point. The second position refers to a mirror line perpendicular (in three dimensions it is the mirror plane parallel) to the axis of rotation and makes an angle $\alpha\left(=\dfrac{2\pi}{n}\right)$ with all other mirror lines (or mirror planes in three dimensions) of the same "form". Since an n-fold axis repeats a mirror into n mirrors, separated by angular intervals α, a two fold symmetry favours the presence of only one mirror line (or plane in three dimensions) in each from because a clockwise or an anticlockwise rotation of $180°$ gives the same result. However, for 3–, 4– and 6-fold symmetry axes there are 3, 2 and 3 mirror lines (or planes in three dimensions) respectively in each form. They are clearly illustrated in Fig. 3.8. The second form of mirror (s) occupy the third position in the full symbol. The relative orientation of the second form of mirror (s) with respect to the first form can be obtained from the equations 6 and $\left(\alpha=\dfrac{2\pi}{n}\right)$ as follows:

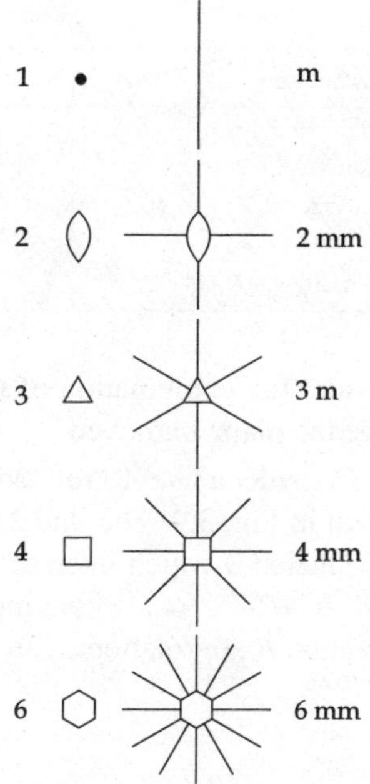

Fig. 3.7 The ten crystallographic point groups in a plane

(i)   Combination of 2-fold and m produces a mirror line of second form at $90°$ away.

(ii) Combination of 4-fold and m produces a pair of mirror lines of second form at $45°$ away.

(iii) Combination of 6-fold and m produces three mirror lines of second form at $30°$ away.

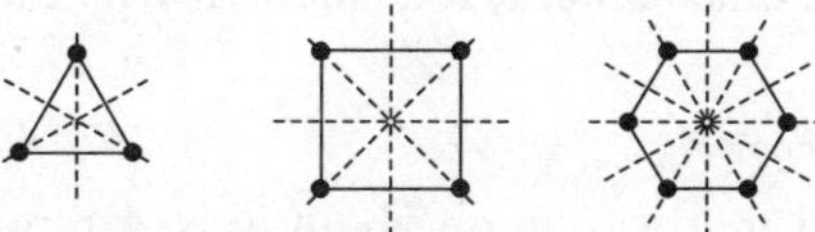

Fig. 3.8 Dashed lines mark the traces of symmetry planes perpendicular to the plane of the regular polygon (axial mirror planes)

Sometimes, it is more convenient to use short symbols to represent the point groups. They are given in the parentheses. In this notation, the principal position (i.e. the rotational symmetry) is left vacant for 1 and 2-folds. Further, in the even-order symbols (except the 2-folds) with the mirror lines, the second m is omitted. However, one can correctly use $2m$ for $2mm$ point group, but a more symmetrical short symbol mm is preferred. A list of two dimensional point groups are provided in Table 3.2.

3.6 PLANE LATTICE CONSISTENT WITH ROTATIONAL SYMMETRY

In general, a combination of proper rotations with translation yields complex symmetry groups which leave space unmoved and these are accordingly called space groups. However, at present we

Table 3.2   point groups in two dimensions

Lattice types	Rotational Symmetry (n-fold)	Point	Groups
Oblique, P	2	1	2
Square, P	4	4	$4mm$ ($4m$)
Hexagonal, P	3	3	$3m$
	6	6	$6mm$ ($6m$)
Rectangular P and C	2	2	$2mm$ (mm)

consider the combination of translation with kind of rotation that (transforms a plane into itself) leave the plane unmoved.

Consider a line AB rotated about an axis A through an angle α to obtain the line AB' labeled 2 as shown in Fig. 3.9. The line 2 is subsequently translated through t symmetrically to obtain the line BA' labeled 3 which intersects with line 1 at B. From simple geometry, it is found that the angle $BAB' = ABA' = \alpha$ also having the same sense. Thus the operations A_α and t is equivalent to the operation B_α (a rotation of an angle α at B) expressed as

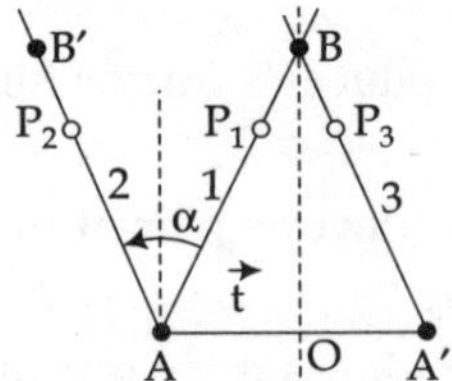

Fig. 3.9   Combination of a rotation, A_α with translation $t_\perp$

$$A_\alpha \cdot t_\perp = B_\alpha \tag{7}$$

Therefore, it can be stated that a rotation about axis A through an angle α followed by a translation t, perpendicular to the axis, is equivalent to a rotation through the same angle α, in the same sense, but about an axis B (parallel to axis A) located at a distance,

$$OB = \left(\frac{t}{2}\right) \cot\left(\frac{\alpha}{2}\right) \tag{8}$$

from the bisector AA'.

The result obtained in eq. 8 can be applied to a plane lattice such that the plane is $\perp$ to the axis of rotation. In doing so, the following principles should be observed:

(i)   For an n-fold axis, each of the rotations $\alpha\left(=\dfrac{2\pi}{n}\right)$, 2α, 3α ... $n\alpha$ ($= 2\pi$) is to be combined with the translation of the plane lattice.

(ii)   For a plane lattice, it is sufficient to combine each rotation with the primitive translations t_1, t_2 and $t_1 + t_2$ of the unit cell (Fig. 3.10).

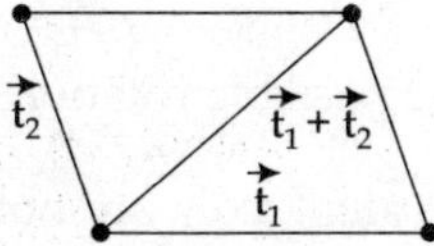

Fig. 3.10 The nonequivalent translations of a plane cell to be combined with rotations of axis *A*

Let us consider the combination of rotation axis of increasing symmetry with the translations of the plane lattice one by one.

The basic procedure is to (i) find the possible rotations for a given (n-fold) rotation axis, (ii) ascertain the possible non-equivalent translations present in the given lattice (iii) combine each rotation with each non-equivalent translation and obtain the corresponding location of perpendicular bisector using eq. 8.

2-fold Axis (*n* = 2)

Under this category, the possible rotations are 1 and π at *A*. The operation 1 (is common to all rotations) is trivial and can be neglected. The rotation of π about *A* is combined with the translation t_1, t_2 and $t_1 + t_2$. As a result, the corresponding rotations of π occur at the locations obtained from eq. 8. They are found to be at points *B*, *C* and *D*, each lying in the middle of their respective translations as shown in Fig. 3.11a. A symbolic representation of the same is shown in Fig. 3.11b. Therefore, a plane lattice consistent with a perpendicular 2-fold rotation axis has four non equivalent (two symmetry elements are said to be non equivalent if no operation of any other symmetry elements carries one of them into other, e.g. neighbouring mirror planes of 4*mm*) 2-fold rotation axes in each of its

primitive cells. Their coordinates are: (0,0), $\left(\frac{1}{2}, 0\right)$, $\left(0, \frac{1}{2}\right)$ and $\left(\frac{1}{2}, \frac{1}{2}\right)$, respectively (Fig. 3.11b).

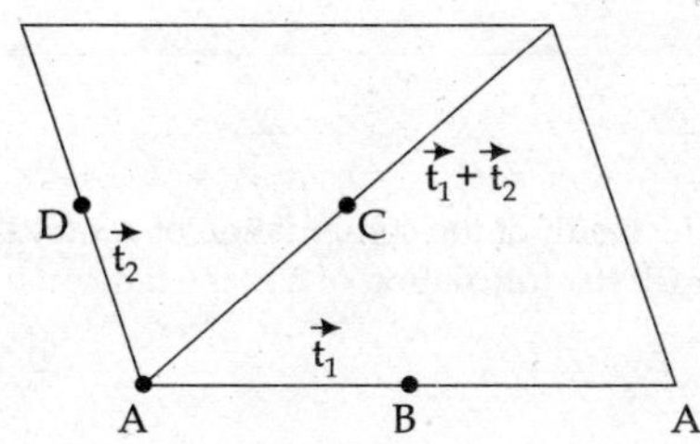

(a) Showing 2-fold symmetry at B, C and D

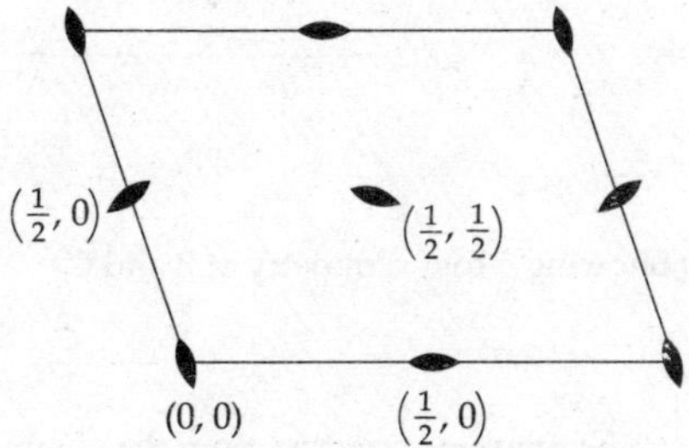

(b) The result of combining the rotations of a 2-fold axis with the translations of a plane lattice

Fig. 3.11

3-fold Axis (*n* = 3)

If a plane lattice consistent with a perpendicular 3-fold axis, then its translations t_1 and t_2 are carried into one another by rotation at *A* and therefore they are equivalent (two symmetry elements are equivalent if an operation of some other symmetry elements carries one of them into the other, e.g. alternate mirror planes of 4*mm*). In this case, it is sufficient to combine each rotation with t_1 and

$t_1 + t_2$ only, at A. Under this category, the possible rotations are: 1, $\dfrac{2\pi}{3}$ and $\dfrac{4\pi}{3}\left(=\dfrac{-2\pi}{3}\right)$ as shown in Fig. 3.12. The positions of non-equivalent axes are obtained from eq. 8. They are found to be located at A, B, and C, in the unit cell as shown in Fig. 3.13a and their respective coordinate are $(0,0)\left(\dfrac{2}{3},\dfrac{1}{3}\right)$ and $\left(\dfrac{1}{3},\dfrac{2}{3}\right)$. A symbolic representation of the same is shown in Fig. 3.13b. The results of the combination of rotations with the translation are provided in Table 3.3.

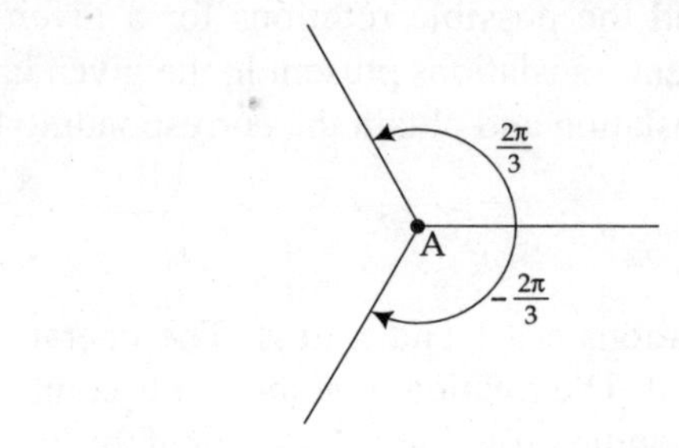

Fig. 3.12

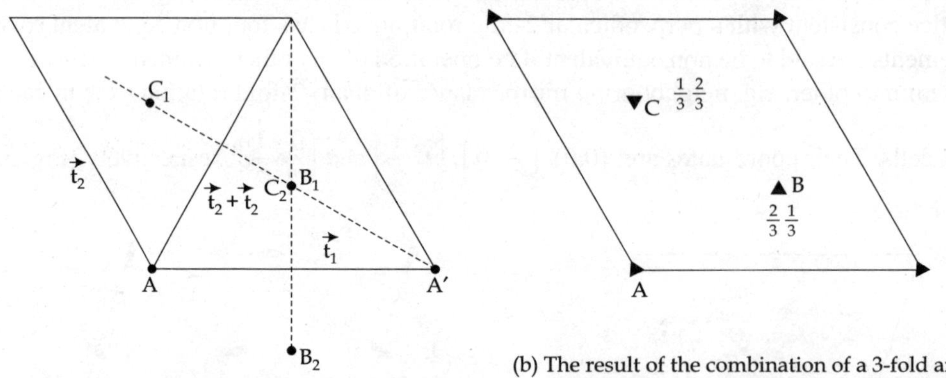

(a) Showing 3-fold symmetry at B and C

(b) The result of the combination of a 3-fold axis with the translation of a plane lattice

Fig. 3.13

Table 3.3   The rotations resulting from the combination of a 3-fold axis with the nonequivalent translations of the cell

Rotations at A	Translations	
	$\vec{t}_1$	$\vec{t}_1 + \vec{t}_2$
1	1	1
$\dfrac{2\pi}{3}$	$\dfrac{2\pi}{3}$ at B_1	$\dfrac{2\pi}{3}$ at C_1
$\dfrac{4\pi}{3}$	$\dfrac{4\pi}{3}$ at $B_2\,(=C_1)$	$\dfrac{4\pi}{3}$ at $C_2\,(=B_1)$

4-fold Axis (*n* = 4)

If a plane lattice is consistent with a 4-fold axis, its translation t_1 and t_2 are equivalent. It is therefore sufficient to combine each rotation with t_1 and $t_1 + t_2$ only, at A. Under this category, the possible rotations are: $1, \dfrac{\pi}{2}, \pi$ and $\dfrac{3\pi}{2}\left(=-\dfrac{\pi}{2}\right)$ as shown in Fig. 3.14. The positions of non equivalent axes are obtained from eq. 8. They are found to be located at $B_1 (\equiv C_2)$, C_1 and C_3 as shown in Fig. 3.15a. Out of these, C_1 and C_3 correspond to the original 4-fold axis at A. Thus the non-equivalent 4-fold axis are located at A and B_1 as shown by the symbolic representation in Fig. 3.15b. The results are provided in Table 3.4.

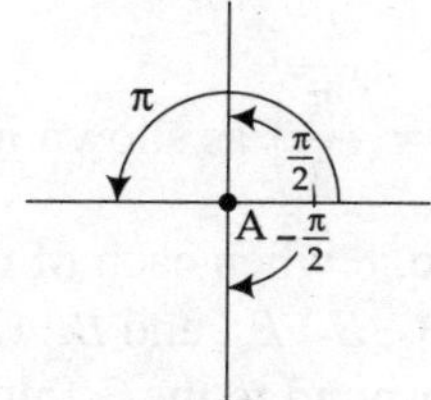

Fig. 3.14

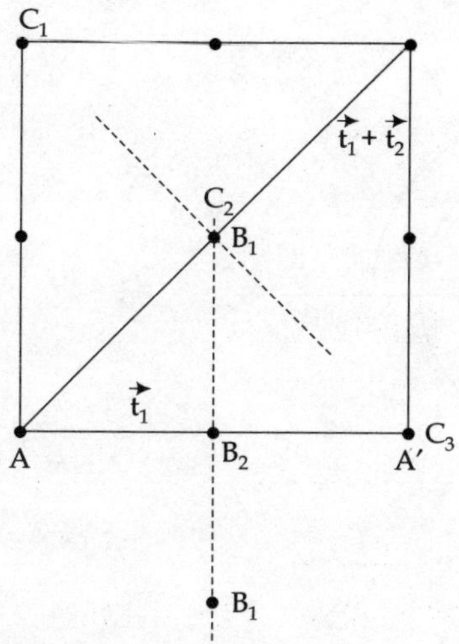

(a) Showing 2-fold and 4-fold Symmetries

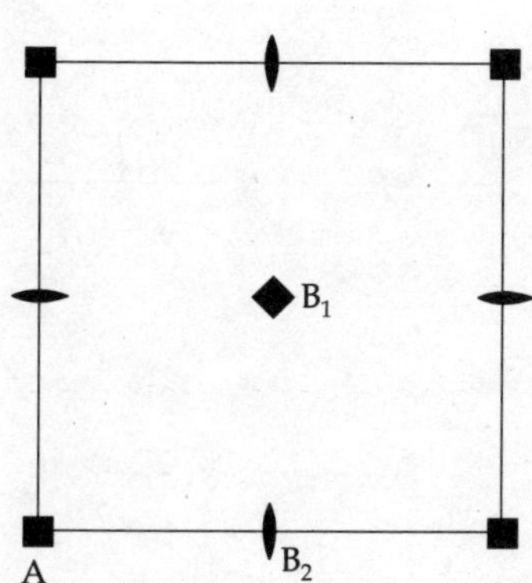

(b) The result of the combination of a 4-fold axis with translation of a plane lattice

Fig. 3.15

6-fold Axis (*n* = 6)

If a plane is consistent with 6-fold axis, then the translation $t_1 = t_1 + t_2 = t_2$. Thus it is sufficient to combine the rotations with the translation t_1 only, at A. Under this category, the possible rotations

Table 3.4 The rotations resulting form the combination of a 4-fold axis with the nonequivalent translations of the cell

Rotations at A	Translations	
	$\bar{t}_1$	$\bar{t}_1 + \bar{t}_2$
1	1	1
$\dfrac{\pi}{2}$	$\dfrac{\pi}{2}$ at B_1	$\dfrac{\pi}{2}$ at $C_1\,(=A)$
π	π at B_2	π at $C_2\,(=B_1)$
$\dfrac{3\pi}{2}$	$\dfrac{3\pi}{2}$ at $B_3\,(=B_2)$	$\dfrac{3\pi}{2}$ at $C_3\,(=A)$

are: $1, \dfrac{\pi}{3}, \dfrac{2\pi}{3}, \pi, \dfrac{4\pi}{3}\left(=-\dfrac{2\pi}{3}\right)$ and $\dfrac{5\pi}{3}\left(=-\dfrac{\pi}{3}\right)$ as shown in Fig. 3.16. The location of the perpen-

dicular bisector can be obtained using eq. 8 when each of these rotations are combined with t_1 and the various positions are marked as B_1, B_2, B_3, B_4, and B_5. Out of these, B_1 and B_5 correspond to the original 6-fold axis at A. B_2 and B_4 correspond to the 3-fold axis and B_3 corresponds to 2-fold axis, respectively as shown in Fig. 3.17a. The complete distribution of symmetry elements is shown in Fig. 3.17b. The results are provided in Table 3.5.

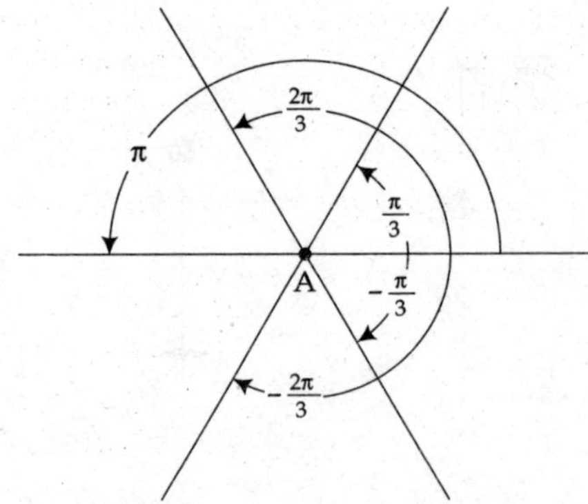

Fig. 3.16

On the basis of above discussions, we conclude that the combination of axial symmetry with the perpendicular translations of the plane lattice produce several nonequivalent axes of the same kind per unit cell. It is interesting to note that the number of such axes decreases with the increase of n (order of rotational symmetry) as shown in Table 3.6. This discussion also suggests that only three kinds of lattice types are required for five proper rotational symmetries. That is, a general parallelogram axes, a 120° rhombus shaped unit cell based on 3 and 6-fold axes, and a square shaped based on 4-fold symmetry, respectively.

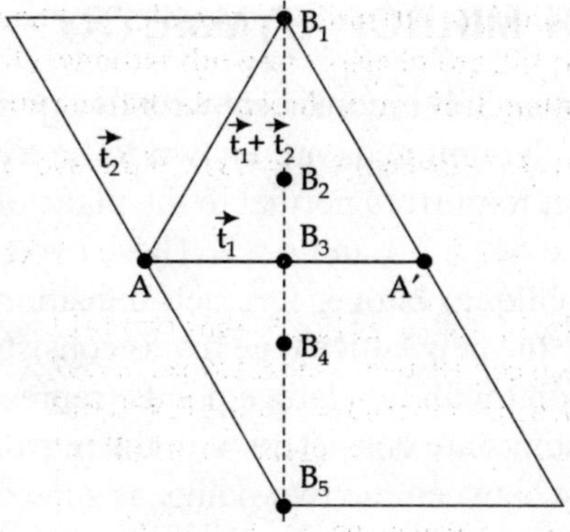

(a) Showing 2, 3 and 6-fold symmetries

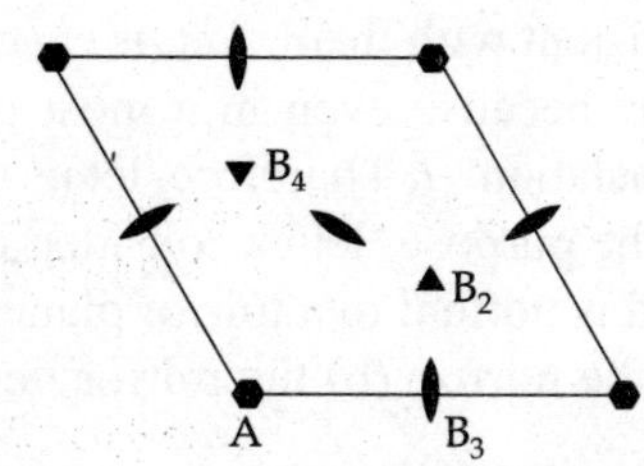

(b) The result of the combination of a 6-fold axis with the translation of a plane lattice

Fig. 3.17

Table 3.5 The rotations resulting from the combination of a 6 fold axis with the nonequivalent translations of the cell

Rotations at A	Translations t_1
1	1
$\dfrac{\pi}{3}$	$\dfrac{\pi}{3}$ at B_1 $(=A)$
$\dfrac{2\pi}{3}$	$\dfrac{2\pi}{3}$ at B_2
π	π at B_3
$\dfrac{4\pi}{3}$	$\dfrac{4\pi}{3}$ at B_4 $(=B_2)$
$\dfrac{5\pi}{3}$	$\dfrac{5\pi}{3}$ at B_5 $(=A)$

Table 3.6 The number and locations of nonequivalent axes

Axis	Number of nonequivalent axes per cell	Coordinates of axes in fractions of translations t_1 and t_2
2	4	$0\,0,\ \dfrac{1}{2}\dfrac{1}{2},\ \dfrac{1}{2}\,0,\ 0\,\dfrac{1}{2}$
3	3	$0\,0,\ \dfrac{2}{3}\dfrac{1}{3},\ \dfrac{1}{3}\dfrac{2}{3}$
4	2	$0\,0,\ \dfrac{1}{2}\dfrac{1}{2}$
6	1	$0\,0$

3.7 PLANE LATTICE CONSISTENT WITH MIRROR SYMMETRY

Combination of lattice translation and rotation at a point (see 3.4) is consistent with five permissible rotational symmetries. Further, for $n > 2$, the shapes of the unit cells are known to be triangular, square and hexagonal. They are symmetrical with respect to mirrors normal to the plane of the net and hence are consistent with them. Let us check for the cases $n = 1$ and $n = 2$. These two cases are inherently the same because even in a most general (oblique) lattice, for each transition t there exists a reverse translation $-t$. Therefore, let us determine the new lattice type that is consistent with the mirror m. For the purpose, let us consider a translation t with two lattice points (representing a row) at its ends and is normal to a mirror plane. These points are consistent with the mirror only if (a) a point lies on the mirror, (b) the mirror lies halfway between the two points as shown in Fig. 3.18.

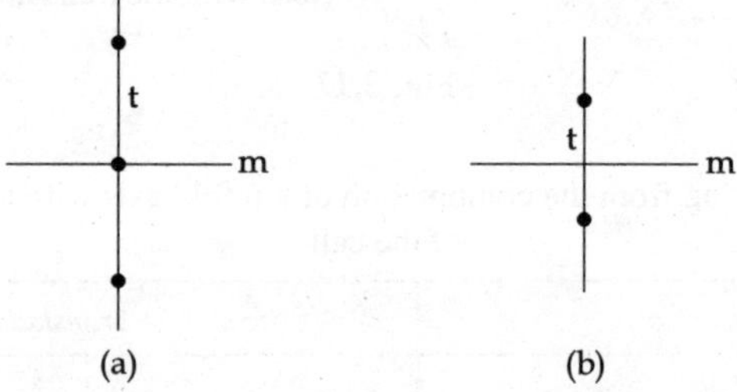

Fig. 3.18 Two situations of mirror plane perpendicular to a row of atoms

Now consider two similar lattice rows of either (a) type or (b) type or dissimilar (a) (b) types as shown in Fig. 3.19. Note that the points in Fig. 3.19 (a) and (b) determine a net with a rectangular mesh (Fig. 3.20a) whereas the points in Fig. 3.19(c) determine a net with the rhombus or diamond mesh (Fig. 3.20b). The resulting rectangular lattice and diamond lattice with the distribution of mirror planes (heavy lines) are shown in Fig. 3.20.

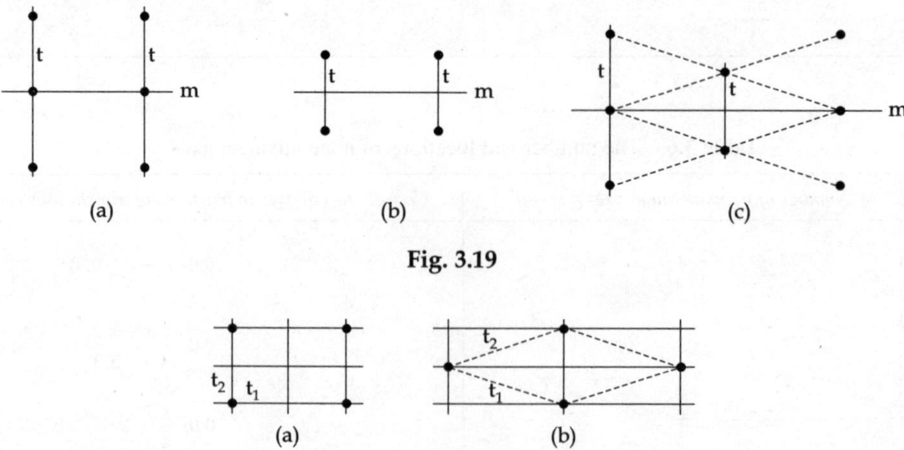

Fig. 3.19

Fig. 3.20 The two plane lattices consistent with symmetry m or 2mm the heavy lines are mirrors

All possible symmetry elements obtained from various consistent combinations for all five plane lattices are distributed in the respective unit cells as shown in Fig. 3.21. They all exhibit the maximum symmetry elements.

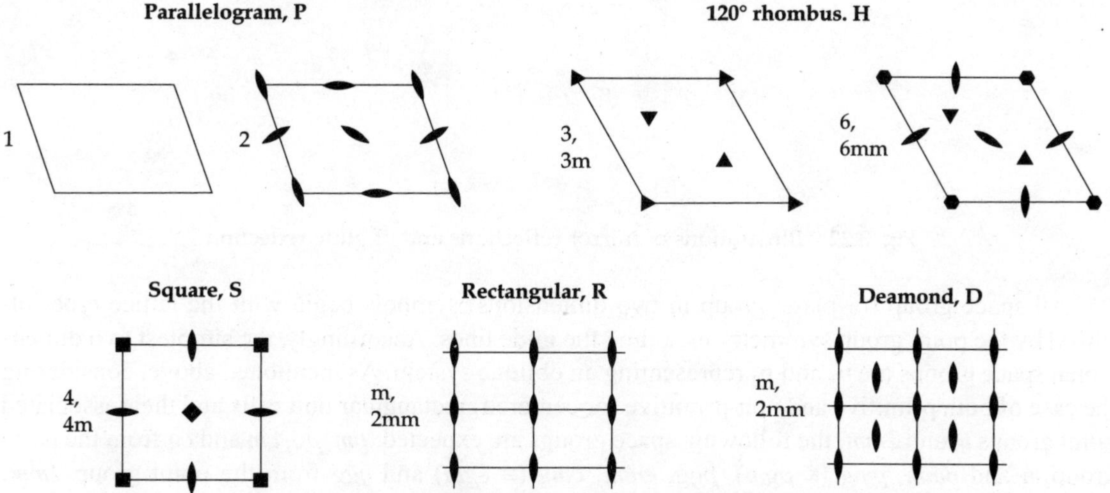

Fig. 3.21   The distribution of rotation axes and mirrors in the five plane lattices type

3.8   COMBINATION OF MICROSCOPIC SYMMETRY OPERATION

Reflection with a parallel translation (glide line) is the only combination possible under microscopic case. Let us consider a one dimensional lattice of right handed number 7. One such motif is marked as (*A*). On reflection, we obtain another lattice row of left-handed motif such as (*B*) as shown in Fig. 3.22a. Now consider the combined effect of reflection and translation, where the left handed motif (*A*) is reflected and subsequently translated to obtain the motif (*B*) as shown in Fig. 3.22b. This is then reflected again and translated to obtain (*A′*) and so on. This combined operation provides a new kind of operator, called the glide line (glide plane in three dimensions). The characteristic feature of the glide line is that the right and left handed forms of the motif are successively repeated on either side of the glide line with a periodicity of half the lattice translation in the direction of the glide. A glide line is symbolically represented as '*g*'.

3.9   SPACE GROUP IN TWO DIMENSIONS (PLANE GROUPS)

Two dimensional space group (or plane groups) can be derived by combining all the ten point group symmetries consistently with the five plane lattices. However, it is important to note that while associating any point group symmetry with a particular lattice type, the relationship between the symmetry and the unit cell shape must be maintained. The combination of reflection and translation discussed in the last section produces a new symmetry element called the glide line. This can either replace or supplement the mirror lines.

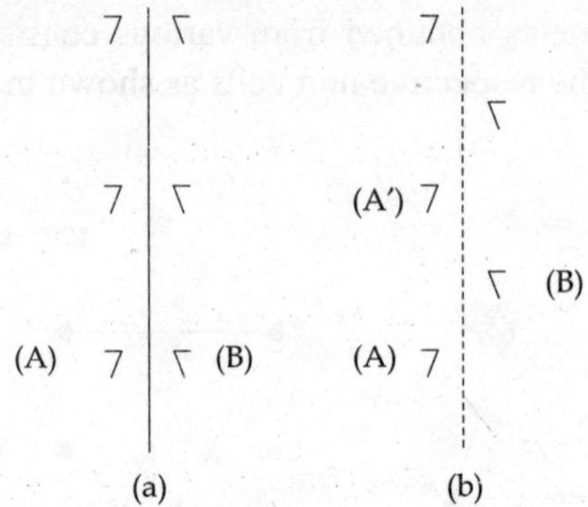

Fig. 3.22   Illustrations of mirror reflections and of glide reflection

All space group (or plane group in two dimensions) symbols begin with the lattice type followed by the point group symmetry including the glide lines. Accordingly, the simplest two dimensional space groups are p_1 and p_2 representing an oblique system. As mentioned above, considering the case of both primitive and non-primitive (or centered) rectangular unit cells and their associated point groups m and $2mm$, the following space groups are expected: pm, pg, cm and cg from the point group m and pmm, pmg ($\equiv pgm$), pgg, cmm, cmg ($\equiv cgm$) and cgg from the point group $2mm$, respectively. However, some of them are found to be equivalent. For example, Fig. 3.23 demonstrates the equivalence between cm and cg; the symmetry elements are the same in both cases, interleaving m and g lines, and the distributions of GEP's (general equivalent positions) can be made identical by a shift in the origin of the unit cell. Similarly, it can be shown that $cmm \equiv cmg \equiv cgg$. Thus the independent plane groups for a rectangular system are: $pm, pg, cm, pmm, pmg, pgg$ and cmm as shown in Table 3.7.

Similar arguments will lead to the derivation of plane groups for square and hexagonal systems. It is observed that there are only 17 different space groups (or plane groups) possible in two dimen-

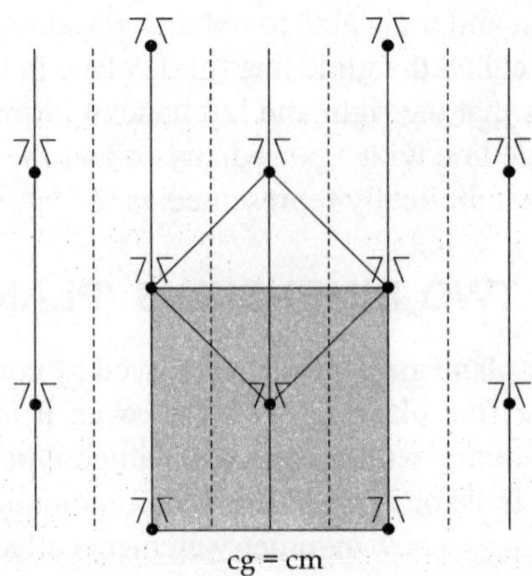

Fig. 3.23   Equivalence between *cm* and *cg*

Table 3.7 Two dimensional lattices, point groups and space groups

Lattice	Point group	Space group symbol	
		Full	*Short*
Oblique p(primitive)	1	p1	p1
	2	p211	p2
Rectangular p and C (Centered)	m	p1m1	pm
		p1g1	pg
		c1m1	cm
	2mm	p2mm	pmm
		p2mg	pmg
		p2gg	pgg
		c2mm	cmm
Square p	4	p4	p4
	4mm	p4mm	p4m
		p4gm	p4g
Hexagonal p	3	p3	p3
	3m	p3m1	p3m1
		p31m	p31m
	6	p6	p6
	6mm	p6mm	p6m

sions. They are listed in Table 3.7 and pictorially shown in Fig. 3.24. It is important to mention that the existence of two distinct space groups $p3m_1$ and $p3_1m$ correspond to two alternative positions for the three reflection lines associated with 3-fold rotation. In $p3m_1$, the initial position of the reflection line bisects γ while in $p3_1m$, it is taken along an edge of the unit cell.

3.10 SUMMARY

1. A two dimensional periodic pattern can have Translation, Rotation and Reflection symmetry elements.

2. There are only five proper rotational symmetry elements possible. They are 1–, 2–, 3–, 4– and 6-folds.

3. A combination of rotation A_α and reflection produces another reflection at an angular separation $\alpha/2$ (for n-even) or coincide (for n-odd).

4. A combination of rotation A_α and perpendicular translation produces several non-equivalent axes of same kind.

5. There are ten point groups and seventeen space groups in two dimensions.

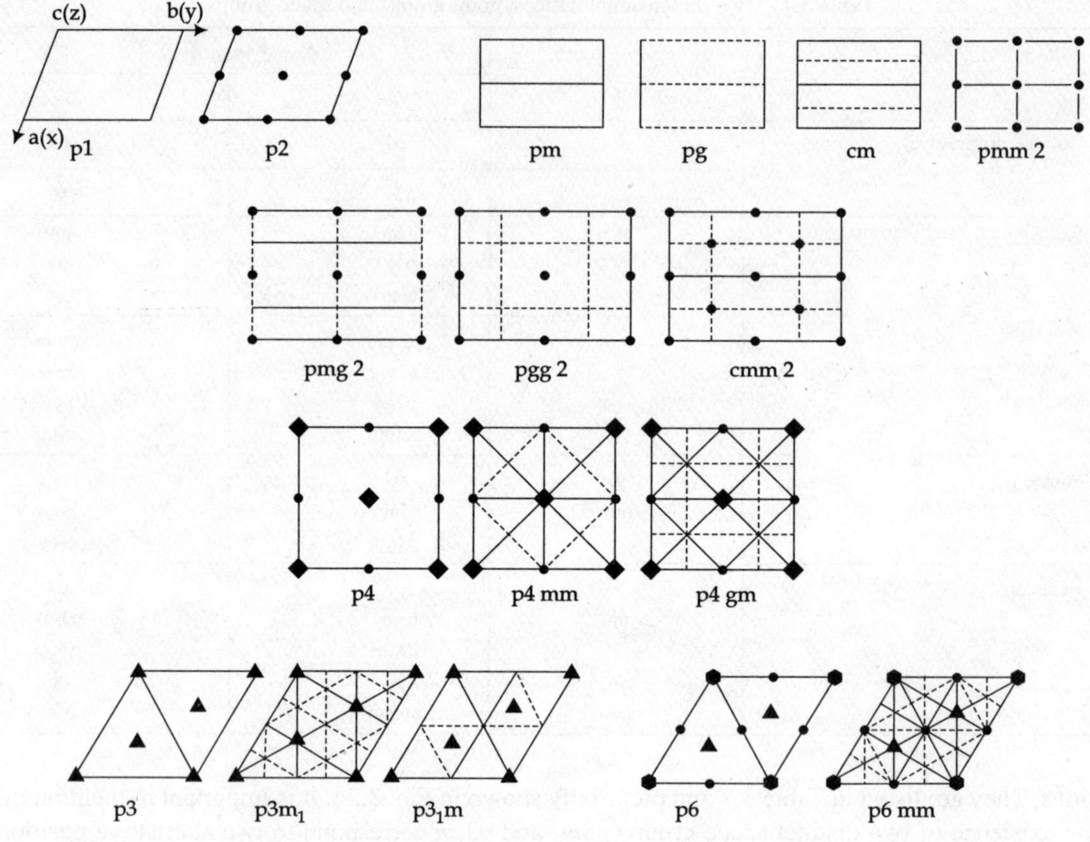

Fig. 3.24   Graphical representation of 17 plane groups

3.11   DEFINITIONS

Glide line: A glide line is a line across which mirror reflection combined with a translation transforms an array of atoms to self coincidence.

Mirror line (Reflection):   A mirror line of symmetry is any line which divides a body into two halves that are mirror images of one another across the line.

Rotational symmetry: A rotational symmetry is about an axis (or line) through which a body may be rotated to one or more positions of self coincidence.

Self coincidence: Two or more positions of a body that are indistinguishable from one another are self coincident positions.

Symmetry element: A symmetry element is any point or line in a body which an appropriate symmetry operation will transform the body into self coincidence.

Symmetry operation: A symmetry operation is any operation that can be performed on a body to transform it to self coincidence.

Translational symmetry: It is the periodic repetetion of lattice points along all its crystallographic directions.

REVIEW QUESTIONS AND PROBLEMS

1. What are symmetry elements and symmetry operations? Describe the principal symmetry operations applicable to two-dimensional lattice.

2. Find the total number of symmetry elements in each of the five plane lattices.

3. A two dimensional lattice possesses 4-fold rotational symmetry. Sketch the net. Position four atoms at the corner of the net and show that the arrangement is consistent with mirror lines parallel to the 4-fold symmetry. Are there two different sets of mirror lines?

4. A unit cell of a two-dimensional lattice has $a = b$, $\gamma = 120°$. Place three atoms about each lattice point in different ways so that the two-dimensional crystal may show the following symmetries in turn: (a) a 6-fold symmetry with mirror lines parallel to it, (b) a 3-fold symmetry, (c) a 2-fold symmetry at the intersection of two sets of mirror lines, (d) a 2-fold and 1-fold symmetries.

5. Show that the permissible rotational symmetries are limited to only five i.e. 1–, 2–, 3–, 4– and 6-folds. Why 5-fold and greater than 6-folds are not possible?

6. What are point groups and plane (space) groups? List them for two-dimensional lattices.

7. Differentiate between non-centrosymmetric and centrosymmetric point groups. Find their numbers, write them separately.   **Ans.** 4,6; 1, *m*, 3, 3*m*, and 2, 2*m*, 4, 4*m*, 6, 6*mm*

8. Find the number of holohedric point groups in two dimensional lattices. Name them.

Ans. 4; 2, 2*m*, 4*mm* and 6*mm*

CHAPTER 4

SYMMETRY ELEMENTS IN THREE DIMENSIONS

4.1 INTRODUCTION

The symmetry elements discussed in the last chapter related to two dimensional lattice is very much relevant to any discussion of the three dimensional periodicity shown by crystalline solids. However, the increase of one more dimension increases the number of point group symmetries and the complexity in their derivation. But the three dimensional unit cell which could be considered as regular solid objects make the visualization of macroscopic symmetry elements relatively simpler.

In this chapter, we shall make a detailed study of macroscopic and microscopic symmetry elements present in crystalline solids. In the process, we shall examine the compatible combinations of various symmetry elements and ultimately derive all thirty two possible point groups found in crystals following simplified conventional approach.

4.2 SYMMETRY ELEMENTS

A symmetry element is defined as a point, a line or a plane in a body (object) about which an appropriate symmetry operation can be performed to transform it to self coincidence.

A three dimensional periodic pattern (a crystal) can have the following symmetry elements

(i) Translation : $T = n_1 a + n_2 b + n_3 c$

(ii) Proper rotation : $\alpha = \dfrac{2\pi}{n}$, $n = 1, 2, 3, 4$ and 6

(iii) Reflection : m

(iv) Inversion : $\bar{1}$

The translation, rotation and reflection operations will have similar functions as discussed in two dimensional case. However, a three dimentional view of proper rotation and reflection are shown in Fig. 4.1. Further, the additional symmetry element, the inversion which has the property to change the character of the motif like reflection and also increases the complexity of the problem to a great extent. Further, like reflections, it also changes a right handed object into a left handed

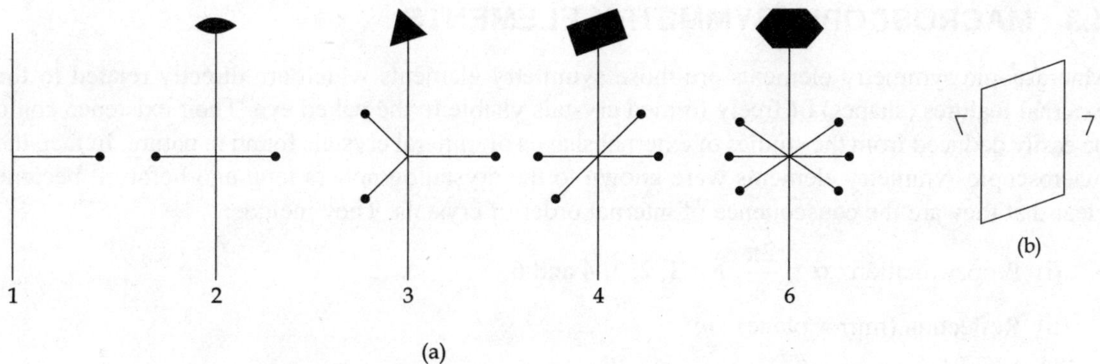

Fig. 4.1 (a) Five proper rotational axes (b) reflection

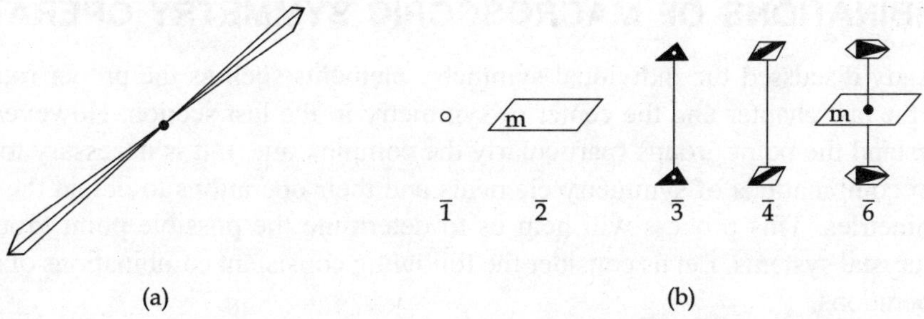

Fig. 4.2 (a) Inversion of an object, (b) five rotoinversion axes

one and vice-versa, i.e. it also produces enantiomorphous pair of objects. But as we know that the reflections occurs in a plane of mirror (Fig. 4.1b), while an inversion is understood to be equivalent to a reflection through a point as shown in Fig. 4.2a. This is called the center of inversion or center of symmetry. In fact, the centre of inversion has the property of inverting all space (at present the number 7) through a point. The international symbol of inversion center is $\bar{1}$. The inversion axes corresponding to five proper rotations are shown Fig. 4.2b.

On a macroscopic scale, crystals are understood to have well defined faces, (i.e. well defined external shapes) sharp edges and precise interfacial angles. On the other hand, on a microscopic scale, they are understood to have atoms or group of atoms arranged in a three dimensional periodic manner. Thus depending upon the external features of a crystal and the internal arrangement of atoms, ions or molecules in them, the symmetry elements in crystalline solids can be broadly classified as:

1. Macroscopic symmetry Elements
2. Microscopic symmetry Elements

In the following, we shall try to understand these symmetry elements and discuss their operations in detail.

4.3 MACROSCOPIC SYMMETRY ELEMENTS

Macroscopic symmetry elements are those symmetry elements which are directly related to the external features (shapes) of freely formed crystals visible to the naked eye. Their existence could be easily deduced from the studies of external shapes of mineral crystals found in nature. In fact, the macroscopic symmetry elements were known to the crystallographers long ago before it became clear that they are the consequence of internal order of crystals. They include:

(i) Proper rotation : $\alpha = \dfrac{2\pi}{n}$, $n = 1, 2, 3, 4$ and 6

(ii) Reflection (mirror plane) : m

(iii) Inversion center : $\bar{1}$
     (Center of symmetry)

4.4 COMBINATIONS OF MACROSCOPIC SYMMETRY OPERATIONS

We have already discussed the individual symmetry elements such as the proper rotaion and the reflection in the last chapter and the center of symmetry in the last section. However, in order to clearly understand the point groups (particularly the complex ones), it is necessary to examine all the consistent combinations of symmetry elements and their operations to obtain the independent possible symmetries. This process will help us to determine the possible point groups for three dimensional crystal systems. Let us consider the following consistent combinations of macroscopic symmetry operations.

1. Rotation at a Point
2. Axial Combinations (two proper rotations)
3. Rotation and Reflection ($m \perp$ to axis of rotations)
4. Rotation and Inversion
5. Proper and Improper Rotations
6. Reflection and Inversion

4.5 ROTATION AT A POINT

In the last chapter, we described the derivation of five proper rotational symmetries using a conventional method normally found in all text books on solid state physics. However, here we shall discuss an alternative approach to derive the same. Let us consider a point $P(x, y, z)$ on $(x - y)$ plane to be rotated through an angle θ in a counterclockwise about the z axis to a point $Q(x', y', z')$ while retaining the same Cartesian coordinate system as shown in Fig. 4.3. Accordingly, let us suppose that i and j are the unit vectors along the OX and OY axes, respectively and the line OP makes an angle ϕ with respect to X-axis. It is to be noted that the vector $|\overrightarrow{OP}| = |\overrightarrow{OQ}|$ and $z = z'$. Then, from the figure, we can write

$$\overrightarrow{OP} = x\vec{i} + y\,\vec{j}$$

$$= |\overrightarrow{OP}| \cos \phi \; \vec{i} + |\overrightarrow{OP}| \sin \phi \; \vec{j}$$

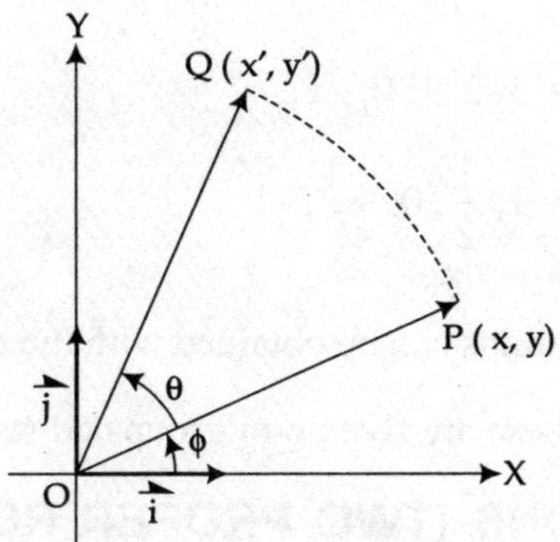

Fig. 4.3 Rotation at a point

$$= |\overrightarrow{OQ}| \cos \phi \, \vec{i} + |\overrightarrow{OQ}| \sin \phi \, \vec{j} \tag{1}$$

and

$$\overrightarrow{OQ} = x'\vec{i} + y'\vec{j}$$

$$= |\overrightarrow{OQ}|\cos(\theta + \phi)\vec{i} + |\overrightarrow{OQ}| \sin (\theta + \phi)\vec{j}$$

$$= |\overrightarrow{OQ}|(\cos\theta . \cos\phi - \sin\theta . \sin\phi)\vec{i} + |\overrightarrow{OQ}| (\sin\theta . \cos\phi + \cos\theta . \sin\phi)\vec{j}$$

$$= (x\cos\theta - y\sin\theta)\vec{i} + (x\sin\theta + y\cos\theta)\vec{j} \tag{2}$$

This gives us

$$x' = x \cos\theta - y \sin\theta$$
$$y' = x \sin\theta + y \cos\theta$$
$$z' = z \tag{3}$$

In the matrix notation eq. 3 can be expressed as

$$\begin{pmatrix} x' \\ y' \\ z' \end{pmatrix} = \begin{pmatrix} \cos\theta & -\sin\theta & 0 \\ \sin\theta & \cos\theta & 0 \\ 0 & 0 & 1 \end{pmatrix} \begin{pmatrix} x \\ y \\ z \end{pmatrix} \tag{4}$$

where the matrix ,

$$A = \begin{pmatrix} \cos\theta & -\sin\theta & 0 \\ \sin\theta & \cos\theta & 0 \\ 0 & 0 & 1 \end{pmatrix}$$

is known as rotation matrix. If we are dealing with a symmetry operation of a crystal, then the traces of such a metrix (sum of diagonal elements) are found to be equal to an integer. Since $|\cos\theta|$ can at most be equal to 1, the trace of the matrix must lie between + 3 and –1 (Streitwolf, 1967).

Therefore, we can write

$$1 + 2 \cos \theta = 3, 2, 1, 0, -1$$

or

$$\cos \theta = 1, \frac{1}{2}, 0, \frac{-1}{2}, -1$$

The corresponding proper rotations can be obtained with the help of the equation $n = \dfrac{2\pi}{\theta}$. They are: 1, 2, 3, 4 and 6 (Table 3.1). These are five point groups corresponding to five proper rotations.

4.6   AXIAL COMBINATIONS (TWO PROPER ROTATIONS)

In general, two proper rotations, one followed by another about a common point can not exitst alone and automatically creates a third rotation equivalent to the combination of the first two. Let us suppose that there are two proper rotation axes OA and OB intersecting at O (Fig. 4.4), rotate an object through angles α and β, respectively. The corresponding rotational operations are called A_α and B_β, respectively. Thus a point marked as P is moved to Q by A_α and then to R by B_β. It is obvious that the motion from P to R could also be achieved directly by a rotation through an angle γ about OC. That is the combination of rotations A_α and B_β is equivalent to a single rotation C_γ. In terms of an operation equation, this can be written as

$$A_\alpha \cdot B_\beta = C_\gamma \tag{5}$$

where the dot in eq. 5 read as "followed by". Equation 5 can be understood as a sequence of operations A_α, followed by B_β followed by C_γ in reversed sense, which restores the original point to itself. This can be expressed as

$$A_\alpha \cdot B_\beta \cdot C_{-\gamma} = 1 \tag{6}$$

Eq. 6 can be represented in a more symmetrical form if the direction of γ (which is accepted as positive) is reversed, i.e.

$$A_\alpha \cdot B_\beta \cdot C_\gamma = 1 \tag{7}$$

where α, β and γ are taken all clockwise or all anticlockwise.

The above discussed form of rotational combination has certain disadvantages when 2-fold axes are involved because they require an axis of $\dfrac{2\pi}{2} \simeq 180°$ each. In order to avoid this problem,

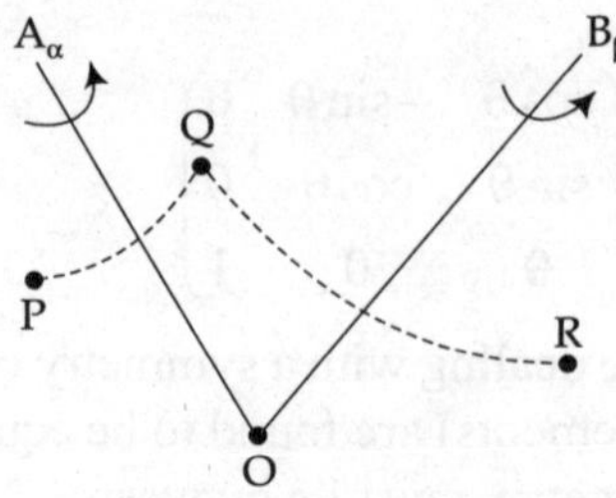

Fig. 4.4   Arrangement of equivalent objects around two intersecting symmetery axes

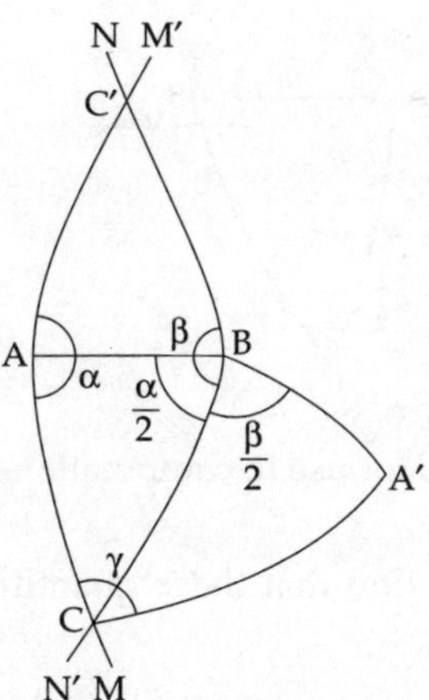

Fig. 4.5 Euler's construction for the combination of a rotation through angle α about *A* with a rotation through angle β about *B*

Eular considered half of the rotation angles for computational work. Euler's construction for the above mentioned rotational motions is illustrated in Fig. 4.5, where the points *A*, *B* and *C* are on the surface of a sphere of radius $OA = OB = OC$. Let us consider the curved line *AM* which is rotated through an angle α about the line *AB* (is the line joining the two rotational axes) to become *AM′*

such that $\angle MAB = \angle M'AB = \dfrac{\alpha}{2}$. Similarly, the line BN is rotated through an angle β about the line

BA to become *BN′* such that $\angle NBA = \angle N'BA = \dfrac{\beta}{2}$. The intersection of *AM′* and *BN* is *C′*, and the

intersection of *BN′* and *AM* is *C*. The rotational motion A_α brings *C* to *C′* but B_β restores *C′* to *C* and hence the combination of rotations A_α and B_β leaves *C* unmoved. Therefore, if there is a motion of points on the sphere due to A_α and B_β it must be a rotation about an axis OC. The amount of rotation can be measured by considering the migration of the point *A*. Since A_α leaves *A* unmoved but B_β moves *A* to *A′* (Fig. 4.5). Now considering the spherical triangles *BA′C*, we have

$$\angle ABC = \angle A'BC = \frac{\beta}{2}$$

also $$AB = A'B$$

Hence Δ *ABC* and Δ *A′BC* are congruent. This implies that

$$\angle ACB = \angle A'CB = \frac{\gamma}{2} \text{ (say)}$$

Therefore, the rotation C_γ will move the point *A* to *A′*.

Now, consider the spherical triangle *UVW* of Fig. 4.5 separately as shown in Fig. 4.6, where the

angles are $U = \dfrac{\alpha}{2}$, $V = \dfrac{\beta}{2}$ and $W = \dfrac{\gamma}{2}$ respectively and the sides opposite to these angles are *u*, *v* and

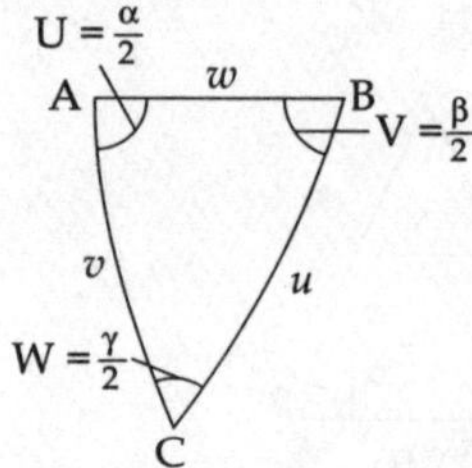

Fig. 4.6 Spherical triangle *ABC* for use in computations based upon Euler's construction

w. From spherical trigonometry, we find that these quantities are related according to the law of cosines, which states that

$$\cos w = \cos u + \cos v + \sin u \sin v \cos w \tag{8}$$

Further, the spherical triangle *UVW* and its polar triangle *uvw* are related through the following equation

$$
\begin{aligned}
u &= 180 - U & U &= 180 - u \\
v &= 180 - U & \quad\text{or}\quad & V &= 180 - v \\
w &= 180 - W & W &= 180 - w
\end{aligned}
\tag{9}
$$

Substituting these in eq. 8 and simplifying the same we obtain the intersection angle,

$$\cos w = \frac{\cos W + \cos U.\cos V}{\sin U \sin V} = \frac{\cos\dfrac{\gamma}{2} + \cos\dfrac{\alpha}{2}.\cos\dfrac{\beta}{2}}{\sin\dfrac{\alpha}{2}\,\sin\dfrac{\beta}{2}} \tag{10}$$

Similarly, we can obtain

$$\cos u = \frac{\cos U + \cos V.\cos W}{\sin V \sin W}$$

and

$$\cos v = \frac{\cos V + \cos W.\cos U}{\sin W \sin U} \tag{11}$$

For general and non-crystallographic cases, any rotation α can be combined with any rotation β at any intersection angle w to produce any third rotation γ. However, for crystallographic cases, the values of α, β and γ are restricted to the permissible values of proper rotations in crystals, namely 1-, 2-, 3-, 4- and 6-fold rotations only. Consequently, the intersection angles are also strictly limited which can be computed using eq. 10 by substituting the permissible values of U, V and W. In Table 4.1, the values of different angles appearing in eq. 10 corresponding to 5 possible proper rotations are provided. These values may further be used to calculate the crystallographically permissible values of intersection angles. They are provided in Table 4.2. In a similar manner, for the intersection angles u, v and w the crystallographically possible rotational combinations can also be obtained. Table 4.3 provides a comprehensive data for all permissible combinations of rotations and

Table 4.1 Spherical angles for 5 proper rotations

Axis at A, B or C	Throw of axis α, β or γ	$U (= \alpha/2)$, $V (= \beta/2)$ or $W (= \gamma/2)$	cos U, V, W	sin U, V, W
1-fold	360°	180°	−1	0
2-fold	180°	90°	0	1
3-fold	120°	60°	$\dfrac{1}{2}$	$\dfrac{\sqrt{3}}{2}$
4-fold	90°	45°	$\dfrac{1}{\sqrt{2}}$	$\dfrac{1}{\sqrt{2}}$
6-fold	60°	30°	$\dfrac{\sqrt{3}}{2}$	$\dfrac{1}{2}$

Table 4.2 Proper polyaxial combinations and corresponding spherical and polar angles

Polyaxial combinations	$U = \alpha/2$	w	$V = \beta/2$	u	$W = \gamma/2$	v
2 2 2	90°	90°	90°	90°	90°	90°
3 2 2	60°	90°	90°	60°	90°	90°
4 2 2	45°	90°	90°	45°	90°	90°
6 2 2	30°	90°	90°	30°	90°	90°
3 3 2	60°	70°32′	60°	54°44′	90°	54°44′
4 3 2	45°	54°44′	60°	35°16′	90°	45°

Table 4.3 Combination of symmetry axes and corresponding angles between them

Combination of symmetry axes	Angles in degrees		
	α	β	γ
2 2 2	90 (2 2)	90 (2 2)	90 (2 2)
3 2 2	90 (2 3)	90 (2 3)	60 (2 2)
4 2 2	90 (2 4)	90 (2 4)	45 (2 2)
6 2 2	90 (2 6)	90 (2 6)	30 (2 2)
2 3 3	54 44′08″ (2 3)	54 44′08″ (2 3)	70 31′44″ (3 3)
4 3 2	35 15′52″ (2 3)	45 (2 4)	54 44′08″ (4 3)

obtained. Table 4.3 provides a comprehensive data for all permissible combinations of rotations and their graphical descriptions are illustrated in Fig. 4.7.

4.7 ROTATION AND REFLECTION (ROTOREFLECTION)

Rotoreflection represents a combined operation of rotation and reflection (where the mirror plane is perpendicular to the axis of rotation). This combined operations repeats an initially right handed object (for example) to a left handed and vice versa. This produces a series of right and left handed objects such that the neighbouring objects are always enantiomorphous, while the alternate objects

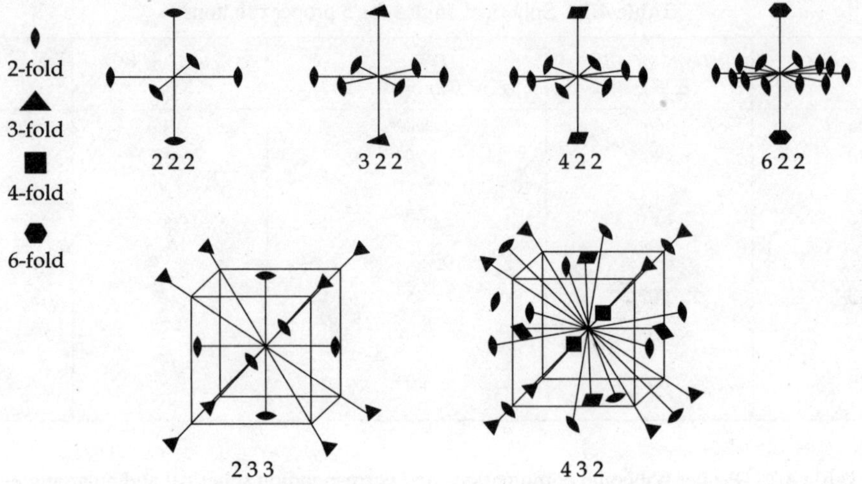

Fig. 4.7 Arrangements of proper symmetry axes for six point groups

are always congruent. There exists rotoreflection axis corresponding to each of the five proper axes. They are symbolically represented as $\tilde{1}$ (read as "one tilde") $\tilde{2}$, $\tilde{3}$ etc. Three dimensional view of $\tilde{1}$ and $\tilde{2}$ are illustrated in Fig. 4.8. A plane projection of the five proper rotation axes and roto reflection axes are shown in Fig. 4.9. In such projections, it is necessary to distinguish between an object lying above or below the mirror plane. A dot/cross is used to indicate an object lying above this plane, and a circle to indicate an object below this plane. The dot/cross and circle are also related to each other enantiomorphically (Fig. 4.9b).

4.8 ROTATION AND INVERSION (ROTOINVERSION)

Like rotoreflection, rotoineversion also represents a combined operation of rotation and inversion. Consequently, these exists a rotoinversion axis corresponding to each of the five proper rotation axes. Rotoinversion axes are symbolically represented as $\bar{1}$ (read as "one bar") $\bar{2}$, $\bar{3}$ etc. A plane

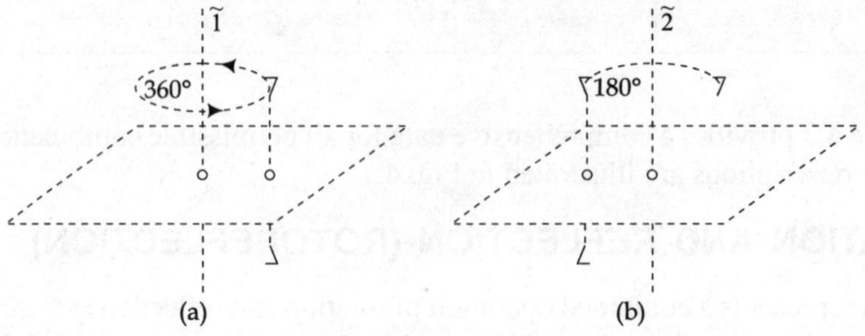

Fig. 4.8 Three dimensional view of $\tilde{1}$ and $\tilde{2}$

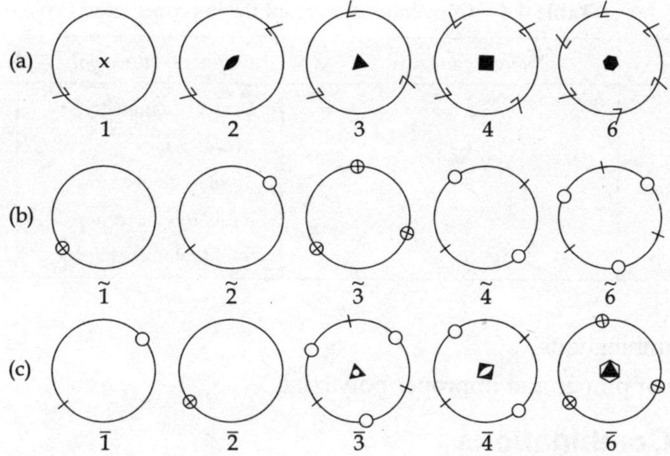

Fig. 4.9 Showing projections of proper rotation axes, rotoreflection axes and rotoinversion axes

projection of the five rotoinversion axes are shown in Fig. 4.9c. The presence of center of symmetry is indicated by placing a white dot at the center (for example in the symbol of $\bar{1}$ and $\bar{3}$ in Fig. 4.9c).

When the five rotoreflection axes are compared with the five rotionversion axes as in Fig. 4.9, it is clear that they are equivalent in pairs. That is, $\tilde{1} = \dfrac{1}{m} = m = \bar{2}$, $\tilde{2} = \bar{1}$, $\tilde{3} = \dfrac{3}{m} = \bar{6}$, $\tilde{4} = \bar{4}$ and $\tilde{6} = \bar{3}$. On the basis of such observations, the following empirical rules can be formulated which can be used for conversion from one system to another, i.e. from rotoreflection to rotoinversion or vice-versa. They are:

$$\tilde{n}_{odd} = \overline{2n} = \frac{n}{m}$$

$$\widetilde{2n} = \bar{n}_{odd} = n.i$$

$$\widetilde{4n} = \overline{4n} \tag{12}$$

where n = 1, 2, 3, 4 and 6, m is a mirror plane and i is the center of inversion, respectively. The International/conventional symbol used to represent these axes are given in Table 4.4.

4.9 PROPER AND IMPROPER ROTATIONS

In the above discussion under preceding sections, we observed that the lattice is consistent with a proper rotation axis n and an improper rotation axis $\bar{n}$ simultaneously. This suggests that the two axes could coincide along a particular direction or they could intersect each other at a point. Accordingly, the combinations of proper and improper rotations are classified as:

Table 4.4   Conventional symbol for improper axes

Rotonversion axes	Rotoreflection axes	Conventional symbol	International symbol
$\bar{1}$	$\tilde{2}$	Center of symmetry	$\bar{1}$
$\bar{2}$	$\tilde{1}$	Mirror plane	m
$\bar{3}$	$\tilde{6}$	3fold rotoineversion	$\bar{3}$
$\bar{4}$	$\tilde{4}$	4 fold rotoineversion	$\bar{4}$
$\bar{6}$	$\tilde{3}$	9 fold rotoineversion	$\bar{6}$

(b) Polyaxial combinations

(c) Coexistence of proper and improper polyaxials

(a) Monoaxial Combinations

The monoaxial combinations of proper axis n and improper axis $\bar{n}$ are customarily written as the fraction $\dfrac{n}{\bar{n}}$. Thus the monoaxial class includes $\bar{n}$ and $\dfrac{n}{\bar{n}}$. The ratio $\dfrac{n}{\bar{n}}$ behaves differently depending whether n is odd, even or a multiple of 4. Agin on the basis of observations and keeping in view the empirical rules can be formulated for the deduction of such ratios, i.e.

$$\frac{n}{\bar{n}} = \frac{n}{n.i} = \bar{n} \quad \text{when n is odd}$$

and
$$\frac{n}{\bar{n}} = \frac{n}{m} \quad \text{when } n \text{ is even and multiple of 4} \tag{13}$$

Making use of above empirical rules, one can easily determine 13 possible combinations of monoaxial symmetry operations (i.e. 13 point groups). They are provided in Table 4.5.

(b) Polyaxial Combinations

Earlier we demonstrated that a proper rotation about an axis A followed by another rotation about an axis B is equivalent to a proper rotation about a third axis C. Complete operation can be expressed symbolically as

$$AB = C \tag{14}$$

Here, we can consider three different cases:

(i) When both the intersecting axes are proper rotations. In this case, eq. 14 will reduce to

$$PP = P \tag{15}$$

where P stands for proper rotations. This case has been considered in sec. 4.6 in detail.

(ii) When one interacting axis is proper and other improper. Under this category, we have two

Table 4.5 Monoaxial combinations of symmetry operations

n	Symmetry operations (Point groups)		
	Proper rotation axis-n	*Improper rotations axis $\bar{n}$*	*Monoaxial combinations $\dfrac{n}{\bar{n}}$*
1	1	$\bar{1}$	$\dfrac{1}{\bar{1}} = \bar{1}$
2	2	$\bar{2} = m$	$\dfrac{2}{\bar{2}} = \dfrac{2}{m}$
3	3	$\bar{3}$	$\dfrac{3}{\bar{3}} = \bar{3}$
4	4	$\bar{4}$	$\dfrac{4}{\bar{4}} = \dfrac{4}{m}$
6	6	$\bar{6} = \dfrac{3}{m}$	$\dfrac{6}{\bar{6}} = \dfrac{6}{m}$
Total	5	5	3 = 13

different possibilities. They are:

$$PI = I$$

and $\qquad\qquad IP = I \qquad\qquad\qquad\qquad\qquad\qquad\qquad\qquad (16)$

(iii) When both the interacting axes are improper rotations. In this case, the first improper rotations will change a right handed object to a left handed (or vice-versa) and the second improper rotation will change the left handed into a right handed object (congruent to the initial object). Thus, the eq. 14 will reduce to

$$II = P \qquad\qquad\qquad\qquad\qquad\qquad\qquad\qquad (17)$$

where I stands for improper rotations. This implies also that $I.I \neq I$.

Equations 14-17 can be used to study the six combinations of these intersecting axes derived earlier. For each of the six polyaxial sets, there is only one possible *PPP* combination. These are given in second column of Table 4.6. However, the combinations under *PII*, *IPI* and *IIP* are not all independent. Distinct possible combinations can be obtained by taking into account the following criteria.

1. If the two of the these axes are symmetry equivalent, they can not be one proper and one improper but they must always be the same. For example, the 3-fold axes in 233 are symmetry equivalent, while diad axes in $n22$ (differ by 45°) are not symmetry equivalent but the proper and improper axes are interchangeable, i.e. $\bar{n}\,2\,\bar{2} \equiv \bar{n}\,\bar{2}\,2$.

Table 4.6   Polyaxial combinations of symmetry operations

Polyaxial combinations	Permissible combination				
	PPP	PII	IPI	IIP	$\left(\dfrac{n_1}{\bar{n}_1}\right)\left(\dfrac{n_2}{\bar{n}_2}\right)\left(\dfrac{n_3}{\bar{n}_3}\right)$
2 2 2	2 2 2	$2\,\bar{2}\,\bar{2}=2\,m\,m$	–	–	$\dfrac{2}{\bar{2}}\dfrac{2}{\bar{2}}\dfrac{2}{\bar{2}}=\dfrac{2}{m}\dfrac{2}{m}\dfrac{2}{m}=m\,m\,m$
3 2 2	3 2 2	$3\,\bar{2}\,(\bar{2})=3m\,(m)$ Called $3m$	–	–	$\dfrac{3}{\bar{3}}\dfrac{2}{\bar{2}}\dfrac{2}{\bar{2}}=\bar{3}\dfrac{2}{m}\dfrac{2}{m}=\bar{3}\dfrac{2}{m}=\bar{3}m$
4 2 2	4 2 2	$4\,\bar{2}\,\bar{2}=4\,m\,m$	$\bar{4}\,2\,\bar{2}=\bar{4}\,2\,m$	–	$\dfrac{4}{\bar{4}}\dfrac{2}{\bar{2}}\dfrac{2}{\bar{2}}=\dfrac{4}{m}\dfrac{2}{m}\dfrac{2}{m}=4/mmm$
6 2 2	6 2 2	$6\,\bar{2}\,\bar{2}=6\,mm$	$\bar{6}\,2\,\bar{2}=\bar{6}\,2\,m$	–	$\dfrac{6}{\bar{6}}\dfrac{2}{\bar{2}}\dfrac{2}{\bar{2}}=\dfrac{6}{m}\dfrac{2}{m}\dfrac{2}{m}=6/mmm$
2 3 3	2 3 3	–	–	–	$\dfrac{2}{\bar{2}}\dfrac{3}{\bar{3}}\dfrac{3}{\bar{3}}=\dfrac{2}{m}\bar{3}\bar{3}=\dfrac{2}{m}\bar{3}=m\,3$
4 3 2	4 3 2	–	$\bar{4}\,3\,\bar{2}=\bar{4}\,3\,m$	–	$\dfrac{4}{\bar{4}}\dfrac{3}{\bar{3}}\dfrac{3}{\bar{2}}=\dfrac{4}{m}\bar{3}=\dfrac{2}{m}=m\,3\,m$
Total	**6**	**4**	**3**	**0**	**6** **= 19**

2. If an even order axis and a $\bar{1}$ axis (or an m plane) co-exist, there will also be an m plane (or a $\bar{1}$ axis) normal to the axis and passing through the intersection point. Conversely, if m and $\bar{1}$ coexist, there will also be a 2-fold (or an even fold) axis passing through center (of inversion) and normal to m.

(c) Coexistence of Proper and Improper Polyaxials

It has been observed that a given lattice is required to be consistent with the different axial symmetries such as n, $\bar{n}$, $\dfrac{n}{\bar{n}}$ etc. and polyaxial combinations of proper and improper axes simultaneously. On a similar ground, the lattice must also be consistent with the polyaxial combinations where the proper and improper axes exist simultaneously, such as $\left(\dfrac{n_1}{\bar{n}_1}\right)\left(\dfrac{n_2}{\bar{n}_2}\right)\left(\dfrac{n_3}{\bar{n}_3}\right)$. Employing the suitable empirical rules and above mentioned criteria, one can easily determine the 19 independent possible polyaxial combinations (Table 4.6). That is, other 19 points groups. From the above discussion of proper and improper rotation axes, it is clear that the axial combinations alone is sufficient to determine the 32 point groups in three dimensions.

4.10 REFLECTION AND INVERSION

The combination of a reflection and an inversion is illustrated in Fig. 4.10. Suppose we start with a right handed object say $1R$. The mirror, m, repeats $1R$ to $2L$ and the inversion center i repeat $2L$ to $3R$. Thus, $1R$ and $3R$ are connected by a 2-fold rotation about the axis A, i.e. A_π. This operation can be described as

$$m.i = A_\pi \tag{18}$$

or $$m.i.A_\pi = 1 \tag{19}$$

This explains the above mentioned criteria (No. 2) which states that the coexistence of mirror and inverse symmetries produce a 2-fold axis passing through the center and normal to m.

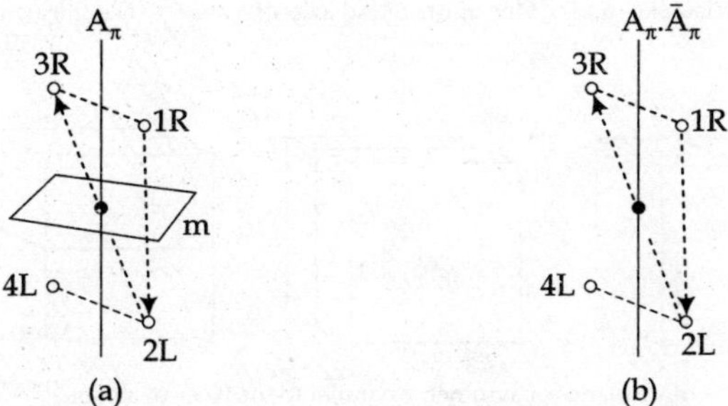

Fig. 4.10 (a) Demonstration that $m.i = A_\pi$ (b) Alternative demonstration of relation in Fig. (a)

Example: Find the total number of symmetry elements that exist in a cube.

Solution: Looking at Fig. 4.11, we find that there exist

(i) 2- fold axes parallel to the face diagonal and passing through the center of the cube. There are six such axes.

(ii) 3- fold rotoinversion axes passing through the body diagonal. There are four such axes.

(iii) 4- fold symmetry about an axis passing through the center of two opposite faces. There are three such symmetry axes.

(iv) Nine (mirror) planes of symmetry, three of them bisecting the parallel faces, and six of them connecting diagonal edges. All of them are passing through the center of the cube.

(v) One center of symmetry at the center of the cube.

Thus, there are:

6 diads + 4 triads + 3 tetrods	= 13 axes
3 surface planes + 6 diagonal planes	= 9 planes
Center of symmetry	= 1

Therefore, the total symmetry elements in a cube is equal to 23. Table 4.7 shows symmetry distribution in each crystal class (point group).

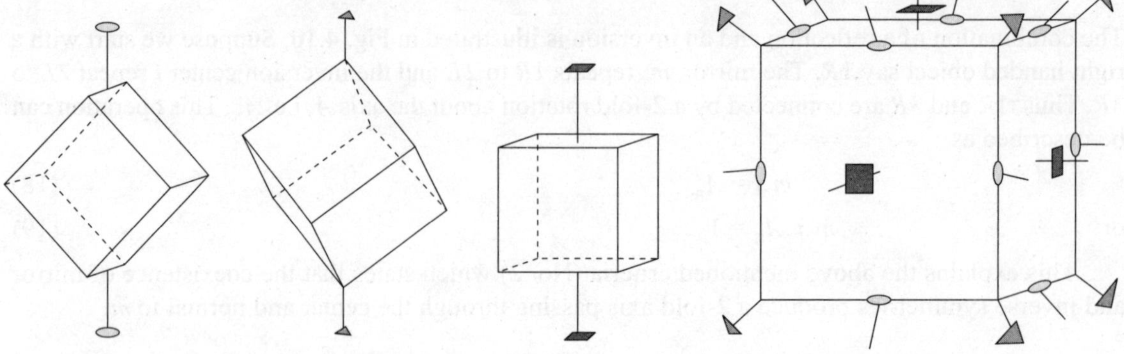

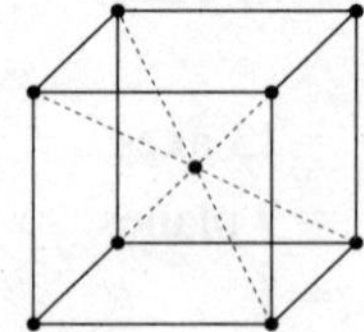

One of the diad-axes   One of the triad axes and   One of the tetrad axes of a cube   The thirteen axes of symmetry shown by a cube

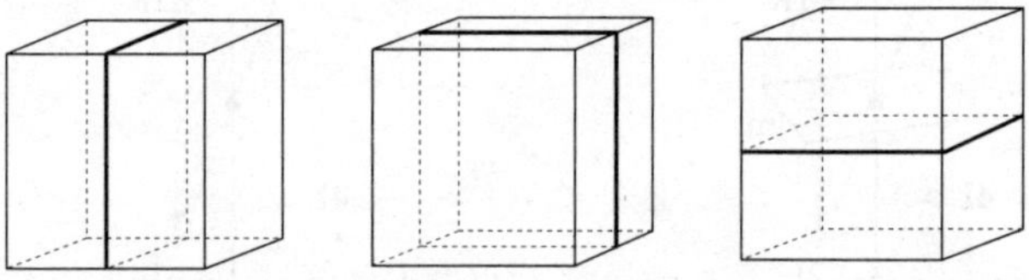

The three planes of symmetry parallel to the faces of a cube.

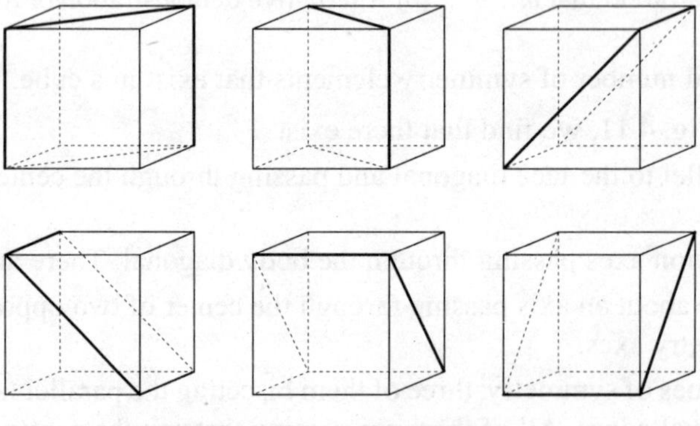

The six diagonal planes of symmetry in a cube

The center of symmetry of a cube

Fig. 4.11   Showing various symmetries of a cube

Table 4.7   Total symmetry of a point group

Crystal system	Point group H-M notation	Symmetry Elements						Total
		Axes				Planes	Center	Symmetry Elements
		2	3	4	6	m	c	Σ
Triclinic	1							0
	$\bar 1$						1	1
Monoclinic	2	1						1
	m					1		1
	$2/m$	1				1	1	3
Orthorhombic	2 2 2	3						3
	$m\,m\,2$	1				2		3
	$2/m\,2/m\,2/m$	3				3	1	7
Trigonal	3		1					1
	$\bar 3$		1				1	2
	3 2	3	1					4
	$3\,m$		1			3		4
	$\bar 3\,2/m$	3	1			3	1	8
Tetragonal	4			1				1
	$\bar 4$			1				1
	$4/m$			1		1	1	3
	4 2 2	4		1				5
	$\bar 4\,m\,m$			1		4	5	5
	$\bar 4\,2\,m$	2		1		2		5
	$4/m\,4/m\,4/m$	4		1		5	1	11
Hexagonal	6				1			1
	$\bar 6$				1	1		2
	$6/m$				1	1	1	3
	6 2 2	6			1			7
	$6\,m\,m$				1	6		7
	$\bar 6\,m\,2$	3			1	3		7
	$6/m\,2/m\,2/m$	6			1	7	1	15
Cubic	2 3	3	4					7
	$2/m\,\bar 3$	3	4			3	1	11
	4 3 2	6	4	3				13
	$\bar 4\,3\,m$		4	3		6		13
	$4/m\,\bar 3\,2/m$	6	4	3		9	1	23

4.11 CLASSIFICATION OF SYMMETRY OPERATIONS

Based on the nature of the final state of an object obtained after some symmetry operations are being performed on it and its relationship with are original state of the object, the symmetry operations can be classified into two broad categories. They are:

(i) The operations of the first sort, and

(ii) The operations of the second sort.

An operation which repeats an object (original state) to become a congruent object (final state) is called an operation of the first sort. Translations and proper rotations are operations of this kind. On the other hand, an operation which repeats an object (original state) to become an enantiomorphous object is called an operation of the second sort. Reflections and inversion are the operations of the second sort. The symmetry operations falling under two categories are provided in Table 4.8.

Table 4.8 Classification of symmetry operations

Symmetry operations		Applicatble to
First sort	*Second sort*	
Proper Rotations	Reflection Inversion Rotoreflection Rotoinversion	Point groups
Translation and Translation + Rotation (Screw axis)	Translation + Reflection (Glide plane)	Space group

4.12 SUMMARY

1. A three dimensional periodic pattern (a crystal) can have the following symmetry elements.

   (i) Translation: $T = n_1 a + n_2 b + n_3 c$, where n_1, n_2, n_3 are integers.

   (ii) Proper rotation: $\alpha = \dfrac{2\pi}{n}$, where $n = 1, 2, 3, 4$ and 6.

   (iii) Reflection: m

   (iv) Inversion center: $\bar{1}$

2. Symmetry elements in crystalline solids can be broadly classified as (i) macroscopic symmetry elements include proper rotations, reflection and inversion, and (ii) microscopic symmetry elements include glide planes and screw axes.

3. In a Cartesian coordinate system, the rotation about z-axis can be represented by a matrix

$$A = \begin{pmatrix} \cos\theta & -\sin\theta & 0 \\ \sin\theta & \cos\theta & 0 \\ 0 & 0 & 1 \end{pmatrix}$$

where θ is the angle of rotation.

4. In three dimensions, there are five proper rotations and six axial combinations possible.

5. Rotoreflection symmetry operations are equivalent in pairs with rotoinversion. The empirical rules for conversion from one system to another are:

$$\tilde{n}_{\text{odd}} = \overline{2n} = \frac{n}{m}$$

$$\widetilde{2n} = \overline{n}_{\text{odd}} = n \cdot i$$

$$\widetilde{4n} = \overline{4n}$$

6. Proper and improper rotations can combine in three different ways. They are:
   (i) Monoaxial combinations
   (ii) Polyaxial combinations
   (iii) Coexistence of proper and improper polyaxials.

4.13 DEFINITIONS

Axial combination: It is a combination of two proper (or improper) rotation axes which intersect each other to form a third axis.

Improper rotation (axes):   They are considered as a combination of symmetry operations of the first kind (e.g. rotation) with any of the second kind (e.g. reflection or inversion).

Inversion center (center of symmetry):   A body has an inversion center if for every point in it there is an identical point equidistant from the center but on the opposite side.

Macroscopic symmetry: It is a symmetry element which is directly related to the external shapes of a freely formed crystal visible to the naked eye.

Rotoinversion symmetry: A body possesses a rotoinversion axis if it can be transformed to self coincidence by the combined effect of rotation and inversion.

Rotoreflection symmetry: A body possesses a rotoreflection axis if it can be transformed to self coincidence by the combined effect of rotation and reflection.

REVIEW QUESTIONS AND PROBLEMS

1. What are the possible symmetry elements in crystalline solid? Explain the existence of characteristic symmetry elements in each crystal system.

2. Find the total number of symmetry elements in each of the fourteen Bravais lattice.

3. What do you understand by macroscopic and microscopic symmetry elements? Point out the compatible combinations of macroscopic symmetry operations.

4. Obtain the expression related to rotation matrix and show that 5-fold and 8-fold rotational symmetries do not exist in a crystal lattice.

5. Describe the theory of axial combinations of two proper rotations and obtain the resulting point group.

6. Describe be rotoreflection and rotoinversion axes. Show that the two sets of improper rotation axes are equivalent in pairs. Deduce empirical rules which can be used for conversion from one system to another.

7. Describe the possible combinations of proper and improper rotations. Show that they follow the basic rules of group theory.

8. Show that the product of two improper rotations is a proper rotations, while the reverse is not true.

9. By means of sketch display the symmetry of space about 5-fold and 8-fold proper rotation axis.

10. Obtains the matrices of 1-, 2-, 3-, 4- and 6-fold rotational symmetries and inversion symmetry.

CHAPTER 5

DERIVATION OF POINT GROUPS

5.1 INTRODUCTION

In the last two chapters, we studied the macroscopic symmetry elements/operations and their compatible combinations in detail. They are found to be related directly with the external shapes of freely formed crystals. Further, the equivalence of rotoreflection and rotoinversion axes in pairs suggests that either of the two improper axes could be used to derive the complete set of point groups in three dimensions. We shall describe both the channels separately in simplified form and obtain the required point groups. In the process, we shall also discuss the two well accepted point group notations based on rotoreflection and rotoinversion axes, respectively.

We shall also develop the concept of matrix representation to understand the simple macroscopic symmetries known as generating elements, with the help of these generating elements, we shall try to understand the derivation of all the point groups in three dimensions. They also help us to understand the equivalence between the number of equivalent positions and the number of symmetry operations in a point group. Further, we shall analyse the existence of Laue symmetry and the relationship among the point groups, crystal classes and crystal systems.

5.2 DERIVATION OF POINT GROUPS (CONVENTIONAL METHOD)

In sections 4.7 and 4.8, we observed that the rotoreflection and rotoinversion axes are equivalent in pairs. This provides us a clue that it is sufficient to consider either rotoreflection axes or rotoinversion axes and not both at the same time to derive the point groups in three dimensions. Accordingly, there are two possible channels available. They are: (i) rotoreflection channel and (ii) rotoinversion channel. Let us consider them one by one

Rotoreflection Channel

In order to derive 32-point groups under rotoreflection channel, let us consider the following symmetry operations:

1. Proper Rotations: X
2. Reflection (mirror plane): m

3. Rotoreflection ($\equiv$ rotoinversion): $\widetilde{X}$

4. Axial combinations: XXX

5. Mirror $\perp$ to Rotation Axis: $\dfrac{X}{m}$

6. Mirror $||$ to Rotation Axis: Xm

7. Mirror $\perp$ to Axial combinations: $\dfrac{XXX}{m}$

8. Mirror $||$ to Axial combinations: $XXX\,m$

Out of the above symmetry operations, we have already discussed many of them in detail earlier in the last chapter. Therefore, their description in the following will be as brief as possible

1. Proper Rotation: X

This symbol includes five proper rotation axes (Fig. 4.1a) i.e. 1, 2, 3, 4 and 6. They are perpendicular to the plane of the paper. They also represent independent point groups.

2. Reflection (Mirror Plane): m

This symbol represents a plane which has the property of dividing the lattice (or object) into two identical halves such that they are mirror images of each other (Fig. 4.1b). This represents an independent point group.

3. Rotoreflection ($\equiv$ Rotoinversion): $\widetilde{X}$

This symbol represents the combined operation of rotation followed by mirror plane perpendicular to the axis i.e. the operations are taking place consecutively. The five rotoreflection axes are:

$$\widetilde{1} = \frac{1}{m} = m = \overline{2},\ \widetilde{2} = \overline{1},\ \widetilde{3} = \frac{3}{m} = \overline{6},\ \widetilde{4} = \overline{4}\ \text{and}\ \widetilde{6} = \overline{3}.$$

This gives us four independent point groups. They are: $\overline{1}$, $\overline{3}$, $\overline{4}$ and $\overline{6}$.

4. Axial Combinations: XXX

This symbol represents the permissible combination of intersecting proper rotation axes. They are: 222, 322 = 32, 422, 622, 233 = 23 and 432 point groups.

5. Mirror $\perp$ to Rotation Axis: $\dfrac{X}{m}$

In this symbol, the combined operation is to place a mirror plane, perpendicular to a given rotation axis i.e. the two operations are taking place concurrently. Thus, the resulting symmetry elements are: $\dfrac{1}{m} = m = \overline{2}$, $\dfrac{2}{m}$, $\dfrac{3}{m} = \overline{6}$, $\dfrac{4}{m}$ and $\dfrac{6}{m}$. This gives us three independent point groups, $\dfrac{2}{m}$, $\dfrac{4}{m}$ and

$\dfrac{6}{m}$. According to the criteria number 2 discussed in section 4.9, all the three symmetry elements will possess center of inversion.

6. Mirror || to Rotation Axis: *Xm*

In this case, a proper rotation axis is combined with a mirror plane parallel to it. The resulting symmetry elements are: $1m$, $2m$, $3m$, $4m$ and $6m$. We have observed earlier in section 3.4 that the combination of a rotation axis of even order (viz. 2, 4 or 6) with a parallel mirror produces additional mirror planes at an angular separation half the order of proper rotation axis. Hence, symmetry elements become $1m$, $2mm$, $3m$, $4mm$, and $6mm$. Excluding the first one, we are left with four independent point groups.

7. Mirror ⊥ to Axial Combinations: $\dfrac{XXX}{m}$

In this symbol, the combined operation is to place a mirror plane perpendicular to each rotation axis involved in the axial combinations. The resulting symmetry elements are: $\dfrac{2}{m}\dfrac{2}{m}\dfrac{2}{m} = mmm$,

$\dfrac{3}{m}m2 = \overline{6}m2$, $\dfrac{4}{m}\dfrac{2}{m}\dfrac{2}{m} = \dfrac{4}{m}mm$, $\dfrac{6}{m}\dfrac{2}{m}\dfrac{2}{m} = \dfrac{6}{m}mm$, $\dfrac{2}{m}\dfrac{3}{m}\dfrac{3}{m} = \dfrac{2}{m}\dfrac{3}{m} = \dfrac{2}{m}\overline{3} = m3$ and

$\dfrac{4}{m}\dfrac{3}{m}\dfrac{2}{m} = \dfrac{4}{m}\overline{3}\dfrac{2}{m} = m3m$. These are all independent members of the $3D$-point groups.

8. Mirror || to Axial combinations: *XXXm*

In this symbol, the combined operation is to place a mirror plane parallel to the combinations 222, 32 and 23. The resulting symmetry elements are: $\overline{4}2m$, $\overline{3}2m$, $\overline{3}m$ and $\overline{4}3m$.

All 32 point groups derived so are presented in Table 5.1 and diagrammatically shown in Fig. 5.1 using Fedorov nomenclature.

Rotoinversion Channel

The derivation of 32 point groups through the rotoinversion channel is somewhat easier and compact than the rotoreflection channel described above. In this case, the simplest way is to determine all the eleven proper rotational point groups first as discussed in section 4.9, where we obtained 5 monoaxial and 6 polyaxial proper rotational point groups. The next step is to multiply each of the eleven point groups by the inversion operator to obtain their corresponding 11centrosymmetric point groups. This has been discussed in section 4.9 under the heading 'coexistence of proper and improper polyaxials' and the results have been summarized in Table 4.5 and 4.6. The concept of multiplication by the inversion operator will be more clear in section 5.6. This way, a total of 22 point groups are obtained.

The eleven centrosymmetric point groups obtained above actually represent the Laue symmetry (section 5.8) and hence they are holohedric. They are also called as the Laue groups because of

Table 5.1   Crystallographic Point Groups Notations

Crystal system	Schoenflies symbol	Hermann Mauguin symbol	Order of group	Laue group
Triclinic	C_1	1	1	
	$C_i(S_2)$	$\bar{1}$	2	$\bar{1}$
Monoclinic	C_2	2	2	
	$C_s(C_h)$	m	2	
	C_{2h}	$2/m$	4	$2/m$
Orthorhombic	D_2	222	4	
	C_{2v}	$mm2$	4	
	$D_{2h}(V_h)$	mmm	8	mmm
Tetragonal	C_4	4	4	
	S_4	$\bar{4}$	4	
	C_{4h}	$4/m$	8	$4/m$
	D_4	422	8	
	C_{4r}	$4mm$	8	
	$D_{2d}(V_d)$	$\bar{4}\,2m$	8	
	D_{4h}	$4/mmm$	16	$4/mmm$
Trigonal	$C3$	3	3	
	$C_{3i}(S_6)$	$\bar{3}$	6	$\bar{3}$
	D_3	32	6	
	C_{3r}	$3m$	6	
	D_{3d}	$\bar{3}\,m$	12	$\bar{3}\,m$
Hexagonal	C_6	6	6	
	C_{3h}	$\bar{6}$	6	
	C_{6h}	$6/m$	12	$6/m$
	D_6	622	12	
	C_{6v}	$6mm$	12	
	D_{3h}	$\bar{6}\,m2$	12	
	D_{6h}	$6/mm$	24	$6/mmm$
Cubic	T	23	12	
	T_h	$m3$	24	
	O	432	24	$m3$
	T_d	$\bar{4}\,3m$	24	
	O_h	$m3m$	48	$m3m$

the presence of inherent inversion axes in them. The presence of 11 centrosymmetric (holohedric) point groups indicates that the remaining 10 point groups will necessarily have the symmetries less than their corresponding holohedries and more than their proper rotational counterparts. That is, the symmetries of the remaining 10 point groups will lie in the intermediate range as shown in Table 5.2.

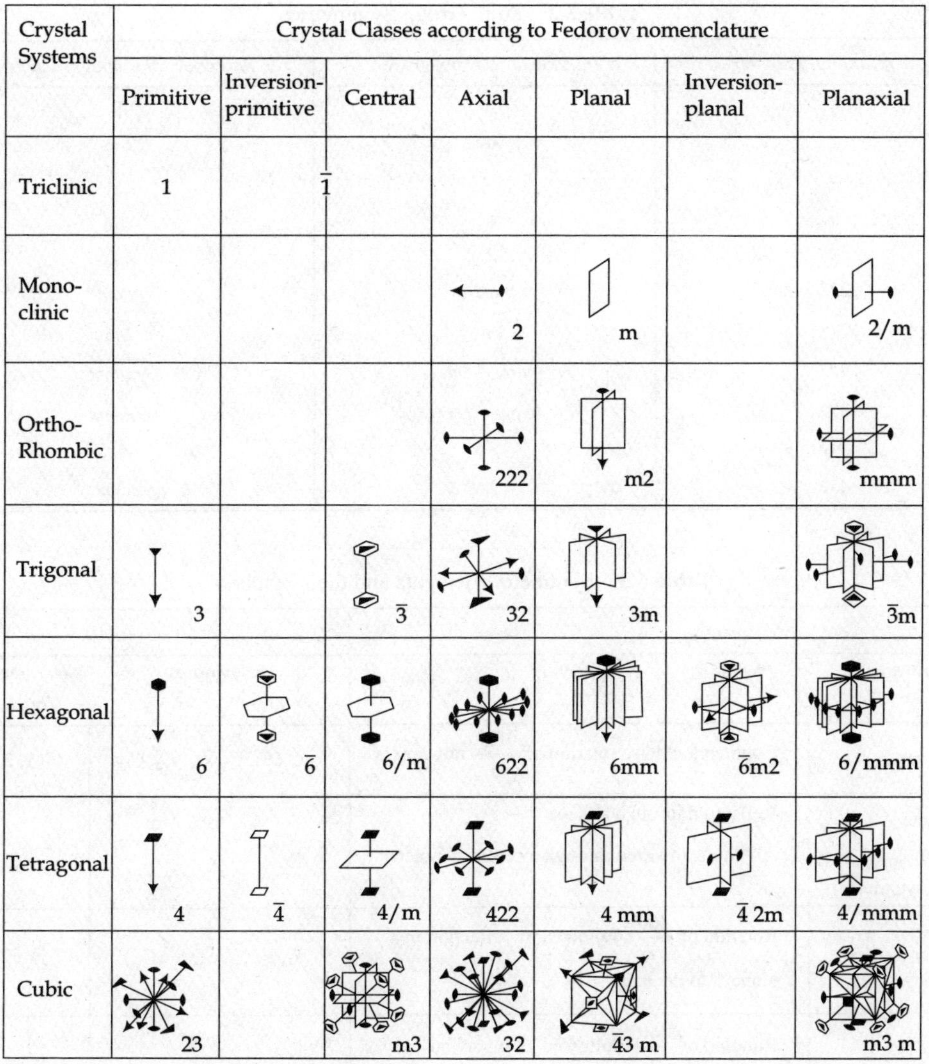

Crystal Systems	Crystal Classes according to Fedorov nomenclature						
	Primitive	Inversion-primitive	Central	Axial	Planal	Inversion-planal	Planaxial
Triclinic	1	$\overline{1}$					
Mono-clinic				2	m		2/m
Ortho-Rhombic				222	m2		mmm
Trigonal	3		$\overline{3}$	32	3m		$\overline{3}$m
Hexagonal	6	$\overline{6}$	6/m	622	6mm	$\overline{6}$m2	6/mmm
Tetragonal	4	$\overline{4}$	4/m	422	4 mm	$\overline{4}$ 2m	4/mmm
Cubic	23		m3	32	$\overline{4}$3 m		m3 m

Fig. 5.1 Three-dimensional view of 32 point groups (crystal classes)

5.3 POINT GROUP NOTATIONS

In the preceding section, we discussed the two channels of point group derivation. They are based on rotoreflection and rotoinversion axes, respectively. The two improper rotation axes are in general represented by two different sets of symbols as provided in Table 5.3. The corresponding point group notations are known as: (i) Schoenflies notation based on rotoreflection axes, and (ii) Hermann Mauguin (also known as 'International') notation based on rotoinversion axes.

Both the point group notations are important on different basis. For example, the chemists and the spectroscopists prefer Schoenflies notation as they find it more systematic to visualize the

Table 5.2 Point Group Symmetries

Proper rotational symmetries	Intermediate symmetries	Improper rotational symmetries
1	–	$\bar{1}$
2	m	$2/m$
3	–	$\bar{3}$
4	$\bar{4}$	$4/m$
6	$\bar{6}$	$6/m$
222	$mm2$	mmm
32(2)	$3m$	$\bar{3}m$
422	$4mm, \bar{4}2m$	$4/mmm$
622	$6mm, \bar{6}2m$	$6/mmm$
23(3)	–	$m3$
432	$\bar{4}3m$	$m3m$

Table 5.3 Symmetry Elements and their Symbols

Symmetry		Symbols	
Element	Operation	Schoenflies	Herman-Mauguin (International)
Rotation axis	Counterclockwise rotation of $\dfrac{360°}{n}$ about axis	$C_1(E), C_2, C_3, C_4, C_6$	1, 2, 3, 4, 6
Mirror plane	Reflection through a plane	σ	m
Centre of Inversion (center of symmetry)	All points inverted through a centre symmetry	i	$\bar{1}$
Rotoreflection axis	Rotation of $\dfrac{360°}{n}$ followed by reflection in a plane $\perp$ to the axis	$S_1, S_2, S_3, S_4, S_5 (C_{3i})$	
Rotoinversion axis	Rotation of $\dfrac{360°}{n}$ followed by inversion through a point on the axis		$\bar{1}, \bar{2}, \bar{3}, \bar{4}, \bar{6}$

molecular symmetries. On the other hand, the solid state physicists and the crystallographers prefer International notation because of the two reasons: (i) It is easily extended to include the translational symmetry elements and (ii) It specifies the directions of the symmetry axes.

Let us briefly describe the general representation of symmetry symbols in the two notations. The symbols used in the cubic case are somewhat specialized.

Schoenflies Notation

1. C_n groups (C = cyclic): A groups is a simple nth order cyclic group of rotation about a single n-fold axis. Its elements are C_n, $C_n^{\,2}$, ... $c_n^{\,n} = E$. For example, the symmetry elements of C_3 group are C_3, C_3^2 and $C_3^3 = E$. They also form an ablian group.

2. Schoenflies represented the rotoreflection (rotation followed by reflection) axes $\tilde{1}$, $\tilde{2}$, $\tilde{3}$, $\tilde{4}$, $\tilde{6}$ by S_1, S_2, S_3, S_4 and S_6. Here $S_1 = C_s$ and $S_2 = C_i$ (centre of inversion). When n is an odd number (e.g. $n = 3$), S_n and C_{nh} are equivalent, but the C_{nh} notation is preferred. The S_6 group is sometime called C_{3i} because its symmetry elements include both, a C_3 axis and a centre of inversion. The S_4 group is a new class whose Hermann-Mauguin equivalent is $\bar{4}$.

3. C_{nh} group (h = horizontal): A C_{nh} group is obtained by placing a mirror plane perpendicular to a C_n axis. For $n = 1$, the symmetry elements are σ and E the group is called C_s. $C_{2h} = C_2 \times C_i$, i.e. the symmetry elements of C_{2h} are: C_2, σ_h, i and E. They also form an abelian group.

4. C_{nv} groups (v = vertical): A C_{nv} group has a C_n rotation axis and $n\sigma_v$ mirror planes containing the rotation axis. For $n = 1$, there is only one mirror plane, and $C_{1v} = C_{1h} = C_s$. A C_{nv} is of order $2n$.

    The above mentioned four groups, i.e. C_n, C_{nh}, C_{nv} and S_n deal with the monoaxial rotational system only, while the remaining groups to be mentioned below deal with the polyaxial rotational systems.

5. D_n group (D = dihedral): A D_n group is obtained by adding a 2-fold axis perpendicular to the principal C_n axis. In analogy to C_{nv}, a D_n group has order $2n$.

6. D_{nh} group: A D_{nh} group is obtained by adding a mirror plane containing all of the 2-fold axes to D_n. This automatically generates n vertical mirror planes. This group can also be obtained as a product of D_n and C_i. That is,

$$D_{nh} = D_n \times C_i \text{ for } n \text{ even}$$
$$= D_n \times C_s \text{ for } n \text{ odd}$$

    This group is of order $4n$

7. D_{nd} group (d = diagonal): A D_{nd} ($n = 2, 3$) group is obtained by adding n vertical mirror planes (σ_d) that bisect the angles between the 2-fold axes in D_n. This group is order $4n$.

    The preceding set of cyclic and dihedral groups are illustrated by particular examples in Fig. 5.2. The remaining point groups that are mentioned below are often referred to collectively as the cubic groups because the symmetry axis and planes occurring in them can be selected from those of a cube. They are illustrated collectively Fig. 5.3 and point group wise in Fig. 5.4.

8. T group (T = tetrahedral): It consists of all the axes of symmetry of a regular tetrahedron. The 2-fold axes pass through the centres of the opposite edges of the cube, whereas the 3-fold axes are formed by the body diagonals of the cube. It is of order 12.

9. T_d group is the complete symmetry group of the regular tetrahedron. It is obtained from T by adding mirror planes, each of which contains one 2-fold axis and one 3-fold axis. Each of these planes contain two diagonally opposite cube edges and the two face diagonals connecting them. It is of order 24.

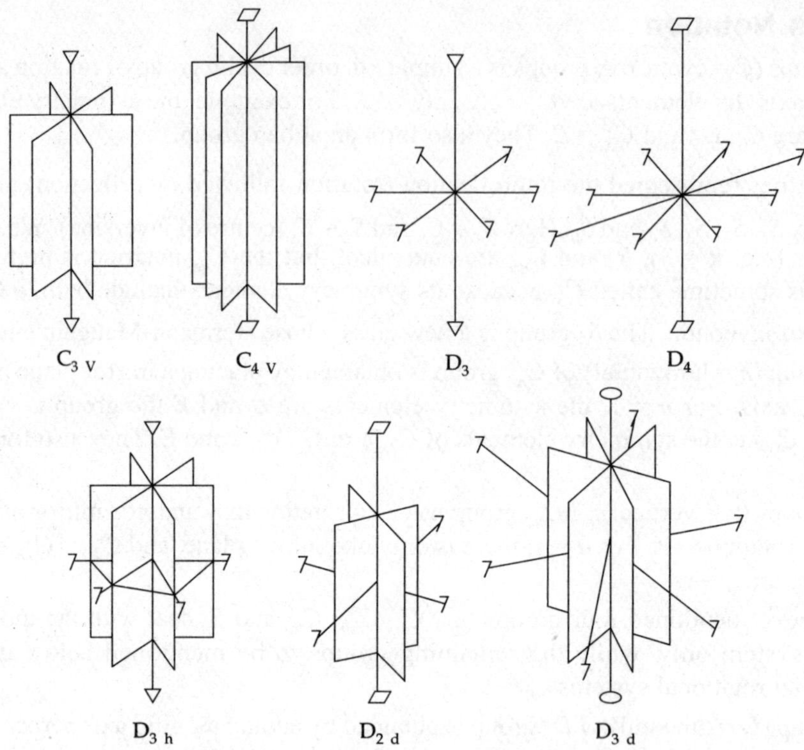

Fig. 5.2 Symmetry axes of some cyclic and dihedral point groups

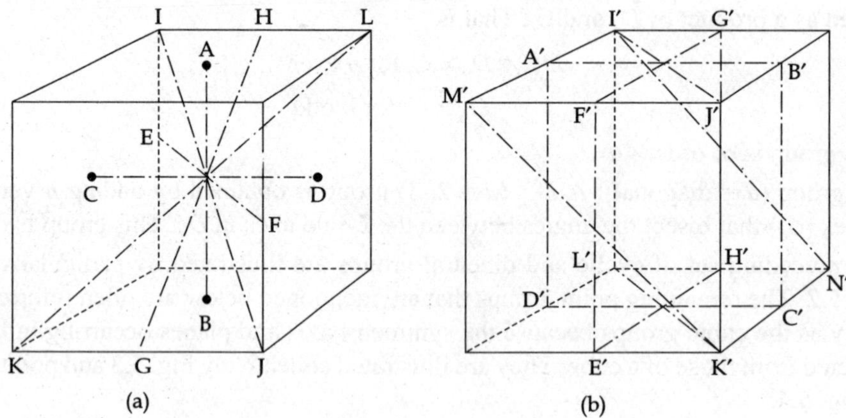

(a)   (b)

Fig. 5.3 Symmetry axes and planes of a cube

10. T_h group is also obtained from T by adding a center of symmetry (C_i). That is $T_h = T \times C_i$. This produces three mirror planes, which divide the cube into the usual octants and turn a proper C_3 axis into an improper C_3 axis. It is of order 24.

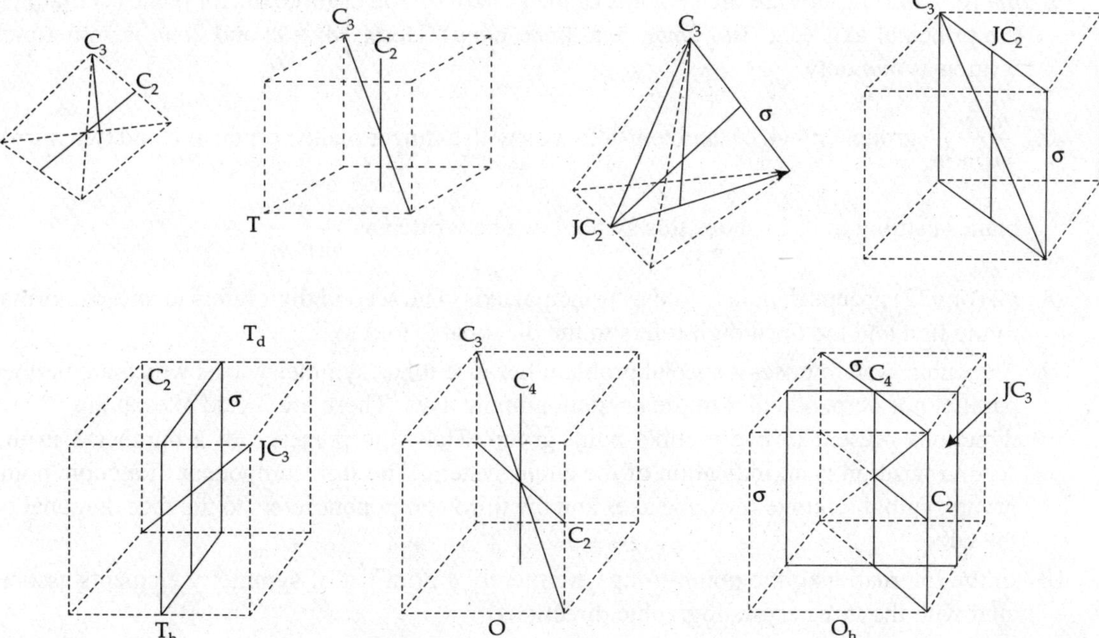

Fig. 5.4 Symmetry planes and axes for the cubic point groups. For T and T_d, the regular tetrahedral are also shown

11. O group (O = octahedral) is the set of proper rotations that carries a regular octahedron into itself. It contains the symmetry axes, C_2, C_3 and C_4 of a cube. It is of order 24.

12. O_h group is the complete symmetry group of a cube. O_h is obtained from O by adding a centre of symmetry (C_i). That is, $O_h = O \times C_i$. It is of order 48.

Hermann-Mauguin (International) Notation

1. Each component of a point group is supposed to refer a direction of its own.

2. n groups (n = 1, 2, 3, 4 and 6) are the cyclic groups, where $n = 1$ is the identity element E. They are known as principal axes.

3. m group (m = mirror plane) is a symbol of mirror plane. The position of m n different point group symbols indicate the directions of the normal to the mirror plane.

4. $\bar{n}$ groups ($\bar{n} = \bar{1}, \bar{2}, \bar{3}, \bar{4}$ and $\bar{6}$) are the rotoinversion axes (rotation followed by inversion) axes, where $\bar{1}$ = center of symmetry, $\bar{2} = m$, etc.

5. $\dfrac{n}{m}$ groups ('n over m') are the n-fold axes with mirror plane normal to them. Here, $\dfrac{1}{m} = m = \bar{2}$ and $\dfrac{3}{m} = \bar{6}$.

6. *nm* (or *nmm*) groups are the symbols of the system of one or more mirror planes containing the principal axis (e.g. 1*m*, 2*mm*, 3*m*, 4*mm*, 6*mm*). Here, 1*m* = *m* and 2*mm* is sometimes written as *mm* only.

7. $\dfrac{n}{m}\dfrac{2}{m}\dfrac{2}{m}$ groups ($n = 4, 6$) denote *n*-fold axis with a mirror plane normal to it and two mirror planes parallel to it. In short, this symbol can be written as $\dfrac{n}{mmm}$.

8. *n2* (or *n22*) groups denote n-fold as principal axis. The second digit refers to an axis normal to the first and the third digit refers to the diagonal 2-fold axis.

9. The cubic system poses a special problem because it has symmetry axes which are neither parallel nor perpendicular to the crystallographic axes. These are 3-fold axes along <111> directions present in every cubic point group. Thus, the presence of a number 3 in the second position is an indication of the cubic system. The first component of a cubic point group symbol refers to the cube axes and the third component refers to the face diagonal of the cube.

10. In the International, the point groups are merely a brief list of symmetry elements associated with the three crystallographic directions.

5.4   LINEAR ORTHOGONAL TRANSFORMATION

We know that the crystallographic coordinate axes x, y, z, are generally used for unambiguous description of planes and directions in crystals. However, the description of physical properties of crystals as well as the analytical representation of their point symmetry groups are based on orthogonal axes say x_1, x_2, x_3. The standard rules for orienting these axes with respect to the crystallographic axes are given in Table 5.4. Both the coordinate systems are always chosen right handed such that if we look them from the end of say x_3 (or z) axis, a rotation from x_1 to x_2 (or x to y) occurs in the counter clockwise direction.

Further, we know that any symmetry operation of a crystal class (a point group) transforms the old set of axes x_1, x_2, x_3, into a new set of axes x'_1, x'_2, x'_3. The angular relationships between the two sets of axes in terms of their directions cosines are provided in Table 5.5, while they have been shown diagramatically in Fig. 5.4. Also coordinates in the two sets of axes are conveniently described by the following linear equations.

$$x'_1 = C_{11}\,x_1 + C_{12}\,x_2 + C_{13}\,x_3$$
$$x'_2 = C_{21}\,x_1 + C_{22}\,x_2 + C_{23}\,x_3 \tag{1}$$
$$x'_3 = C_{31}\,x_1 + C_{32}\,x_2 + C_{33}\,x_3$$

The set of equations can be written in the matrix form as,

$$\begin{pmatrix} x'_1 \\ x'_2 \\ x'_3 \end{pmatrix} = \begin{pmatrix} C_{11} & C_{12} & C_{13} \\ C_{21} & C_{22} & C_{23} \\ C_{31} & C_{32} & C_{33} \end{pmatrix} \begin{pmatrix} x_1 \\ y_1 \\ z_1 \end{pmatrix} \tag{2}$$

Table 5.4 Standard Rules for Axes Orientation

Crystal system	Orthogonal Axes		
	X_3	X_2	X_1
Triclinic	[001]	In the plane $\perp$ to [001]	
Monoclinic	[001]	[010]	In the plane (010)
Orthorhombic	[001]	[010]	[100]
Tetragonal	[001]	[010]	[100]
Trigonal and Hexagonal	[0001]	$[01\bar{1}0] \equiv [120]$	$[\bar{2}1\bar{1}0] \equiv [100]$
Cubic	[001]	[010]	[100]

Table 5.5 Relationships between two sets of axes

Axes Old $\rightarrow$ $\downarrow$New	x_1	x_2	x_3
x_1'	C_{11}	C_{12}	C_{13}
x_2'	C_{21}	C_{22}	C_{23}
x_3'	C_{31}	C_{32}	C_{33}

The eq. 2 in turn can be written in short as

$$x'_i = \sum_{j=1}^{3} C_{ij}\, x_j \; (i = 1, 2, 3) \tag{3}$$

where the nine C_{ij} coefficients are not independent of one another and can be written as

$$C_{ij} = \begin{cases} 1 \text{ for } i = j \\ 0 \text{ for } i \neq j \end{cases} \tag{4}$$

The first scripts in the symbol C_{ij} refers to the new axes and the second to the old ones. The matrix elements C_{ij} may be conveniently recognized as the direction cosines between new and old axes in the right handed coordinate system taken in the order old $\rightarrow$ new as shown in Fig. 5.5. Accordingly, the elements of the matrix C_{ij} in eq. 2 can be written in terms of direction cosines as

$$C_{ij} = \begin{pmatrix} \cos(x_1\, x_1') & \cos(x_1\, x_2') & \cos(x_1\, x_3') \\ \cos(x_2\, x_1') & \cos(x_2\, x_2') & \cos(x_2\, x_3') \\ \cos(x_3\, x_1') & \cos(x_3\, x_2') & \cos(x_3\, x_3') \end{pmatrix} \tag{5}$$

Since we know that an inverse transformation is equally possible, the corresponding matrix C_{ji} is obtained by transposing the elements of the matrix in eq. 2. The values of C_{ij} satisfy the orthogonality relationships.

$$C_{ij}\, C_{kj} = \delta_{ik} = \begin{cases} 1, i = k \\ 0, i \neq k \end{cases} \tag{6}$$

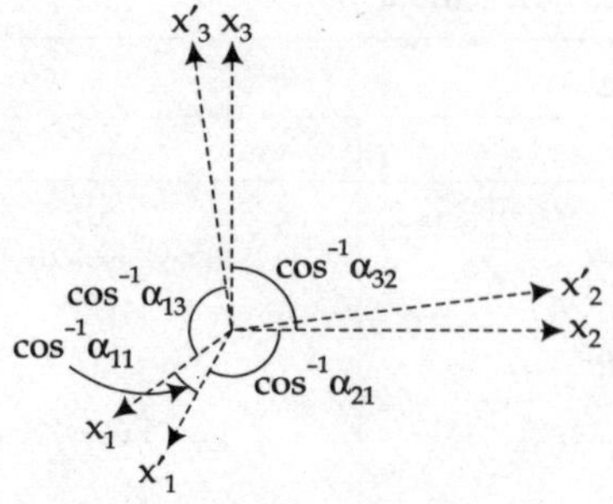

Fig. 5.5 Direction cosines

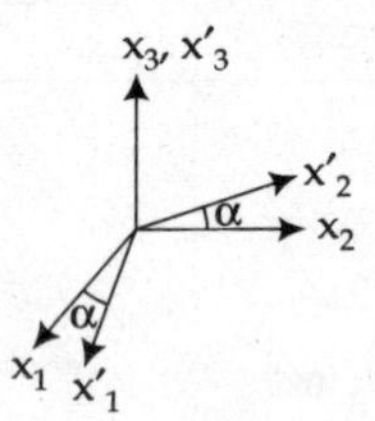

Fig. 5.6 Rotation of axes about x_3 (x'_3) axis

where the symbol δ_{ik} is known as kronecker delta. It is interesting to note that the determinant of the direction cosine matrix, $|C_{ij}|$ is equal to $+1$ for the transformations of the first kind (where the hand of the axes remains unchanged), and to -1 for the transformations of the second kind (where the hand of the axes changes). Making use of the eq. 5, a rotation of an angle α (Fig. 5.6) about any principal axis (say x_3-axis) can be written as

$$C_{ij} = \begin{pmatrix} \cos\alpha & \cos(90+\alpha) & \cos90 \\ \cos(90-\alpha) & \cos\alpha & \cos90 \\ \cos90 & \cos90 & \cos0 \end{pmatrix}$$

which on simplification gives us

$$C_{ij} = \begin{pmatrix} \cos\alpha & -\sin\alpha & 0 \\ \sin\alpha & \cos\alpha & 0 \\ 0 & 0 & 1 \end{pmatrix} \tag{7}$$

From eq.7, one can easily determine the five possible proper rotational symmetries by replacing $\alpha = \dfrac{2\pi}{n}$ (where $n = 1, 2, 3, 4$ and 6).

5.5 SYMMETRY OPERATIONS AND GROUP THEORY

Although we do not necessarily require the group theoretical concept to derive 32 point groups exhibited by crystals, it is useful to appreciate the intimate connection between the symmetry operations (point groups) and the mathematical groups. Here, we shall discuss only certain elementary aspects of the group theory.

Group

A set of symmetry operations forms a group if and only if the following conditions are satisfied.

1. A product of two symmetry operations A and B in a symmetry group is equivalent to a symmetry operation C also to the same group:

$$AB = C$$

where the product AB means the operations B followed by the operation A.

Let us consider a product of symmetry transformations (a transformation which leaves a physical system invariant) with respect to a coordinate system fixed to a crystal. This corresponds to the product of the corresponding matrices just like the standard matrix multiplication. However, the result of the consecutive symmetry transformations will depend on the order in which the operations are applied. In general, $AB \neq BA$. However, if $AB = BA$, the group is said to be commutative or abelian.

2. The multiplication operation is associative

$$AB(C) = A(BC)$$

3. Among the symmetry operations, there exists an operation E called the identity operation such that

$$AE = EA = A$$

The identify operation in terms of symmetry transformation means either no rotation or a rotation of 360° around any arbitrary direction in a crystal.

4. For each operation A, there exists an inverse operation A^{-1} belonging to the same symmetry group such that

$$AA^{-1} = A^{-1}A = E$$

For example, if the operation A is a clockwise rotation by an angle α, then A^{-1} represents the counterclockwise rotation by the angle $(360 - \alpha)$.

Order of the Group

The number of nonequivalent symmetry operations (symmetry operations are equivalent if they give to an identical results; for example the operation of clockwise rotation by 60° and the operation of anticlockwise rotation by 300° are equivalent) forming a symmetry group is called the order of the group. For example, the point group $\dfrac{2}{m}$ has four nonequivalent symmetry operations E, m, 2 and $\bar{1}$. Hence, the order of the point group $\dfrac{2}{m}$ is 4. The character table for the point group $\dfrac{2}{m}$ is provided in Table 5.6.

Table 5.6 The character table of point group $\dfrac{2}{m}$

	E	2	m	$\bar{1}$
E	E	2	m	$\bar{1}$
2	2	E	$\bar{1}$	m
m	m	$\bar{1}$	E	2
$\bar{1}$	$\bar{1}$	m	2	E

Table 5.7 The character table of point group 4

	4	$4^2 (= 2)$	4^3	$4^4 (= E)$
4	2	4^3	E	4
4^2	4^3	E	4	2
4^3	E	4	2	4^3
E	4	2	4^3	E

Cyclic Group

If an operation A belonging to a group transforms a crystal polyhedron into itself, then a repetition of this operation will again transform it into itself. The result of a sequence of symmetry operations is denoted as the power of this operation, such as A^2, A^3, ..., also belong to the same group. Thus, a group generated by a symmetry element is a cyclic group and the element is called the generator of the group. Cyclic groups are abelian but the converse is not true. For a finite group, there exists a finite integer n such that $A^n = E$, called the identity operation. The integer n represents the order of the group. For example, the character table for the point group 4 is provided in Table 5.7.

Generators of a Finite Group

It is possible to generate all the elements of a group by starting from a certain set of elements (at the most three) and taking their powers and products. However, it is to be noted that the definition of the generators is not always unique. For example, in the point group 222 the possible generators are: 2[100] and 2[010], or 2[100] and 2[001] or 2[010] and 2[001].

Subgroups and Super Groups

A set of symmetry elements is said to be a subgroup of a bigger group (is considered as a super group of the group) if the set itself form a group and satisfy the group conditions. In general, every group has two trivial subgroups; the identity element and the group itself. However, in the simplest term it can be said that the addition of symmetry elements to a point group produce super groups while the suppression of symmetry elements from a point group produce subgroups. For example, we know that the point group 1 is the least symmetric and is the sub-group of all other 31 point groups. On the other hand, the point groups $\dfrac{6}{mmm}$ and $m3m$ can have no super groups because it is not possible to obtain any new point group by adding symmetry elements to them.

A subgroup is called proper subgroup if there are symmetry elements of the super group not contained in the subgroup. For example, the point groups 1, 2, $\bar{1}$, m are subgroups of the point group $2/m$.

5.6 MATRIX REPRESENTATION OF SYMMETRY OPERATIONS

The matrix elements in terms of direction cosines obtained in eq. 5 is the basis to understand the matrix representation of any given symmetry operation. The matrices of some fundamental symmetry operations with the help of which all the 32 point groups in three dimensions can be determined are termed as the generating elements. They are ten in numbers including five proper rotation axes, i.e., 1, 2, 3, 4 and 6; one mirror symmetry, *m* and four rotoinversion axes, i.e., $\bar{1}$, $\bar{3}$, $\bar{4}$ and $\bar{6}$.

Let us first obtain the matrices of these generating elements one by one for the two system of axes and then we shall derive all point groups with their help later in the chapter.

(i) Orthogonal Axes

1. 1($\bar{1}$)-fold operation

1-fold operation is equivalent to a no rotation or a rotation of 360° around any direction in a crystal. This operation does not bring about any change in the axes. This gives us $x'_1 = x_1$, $x'_2 = x_2$ and $x'_3 = x_3$ (Fig. 5.7). The matrix corresponding to this operation is obtained either from eq. 5 by substituting the value of direction cosines or directly from eq. 7 with $\alpha = 0$. This is known as identity matrix and is represented as

$$1 \equiv \begin{pmatrix} 1 & 0 & 0 \\ 0 & 1 & 0 \\ 0 & 0 & 1 \end{pmatrix} \tag{8}$$

On the other hand, $\bar{1}$-operation inverts the whole space through a point called the center of symmetry. This operation gives us $x'_1 = -x_1$, $x'_2 = -x_2$ and $x'_3 = -x_3$ as shown in Fig. 5.8. The corresponding matrix is represented as

$$\bar{1} = \begin{pmatrix} -1 & 0 & 0 \\ 0 & -1 & 0 \\ 0 & 0 & -1 \end{pmatrix} \tag{9}$$

This matrix can also be obtained by simply changing the sign of the digits of the matrix in eq. 8.

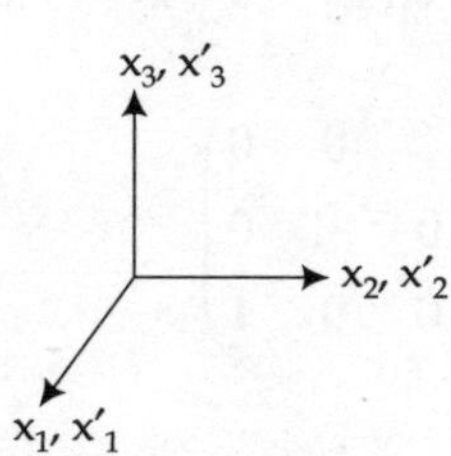

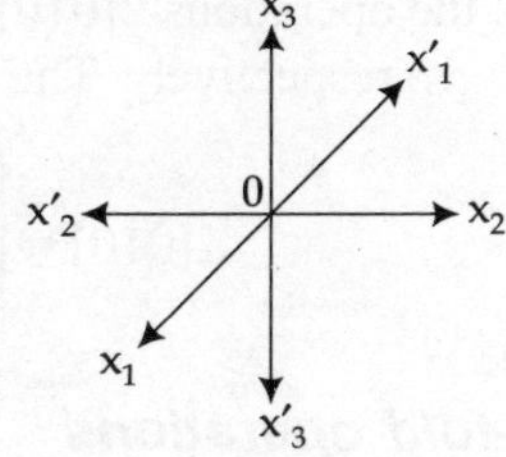

Fig. 5.7

Fig. 5.8 Transformation of the crystallophysical coordinate axes by a center of inversion

2. 2($\bar{2}$)-fold operation

In general, 2 and ($\bar{2}$) fold operations are common to all principal crystallographic directions except [111].

A 2-fold proper rotation along x_3(or z) axis is shown in Fig. 5.9. This suggests that $x'_1 = -x_1$, $x'_2 = -x_2$ and $x'_3 = x_3$. In Miller index notation, this operation is denoted as 2[001]. The corresponding

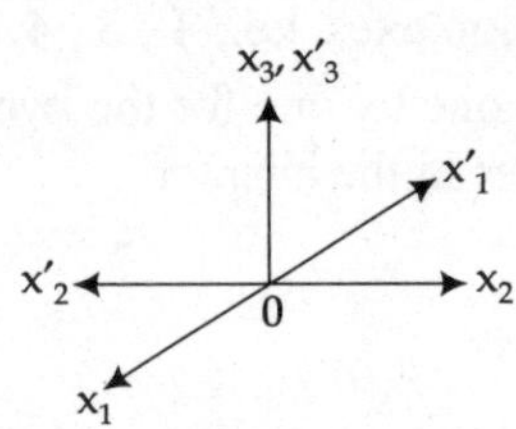

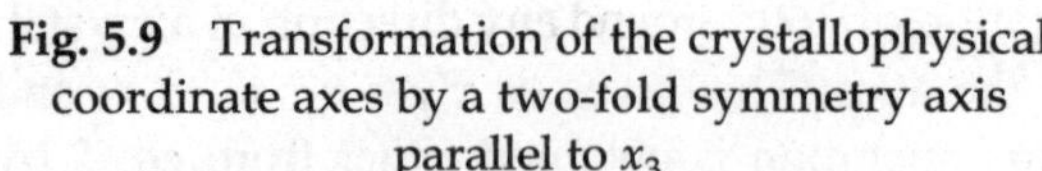

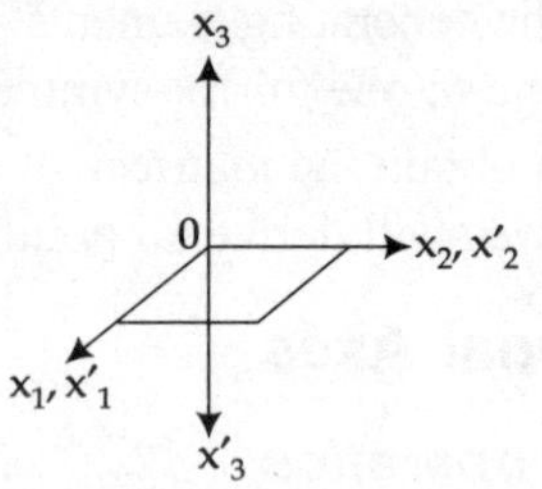

Fig. 5.9 Transformation of the crystallophysical coordinate axes by a two-fold symmetry axis parallel to x_3

Fig. 5.10 Transformation of the crystallophysical coordinate axes by a symmetry plane perpendicular to the axis x_3

matrix can be obtained either from eq. 5 or from eq. 8 directly by changing the sign of the first two axes. This operation is represented as

$$2[001] \equiv \begin{pmatrix} -1 & 0 & 0 \\ 0 & -1 & 0 \\ 0 & 0 & 1 \end{pmatrix} \tag{10}$$

On the other hand, a $\bar{2}$-operation is an improper rotation equivalent to a mirror plane, i.e. $\bar{2} \equiv m$. A mirror plane perpendicular to the $x_3(z)$ axis is shown in Fig. 5.10. This suggests that $x_1' = x_1$, $x_2' = x_2$ and $x_3' = -x_3$.

The corresponding matrix is represented as

$$\bar{2}[001] \equiv m[001] \equiv \begin{pmatrix} 1 & 0 & 0 \\ 0 & 1 & 0 \\ 0 & 0 & -1 \end{pmatrix} \tag{11}$$

Similarly, the operations 2[010] and m [010] give us $x_1' = -x_1$, $x_2' = x_2$, $x_3' = -x_3$ and $x_1' = x_1$, $x_2' = -x_2$, $x_3' = x_3$, respectively. The corresponding matrices are

$$2[010] \equiv \begin{pmatrix} -1 & 0 & 0 \\ 0 & 1 & 0 \\ 0 & 0 & -1 \end{pmatrix} \text{ and } m[010] \equiv \begin{pmatrix} 1 & 0 & 0 \\ 0 & -1 & 0 \\ 0 & 0 & 1 \end{pmatrix} \tag{12}$$

3. 3 and 6-fold operations

A 3-fold operation coupled with the orthogonal axes x_1, x_2, x_3 is shown in Fig. 5.11. A proper rotation about x_3-axis transforms x_1 into x_1' and x_2 into x_2', each of which is thrown 120° away from its initial position. Substituting the values α (= 120°) in eq. 7, the corresponding matrix can be obtained as

$$3[001] = \begin{pmatrix} \cos 120° & -\sin 120° & 0 \\ \sin 120° & \cos 120° & 0 \\ 0 & 0 & 1 \end{pmatrix} = \begin{pmatrix} -\dfrac{1}{2} & -\dfrac{\sqrt{3}}{2} & 0 \\ \dfrac{\sqrt{3}}{2} & -\dfrac{1}{2} & 0 \\ 0 & 0 & 1 \end{pmatrix} \tag{13}$$

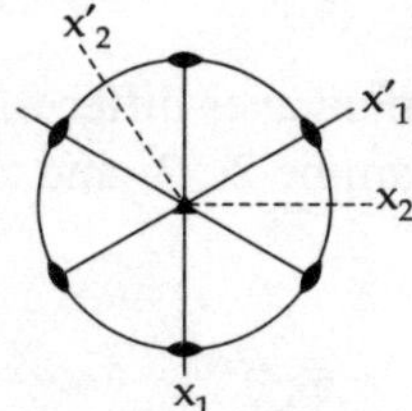

Fig. 5.11   Crystallophysical axes chosen for an equilateral triangle

In a similar manner, the matrix corresponding to a 6-fold rotation is obtained as

$$6[001] = \begin{pmatrix} \cos 60° & -\sin 60° & 0 \\ \sin 60° & \cos 60° & 0 \\ 0 & 0 & 1 \end{pmatrix} = \begin{pmatrix} \dfrac{1}{2} & -\dfrac{\sqrt{3}}{3} & 0 \\ \dfrac{\sqrt{3}}{3} & \dfrac{1}{2} & 0 \\ 0 & 0 & 1 \end{pmatrix} \tag{14}$$

However, the matrix corresponding to a 3-fold rotation along the body diagonal of the cube, i.e. 3[111] can be obtained using eq. 5 as

$$3[111] = \begin{pmatrix} \cos 90° & \cos 90° & \cos 0° \\ \cos 0° & \cos 90° & \cos 90° \\ \cos 90° & \cos 0° & \cos 90° \end{pmatrix} = \begin{pmatrix} 0 & 0 & 1 \\ 1 & 0 & 0 \\ 0 & 1 & 0 \end{pmatrix} \tag{15}$$

4. 4($\bar{4}$)-fold operation

A 4($\bar{4}$)-fold operation is possible along the three crystallographic directions [100], [010] and [001]. These directions are equivalent to orthogonal axes. A 4-fold rotation along [001] transforms the axis x_1 into x_1' and x_2 into x_2', each of which is thrown 90° away from its initial position. This gives us $x_1' = x_2$ and $x_2' = -x_1$. The corresponding matrix can be represented as

$$4[001] = \begin{pmatrix} 0 & -1 & 0 \\ 1 & 0 & 0 \\ 0 & 0 & 1 \end{pmatrix} \tag{16}$$

This matrix can also be obtained from eq. 7 by substituting $\alpha = 90°$. On the other hand, $\bar{4}\,[001]$ can be obtained from eq. 16 by changing the sign of each digit in the matrix, i.e.

$$\bar{4}\,[001] = \begin{pmatrix} 0 & 1 & 0 \\ -1 & 0 & 0 \\ 0 & 0 & -1 \end{pmatrix} \tag{17}$$

(ii) Crystallographic Axes

The orthogonal axes and the crystallographic axes differ for some crystal systems. Let us consider trigonal/hexagonal crystal system and examine 2-, 3- and 6-fold operations under crystallographic axes (Fig. 5.12) separately.

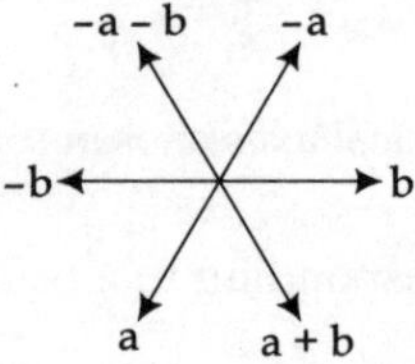

Fig. 5.12   The hexagonal system of axes

1. 2-fold operation

2-fold operations are possible along the three crystallographic axes [100], [010] and [001] as shown in Fig.5.12. A rotation of 180° (π) along [100] gives us $a' = a$, $b' = -a - b$ and $c' = -c$. Therefore, the radius vector

$$
\begin{aligned}
r' &= x_1 a' + x_2 b' + x_3 c' \\
&= x_1 a + x_2(-a - b) + x_3(-c) \\
&= (x_1 - x_2)a + x_2(-b) + x_3(-c)
\end{aligned}
$$

The corresponding matrix can be represented as

$$
2[100] = \begin{pmatrix} 1 & -1 & 0 \\ 0 & -1 & 0 \\ 0 & 0 & -1 \end{pmatrix} \tag{18}
$$

Similarly, we can obtain other matrices. They are given as

$$
2[010] = \begin{pmatrix} -1 & 0 & 0 \\ -1 & 1 & 0 \\ 0 & 0 & -1 \end{pmatrix} \tag{19}
$$

and

$$
2[001] = \begin{pmatrix} -1 & 0 & 0 \\ 0 & -1 & 0 \\ 0 & 0 & 1 \end{pmatrix} \tag{20}
$$

2. 3- and 6-fold operations

In a hexagonal system of axes (Fig. 5.12), a 3-fold axis is possible only along [001] axis. A rotation of 120° $\left(\dfrac{2\pi}{3} \right)$ along [001] direction gives us $a' = b$ and $b' = -a - b$ and $c' = c$.

Therefore, the radius vector

$$r' = x_1 a' + x_2 b' + x_3 c'$$
$$= x_1 b + x_2(-a - b) + x_3 c$$
$$= -x_2 a + (x_1 - x_2) b + x_3 c$$

and the corresponding matrix can be represented as

$$3[001] = \begin{pmatrix} 0 & -1 & 0 \\ 1 & -1 & 0 \\ 0 & 0 & 1 \end{pmatrix} \qquad (16)$$

Like a 3-fold axis, a 6-fold is also possible only along [001] axis. Proceeding in a similar manner, a 6-fold rotation, i.e. a rotation of $60°\left(\dfrac{\pi}{3}\right)$ along [001] will lead us to obtain the matrix

$$6[001] = \begin{pmatrix} 1 & -1 & 0 \\ 1 & 0 & 0 \\ 0 & 0 & 1 \end{pmatrix} \qquad (17)$$

5.7 DERIVATION OF POINT GROUP (MATRIX METHOD)

With the help of generating elements listed in Table 5.8 (for orthogonal axes) or 5.9 (for crystallographic axes) and taking into account the conditions imposed by a group on symmetry operations, we proceed to derive 32 point groups for seven crystal systems, where each crystal system has a particular unit cell shape. We shall start the process with the lowest symmetry crystal system (triclinic) and then progressively proceed on to the highest symmetry crystal system (cubic).

We know that the cell (or cells) of a given crystal system displays a minimum symmetry, which is inherently related to the unit cell shape. This is known as essential symmetry of the unit cell. Similarly, the same cell (cells) has its own characteristic symmetry, which represents the maximum symmetry in a given crystal system. This is referred to as holosymmetry. Therefore the possible symmetry elements of a given crystal system will lie in the range

$$\text{essential symmetry} \leq \text{intermediate symmetry} \leq \text{holosymmetry}$$

Thus in order to derive the possible point groups for a given crystal system, we shall start with the essential (inherent) symmetry element and then go on adding symmetry elements which is compatible with the unit cell shape till the maximum symmetry (holosymmetry) corresponding to the given crystal system is obtained. The results obtained can be verified from Appendix 2 given at the end of this chapter.

1. Triclinic Crystal System

Crystal Classes: $1, \bar{1}$

Both the point groups 1 and $\bar{1}$ of this crystal system are the members of generating elements.

Table 5.8 Generating elements and their matrices (orthogonal axes)

$\text{Identity},\ 1 \equiv \begin{pmatrix} 1 & 0 & 0 \\ 0 & 1 & 0 \\ 0 & 0 & 1 \end{pmatrix}$	$\text{Inversion},\ \bar{1} \equiv \begin{pmatrix} -1 & 0 & 0 \\ 0 & -1 & 0 \\ 0 & 0 & -1 \end{pmatrix}$
$2[001] \equiv \begin{pmatrix} -1 & 0 & 0 \\ 0 & -1 & 0 \\ 0 & 0 & 1 \end{pmatrix}$	$\bar{2}\,[001] = m[001] \equiv \begin{pmatrix} 1 & 0 & 0 \\ 0 & 1 & 0 \\ 0 & 0 & -1 \end{pmatrix}$
$2^{*}[010] \equiv \begin{pmatrix} -1 & 0 & 0 \\ 0 & 1 & 0 \\ 0 & 0 & -1 \end{pmatrix}$	$\bar{2}^{*}[010] = m\,[010] \equiv \begin{pmatrix} 1 & 0 & 0 \\ 0 & -1 & 0 \\ 0 & 0 & 1 \end{pmatrix}$
$3[001] \equiv \begin{pmatrix} -\dfrac{1}{2} & -\dfrac{\sqrt{3}}{2} & 0 \\ \dfrac{\sqrt{3}}{2} & -\dfrac{1}{2} & 0 \\ 0 & 0 & 1 \end{pmatrix},\ \bar{3}\,[001] = \begin{pmatrix} \dfrac{1}{2} & \dfrac{\sqrt{3}}{2} & 0 \\ -\dfrac{\sqrt{3}}{2} & \dfrac{1}{2} & 0 \\ 0 & 0 & 1 \end{pmatrix}$	$3^{*}[111] = \begin{pmatrix} 0 & 0 & 1 \\ 1 & 0 & 0 \\ 0 & 1 & 0 \end{pmatrix}$
$4[001] \equiv \begin{pmatrix} 0 & -1 & 0 \\ 1 & 0 & 0 \\ 0 & 0 & 1 \end{pmatrix}$	$\bar{4}\,[001] \equiv \begin{pmatrix} 0 & 1 & 0 \\ -1 & 0 & 0 \\ 0 & 0 & -1 \end{pmatrix}$
$6[001] \equiv \begin{pmatrix} \dfrac{1}{2} & -\dfrac{\sqrt{3}}{2} & 0 \\ \dfrac{\sqrt{3}}{2} & \dfrac{1}{2} & 0 \\ 0 & 0 & 1 \end{pmatrix}$	$\bar{6}\,[001] \equiv \begin{pmatrix} -\dfrac{1}{2} & \dfrac{\sqrt{3}}{2} & 0 \\ -\dfrac{\sqrt{3}}{2} & -\dfrac{1}{2} & 0 \\ 0 & 0 & -1 \end{pmatrix}$

*Supplementary generating elements

2. Monoclinic Crystal System

Crystal Classes: $2,\ m,\ \dfrac{2}{m}$

Under monoclinic crystal system, the point groups 2 and $\bar{2}$ ($\equiv m$) are the members of the generating elements. However, the third point group $\dfrac{2}{m}$ can be obtained by placing a mirror plane perpendicular to the 2-fold symmetry axis. In the matrix representation, this can be written as

Table 5.9 Generating elements and their matrices (crystallographic axes)

Identity, $1 \equiv \begin{pmatrix} 1 & 0 & 0 \\ 0 & 1 & 0 \\ 0 & 0 & 1 \end{pmatrix}$	Inversion $\bar{1} \equiv \begin{pmatrix} -1 & 0 & 0 \\ 0 & -1 & 0 \\ 0 & 0 & -1 \end{pmatrix}$
$2[001] \equiv \begin{pmatrix} -1 & 0 & 0 \\ 0 & -1 & 0 \\ 0 & 0 & 1 \end{pmatrix}$	$\bar{2}\,[001] = m[001] \equiv \begin{pmatrix} 1 & 0 & 0 \\ 0 & 1 & 0 \\ 0 & 0 & -1 \end{pmatrix}$
$3\,[001] \equiv \begin{pmatrix} 0 & -1 & 0 \\ 1 & -1 & 0 \\ 0 & 0 & 1 \end{pmatrix},\ \bar{3}\,[001] \equiv \begin{pmatrix} 0 & 1 & 0 \\ -1 & 1 & 0 \\ 0 & 0 & -1 \end{pmatrix}$	$3^{*}[111] \equiv \begin{pmatrix} 0 & 0 & 1 \\ 1 & 0 & 0 \\ 0 & 1 & 0 \end{pmatrix}$
$4[001] \equiv \begin{pmatrix} 0 & -1 & 0 \\ 1 & 0 & 0 \\ 0 & 0 & 1 \end{pmatrix}$	$\bar{4}\,[001] \equiv \begin{pmatrix} 0 & 1 & 0 \\ -1 & 0 & 0 \\ 0 & 0 & -1 \end{pmatrix}$
$H \rightarrow 2^{*}[010] \equiv \begin{pmatrix} -1 & 0 & 0 \\ -1 & 1 & 0 \\ 0 & 0 & -1 \end{pmatrix}$	$H \rightarrow \bar{2}^{*}[010] = m\,[010] = \begin{pmatrix} 1 & 0 & 0 \\ 1 & -1 & 0 \\ 0 & 0 & 1 \end{pmatrix}$
$6[001] \equiv \begin{pmatrix} 1 & -1 & 0 \\ 1 & 0 & 0 \\ 0 & 0 & 1 \end{pmatrix}$	$\bar{6}\,[001] = \begin{pmatrix} -1 & 1 & 0 \\ -1 & 0 & 0 \\ 0 & 0 & -1 \end{pmatrix}$

*Supplementary generating elements

$$\frac{2}{m} = 2[010] \times m[010] = \begin{pmatrix} -1 & 0 & 0 \\ 0 & 1 & 0 \\ 0 & 0 & -1 \end{pmatrix} \begin{pmatrix} 1 & 0 & 0 \\ 0 & -1 & 0 \\ 0 & 0 & 1 \end{pmatrix}$$

$$= \begin{pmatrix} -1 & 0 & 0 \\ 0 & -1 & 0 \\ 0 & 0 & -1 \end{pmatrix} = I(\text{inversion})$$

3. Orthorhombic Crystal System

Crystal classes: 222, *mm2*, *mmm*

(i) Crystal class: 222

The crystal class 222 is an essential symmetry of this crystal system. In this case, all the 2-fold axes are mutually perpendicular to each other. Let us consider the matrix multiplication of two mutually perpendicular 2-fold axes.

$$2[001] \times 2[010] = \begin{pmatrix} -1 & 0 & 0 \\ 0 & -1 & 0 \\ 0 & 0 & 1 \end{pmatrix} \begin{pmatrix} -1 & 0 & 0 \\ 0 & 1 & 0 \\ 0 & 0 & -1 \end{pmatrix}$$

$$= \begin{pmatrix} 1 & 0 & 0 \\ 0 & -1 & 0 \\ 0 & 0 & -1 \end{pmatrix} = 2[100]$$

The resulting 2-fold axis is perpendicular to the first two 2-fold axes.

(ii) Crystal class: mm2

In this case, let us place one mirror plane perpendicular to either 2[100] or 2[010] axis without affecting the unit cell geometry. Therefore, let us consider the product

$$2[001] \times m[100] = \begin{pmatrix} -1 & 0 & 0 \\ 0 & -1 & 0 \\ 0 & 0 & -1 \end{pmatrix} \begin{pmatrix} -1 & 0 & 0 \\ 0 & 1 & 0 \\ 0 & 0 & 1 \end{pmatrix}$$

$$= \begin{pmatrix} 1 & 0 & 0 \\ 0 & -1 & 0 \\ 0 & 0 & 1 \end{pmatrix} = m[010]$$

The resulting mirror plane is parallel to 2[100] and 2[001] while perpendicular to m[100]

(iii) Crystal class: mmm

Like monoclinic case, let us place one mirror plane perpendicular to each of the three (mutually perpendicular) 2-fold axes to obtain the point group $\dfrac{2}{m}\dfrac{2}{m}\dfrac{2}{m}$. Similarly, the matrix multiplications of 2[100] × m[100] and 2[001] × m[001] provide us the same result as given by the product 2[010] × m[010] worked out earlier in monoclinic case. Also, the product of all the three operations does not bring about any change. In short, this point group is written as mmm indicating that three mirror planes are mutually perpendicular to each other.

4. Tetragonal Crystal System

Crystal classes: 4, $\bar{4}$, $\dfrac{4}{m}$, 4mm, 422, $\bar{4}2m$, $\dfrac{4}{mmm}$

The point group 4 and $\bar{4}$ are essential symmetry elements and also the members of the generating elements. Other point groups can be derived with their help.

(i) Crystal class: $\dfrac{4}{m}$

Let us place a mirror plane perpendicular to the 4-fold axis to get the point group $4/m$. The corresponding matrix representation is given as

$$4[001] \times m[001] = \begin{pmatrix} 0 & -1 & 0 \\ 1 & 0 & 0 \\ 0 & 0 & 1 \end{pmatrix} \begin{pmatrix} 1 & 0 & 0 \\ 0 & 1 & 0 \\ 0 & 0 & -1 \end{pmatrix}$$

$$= \begin{pmatrix} 0 & -1 & 0 \\ 1 & 0 & 0 \\ 0 & 0 & -1 \end{pmatrix} = \bar{4}^{3}[001]$$

The resulting symmetry element clearly shows inversion effect.

(ii) Crystal class: 4mm

This point group can be obtained by placing a mirror plane parallel to the 4-fold axis (or equivalently perpendicular to either 2[100] or 2[010] axis). We can write the corresponding matrix multiplication as

$$4[001] \times m[010] = \begin{pmatrix} 0 & -1 & 0 \\ 1 & 0 & 0 \\ 0 & 0 & 1 \end{pmatrix} \begin{pmatrix} 1 & 0 & 0 \\ 0 & -1 & 0 \\ 0 & 0 & 1 \end{pmatrix}$$

$$= \begin{pmatrix} 0 & 1 & 0 \\ 1 & 0 & 0 \\ 0 & 0 & 1 \end{pmatrix} = m[1\bar{1}0]$$

This operation provides another set of mirror plane 45° away from the first set.

(ii) Crystal class: 422

This point group is obtained by placing a 2-fold axis perpendicular to 4-fold axis either in the [010] or [100] direction. The corresponding matrix multiplication is written as

$$4[001] \times 2[010] = \begin{pmatrix} 0 & -1 & 0 \\ 1 & 0 & 0 \\ 0 & 0 & 1 \end{pmatrix} \begin{pmatrix} -1 & 0 & 0 \\ 0 & 1 & 0 \\ 0 & 0 & -1 \end{pmatrix}$$

$$= \begin{pmatrix} 0 & -1 & 0 \\ -1 & 0 & 0 \\ 0 & 0 & -1 \end{pmatrix} = 2[1\bar{1}0]$$

This creates a 2-fold symmetry along the face diagonal, i.e. along $[1\bar{1}0]$ direction.

(iii) Crystal class: $\bar{4}\,2m$

This point group can be obtained by placing a 2-fold axis perpendicular to the $\bar{4}$-fold axis. In matrix notation, this can be written as

$$\bar{4}\,[001] \times 2[010] = \begin{pmatrix} 0 & 1 & 0 \\ -1 & 0 & 0 \\ 0 & 0 & -1 \end{pmatrix} \begin{pmatrix} -1 & 0 & 0 \\ 0 & 1 & 0 \\ 0 & 0 & -1 \end{pmatrix}$$

$$= \begin{pmatrix} 0 & 1 & 0 \\ 1 & 0 & 0 \\ 0 & 0 & 1 \end{pmatrix} = m[1\,\bar{1}\,0]$$

This introduces a mirror plane perpendicular to $[1\,\bar{1}\,0]$ direction.

(iv) Crystal class: $\dfrac{4}{m\,m\,m}$

This point group can be obtained when we place a mirror plane perpendicular to each of the 4-, 2-, and 2-fold axes in the point group 422. The corresponding point group is normally represented as

$\dfrac{4}{m}\,\dfrac{2}{m}\,\dfrac{2}{m}$. In short, this point group is usually written as $\dfrac{4}{m\,m\,m}$.

5. Trigonal Crystal System

Crystal classes: 3, $\bar{3}$, 32, 3m, $\bar{3}\,m$

The point groups 3 and $\bar{3}$ are the essential symmetry elements and also the members of the generating elements. With their help, we can derive rest of the point groups of this crystal system. It is to be mentioned here that separate transformation matrix exits for crystallographic and orthogonal axes. We shall use the transformation matrix corresponding to crystallographic axes. However, the use of orthogonal transformation matrix will also lead to the same result.

(i) Crystal class: 32

This point group can be obtained by placing a 2-fold axis perpendicular to a 3-fold axis either in the [010] or [100] direction. The corresponding matrix multiplication in terms of crystallographic axes is written as

$$3[001] \times 2[010] = \begin{pmatrix} 0 & -1 & 0 \\ 1 & -1 & 0 \\ 0 & 0 & 1 \end{pmatrix} \begin{pmatrix} -1 & 0 & 0 \\ -1 & 1 & 0 \\ 0 & 0 & -1 \end{pmatrix}$$

$$= \begin{pmatrix} 1 & -1 & 0 \\ 0 & -1 & 0 \\ 0 & 0 & -1 \end{pmatrix} = 2[100]$$

This produces a 2-fold axis perpendicular to [010] direction. A similar result can be obtained when the matrices of orthogonal axes are considered. Here, it is to be pointed out that a reverse matrix multiplication of two 2-fold crystallographic axes will produce a 3-fold axis according to the relation

$$2[100] \times 2[010] = 3[001]$$

(ii) Crystal class: 3m

Similar to the crystal class 32, this can be obtained by placing a mirror plane instead of a 2-fold axis. However, the mirror plane is parallel to 3-fold axis and perpendicular to either [010] or [100] axis. The corresponding matrix multiplication in terms of crystallographic axes is given by

$$3[001] \times m[010] = \begin{pmatrix} 0 & -1 & 0 \\ 1 & -1 & 0 \\ 0 & 0 & 1 \end{pmatrix} \begin{pmatrix} 1 & 0 & 0 \\ 1 & -1 & 0 \\ 0 & 0 & 1 \end{pmatrix}$$

$$= \begin{pmatrix} -1 & 1 & 0 \\ 0 & 1 & 0 \\ 0 & 0 & 1 \end{pmatrix} = m[100]$$

This introduces a mirror plane perpendicular to [100] axis.

(iii) Crystal class: $\bar{3}m$

This point group can be obtained by placing a 2-fold axis perpendicular to $\bar{3}[001]$ axis and is directed along [010] axis. The resulting point group is expressed as $\bar{3}\dfrac{2}{m}$, which in short is usually written as $\bar{3}m$. The corresponding matrix multiplication in terms of crystallographic axes is written as

$$\bar{3}[001] \times 2[010] = \begin{pmatrix} 0 & 1 & 0 \\ -1 & 1 & 0 \\ 0 & 0 & -1 \end{pmatrix} \begin{pmatrix} -1 & 0 & 0 \\ -1 & 1 & 0 \\ 0 & 0 & -1 \end{pmatrix}$$

$$= \begin{pmatrix} -1 & 1 & 0 \\ 0 & 1 & 0 \\ 0 & 0 & 1 \end{pmatrix} = m[100]$$

6. Hexagonal Crystal System

Crystal class: 6, $\bar{6}$, $\dfrac{6}{m}$, $6mm$, 622, $\bar{6}2m$, $\dfrac{6}{mmm}$.

The point groups 6 and $\bar{6}$ are essential symmetry elements and also the members of the generating elements. Other point groups of this crystal system can be derived with their help. It is to be pointed out that the point groups of the hexagonal crystal system can also be derived exactly on a similar line as the point groups of tetragonal crystal system by simply replacing the rotational symmetry 4 by 6.

7. Cubic Crystal System

Crystal classes: 23, $m3$, $\bar{4}3m$, 43, $m3m$

The cubic crystal system is relatively more difficult to treat because of the absence of any principal axis in it. However, this crystal system is defined by the presence of four 3-fold axes (inclined at an angle of $109° 28'$ with respect to each other) along the cube diagonals. Therefore, the additional symmetry operations could be combined to these axes in such a manner that the basic nature of the cubic unit cell should not be disturbed.

(i) Crystal class: 23

This point group can be obtained when a 2-fold axis about a-, b- or c-axis is combined with any of the four 3-fold axes. For example,

$$2[100] \times 3[111] = \begin{pmatrix} 1 & 0 & 0 \\ 0 & -1 & 0 \\ 0 & 0 & -1 \end{pmatrix} \begin{pmatrix} 0 & 0 & 1 \\ 1 & 0 & 0 \\ 0 & 1 & 0 \end{pmatrix}$$

$$= \begin{pmatrix} 0 & 0 & 1 \\ -1 & 0 & 0 \\ 0 & -1 & 0 \end{pmatrix} = 3^2[1\,\bar{1}\,1]$$

The resulting point group is expressed as 233, which in short is usually written as 23. It is to be pointed out that one should not get confused this point group with the trigonal point group 32 in the international notation.

(ii) Crystal class: m3

This point group can be obtained by placing a mirror plane perpendicular to any of the three 2-fold axes. The resulting point group is expressed as $\dfrac{2}{m}\bar{3}$, which in short is usually written as $m3$. One such matrix multiplication is given by

$$m[100] \times 3^2[1\,\bar{1}\,1] = \begin{pmatrix} -1 & 0 & 0 \\ 0 & 1 & 0 \\ 0 & 0 & 1 \end{pmatrix} \begin{pmatrix} 0 & 0 & 1 \\ -1 & 0 & 0 \\ 0 & -1 & 0 \end{pmatrix}$$

$$= \begin{pmatrix} 0 & 0 & -1 \\ -1 & 0 & 0 \\ 0 & -1 & 0 \end{pmatrix} = \overline{3}\,[111]$$

Again, it is to be pointed out that one should not get confused the cubic point $m3$ with the trigonal point group $3m$.

(iii) Crystal class: $\overline{4}3m$

This point group is the result of the combination of $\overline{4}\,[001]$ with any of the four $3[111]$ axes. In the matrix notation, one such combination can be represented as

$$\overline{4}\,[001] \times 3[111] = \begin{pmatrix} 0 & -1 & 0 \\ 1 & 0 & 0 \\ 0 & 0 & -1 \end{pmatrix} \begin{pmatrix} 0 & 1 & 0 \\ 0 & 0 & 1 \\ 1 & 0 & 0 \end{pmatrix}$$

$$= \begin{pmatrix} 0 & 0 & -1 \\ 0 & 1 & 0 \\ -1 & 0 & 0 \end{pmatrix} = m[101]$$

(iv) Crystal class: 43

This point group is obtained when the $4[001]$ axis is combined with any of the four $3[111]$ axes. One such combination in matrix notation is represented as

$$4[001] \times 3[111] = \begin{pmatrix} 0 & 1 & 0 \\ -1 & 0 & 0 \\ 0 & 0 & 1 \end{pmatrix} \begin{pmatrix} 0 & 1 & 0 \\ 0 & 0 & 1 \\ 1 & 0 & 0 \end{pmatrix}$$

$$= \begin{pmatrix} 0 & 0 & 1 \\ 0 & -1 & 0 \\ 1 & 0 & 0 \end{pmatrix} = 2[101]$$

The resulting point group is expressed as 432, which in short is usually written as 32.

(v) Crystal class: $m3m$

This point group is obtained by placing mirror planes perpendicular to the 4-fold axes. The resulting point group is expressed as $\dfrac{4}{m}\,3\,\dfrac{2}{m}$, which in short is usually written as $m3m$.

All the 32 point group along with their generating elements are provided in Table 5.10.

5.8 EQUIVALENT POSITIONS IN POINT GROUPS

In the field of crystallography, we are very much familiar with the terms like crystallographically equivalent planes and crystallographically equivalent directions, especially in a symmetric cubic

Table 5.10 Point groups and their generating elements

Triclinic	(1)	($\bar{1}$)					
	1	$\bar{1}$					
Monoclinic	(2)	(*m*)	(2/*m*)				
	2[010]	*m*[010]	$\bar{1}$, 2[010]				
Orthorhombic	(222)	(*mm*2)	(*mmm*)				
	2[001], 2[010]	*m*[100], 2[001]	2[001], $\bar{1}$ 2[010]				
Rhombohedral	(3)	($\bar{3}$)	(32)	(3*m*)	($\bar{3}m$)		
	3[001]	$\bar{3}$[001]	3[001], 2[010]	3[001] *m*[010]	$\bar{3}$[001] 2[010]		
Tetragonal	(4)	($\bar{4}$)	(4/*m*)	(42)	(4*mm*)	4/*mmm*	($\bar{4}2m$)
	4[001]	$\bar{4}$[001]	4[001], $\bar{1}$	4[001], 2[010]	4[001], *m*[010]	4[001] $\bar{1}$, 2[010]	$\bar{4}$[001], 2[010]
Hexagonal	(6)	($\bar{6}$)	(6/*m*)	(62)	(6*mm*)	(6/*mmm*)	($\bar{6}2m$)
	6[001]	$\bar{6}$[001]	6[001], $\bar{1}$	6[001], 2[010]	6[001], *m*[010]	6[001], $\bar{1}$, 2[010]	$\bar{6}$[001], 2[010]
Cubic	(23)	(*m*3)	(43)	($\bar{4}3m$)	(*m*3*m*)		
	2[001] 3[111]	2[001] 3[111], $\bar{1}$	4[001] 3[111]	$\bar{4}$[001] 3[111]	4[001], 3[111], $\bar{1}$		

crystal system. In a similar manner, there exist crystallographically equivalent positions (or points) which are obtained from the successive application of symmetry operations of the point groups. The number of equivalent positions are found to vary from 1 to 48 for 32 point groups. They correspond to the least symmetric point group 1 and the most symmetric point group *m*3*m*

$$\left(\equiv \frac{4}{3}\bar{3}\frac{2}{m} \right),$$ respectively.

The point group 1 provides no equivalent position. However, for a point (x, y, z) in space, the point group 2 (taken along z-axis) provides an equivalent point group at $(-x, -y, z)$, the point group $m(\perp$ to $z)$ at $(x, y, -z)$, the point group $\bar{1}$ at $(-x, -y, -z)$ as shown in Fig. 5.13. Similarly, applying various symmetry operations of a points group, the corresponding total number of equivalent positions can be easily obtained. The number of equivalent positions (or equivalently the number of non-equivalent symmetry operations) corresponding to a given point group is termed as the 'order' of the group. This is provided in Table 5.1.

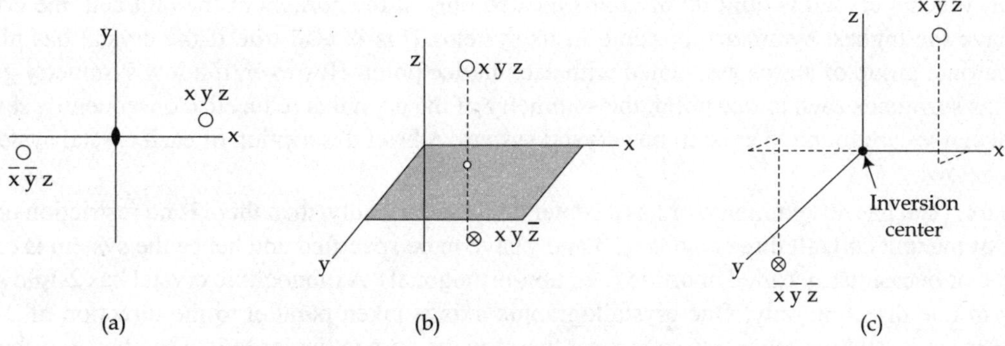

Fig. 5.13 Equivalent positions for point groups 2, *m* and $\bar{1}$

5.9 LAUE SYMMETRY

It is interesting to know that the diffraction pattern obtained from a single crystal (in the absence of anomalous diffraction) is centrosymmetric even if the crystal belongs to a non-centrosymmetric class. This is because of the fact that the *X*-ray (electron/neutron) diffraction automatically introduces a center of symmetry to the point group of the crystal. The diffraction symmetry thus help us in classifying the 32 crystal classes (point groups) in to 11 centrosymmetric classes which for historical reason are known as 11 Laue groups, Laue classes or Laue symmetry as enlisted in Table 5.1. Consequently, it means that each Laue class is inherently associated with its non-centrosymmetric partners of a crystal system. The diffraction symmetry also helps us in identifying a crystal system.

The classification of 32 point groups in terms of monoaxial and poly axial axes makes it easy to visualize the existence of 11 Laue classes, because we observe that a given crystal system possesses either one Laue class contributed by monoaxial point groups only or two Laue classes, one each contributed by monoaxial and polyaxial point groups.

5.10 POINT GROUPS, CRYSTAL CLASSES AND CRYSTAL SYSTEMS

We know that a point group can be defined as a group of symmetry operations which when applied to a crystal lattice, it leaves at least one lattice point unmoved. Earlier, we obtained 32 point groups using different methods. On the other hand, a crystal calss in geometrical sense is defined as the set of all crystals having the same point group symmetry. Interestingly, from a different approach altogether, the number of crystal classes also have been found to be 32.

Historically, the characterization of crystals began with the study of their external faces. Further, drawing normals to each face from a common point it has been shown that all crystals could be classified as belonging to one of the 32 crystal classes, where each of these crystal classes correspond to one of the 32 point groups. Therefore in crystallography, the terms like crystal classes and point groups are used synonymously.

As far as the applications of point groups are concerned, they are not merely limited to the morphology and other macroscopic studies of crystals (as prima-facie appears to be) but they also describe the symmetry of the group of atoms (or molecules) that surrounds each lattice point in a crystal. When a crystal is built up of atoms located only at the corners of the unit cell, the crystal will have the highest symmetry possible in its system. This is also true if the crystal has highly symmetrical group of atoms associated with each lattice point. However, if a low symmetry group of atoms surrounds each lattice point, the symmetry of the crystal is reduced. Consequently, several crystal classes are found to exist in one crystal system. A brief description of each crystal system is given below:

If a crystal has no symmetry, or has a center of symmetry only, then there is no restriction on the shape of the unit cell; all three angles α, β and γ have to be specified and hence the system is called triclinic or occasionally called anorthic (i.e., non-orthogonal). A monoclinic crystal has 2-fold symmetry in one direction only. One crystallographic axis is taken parallel to the direction of 2-fold symmetry axis, and the other two axes must therefore be perpendicular to it. The choice of the cell is thus restricted to a parallelepiped having two right angles; one angle has to be specified, hence the name monoclinic.

An orthorhombic crystal has 2-fold symmetry in three mutually perpendicular directions. The crystallographic axes are taken parallel to the symmetry directions, so that the unit cell is orthogonal. Higher symmetry involving axes of order greater than 2, impose more stringent restrictions on the shapes of the unit cell. The trigonal, tetragonal or hexagonal systems are characterized by a single 3-, 4-. or 6-fold axis, respectively. One crystallographic axis, always designated c, is taken parallel to this unique axis (except for primitive setting of rhombohedral cells), the other two must be perpendicular to it and are equal.

The highest symmetry among all is that of cubic system. The three crystallographic axes are orthogonal and equivalent. In the cubic system there are three lattices; the simple cubic (sc) lattice, the body centred cubic (*bcc*) lattice and the face centred cubic (*fcc*) lattice. A detail description of the three dimensional Bravais lattice is given in Table 5.11.

5.11 SUMMARY

1. The rotoreflection axes and the rotoinversion axes (two improper rotation axes) are found to be equivalent in pairs. This indicates that either of the two improper axes could be used to derive the complete set of point groups in three dimensions.

2. Two improper rotation axes given rise to two different point group notations. They are Schoenflies notation based on rotoreflection axes and Harmann-Mauguin (also known as 'International') notation based on rotoinversion axes.

3. The Schoenflies notation is preferred by the chemists and the spectroscopists while International notation is preferred by solid state physicists and the crystallographers.

4. There are two different set of axes used in crystallography. They are the orthogonal axes and the crystallographic axes. However, both the system of axes are always chosen in right handed form.

5. Any symmetry operation (of point groups) changes the coordinate axes and gives rise to a transformation matrix whose determinant is equal to +1or –1.

6. Matrix representation of symmetry elements makes it easy to derive the point groups purely in terms of matrix multiplications of symmetry operations.

Table 5.11 Bravais lattice in three dimensions

Crystal system	Restriction on conventional cell; axes and angles	Associated lattice		Characteristic symmetry elements	To be specified		
		Number	Symbol*		Axes	Angle	Total parameter
Triclinic	$a \neq b \neq c$ $\alpha \neq \beta \neq \gamma \neq 90°$	1	P	None	a, b, c	α, β, γ	6
Monoclinic	$a \neq b \neq c$ $\alpha = \gamma = 90°$ $\neq \beta$	2	P, C	One 2-fold rotation axis or m ($\equiv \bar{2}$)	a, b, c	γ	4
Orthorhombic	$a \neq b \neq c$ $\alpha = \beta = \gamma = 90°$	4	P, C, F, I	Three 2-fold rotation axis or m (mutually) perpendicular)	a, b, c	–	3
Tetragonal	$a = b \neq c$ $\alpha = \beta = \gamma = 90°$	2	P, I	One 4 fold rotation axis or $\bar{4}$	a, c	–	2
Trigonal (Rhombohedral)	$a = b = c$ $\alpha = \beta = \gamma < 120°$ $\neq 90°$	1	P	one 3-fold rotation axis or $\bar{3}$	a	α	2
Hexagonal	$a = b \neq c$ $\alpha = \beta = 90°$ $\gamma = 120°$	1	P	One 6-fold rotation axis or $\bar{6}$	a, c	–	2
Cubic	$a = b = c$ $\alpha = \beta = \gamma = 90°$	3	P or sc I or bcc F or fcc	Four 3 fold rotation axis or $\bar{3}$ (parallel 1 to cube diagonal)	a	–	1

*Lattice symbol:

P – Primitive (Lattice point are the corners of the unit only)

C – Side centred or Base centred (Lattice point are at the corners and at two face centres opposite to each other)

I – Body centred (Lattice point are at the corners and at the body center)

F – Face centred (Lattice points are at the corners and at the six face centres)

7. Successive application of symmetry operations of a point group gives rise to a number of equivalent positions, termed as the order of the group.

8. There are seven crystal systems, eleven Laue classes and thirty two crystal classes (point groups) in three dimensions.

5.12 DEFINITIONS

Abelian group: If all the elements of a group commute with each other, group is called the abelian group.

Crystal class: In geometrical sense, it is defined as a set of all crystals having the same point-group symmetry.

Crystal system: It is the convenient grouping of common features of some crystal classes together in such a way that the crystals belonging to these classes can be described by the unit cells of the same type.

Cyclic group: It is a group in which all the elements are the powers of a single generating element like $A, A^2, A^3, ... A^n = E$. All cyclic groups are abelians.

Direction cosines: In a general system of axes with the angle α, β and γ between them, the sum of the squares of cosine of the angles are unity, i.e., $\cos^2 \alpha + \cos^2 \beta + \cos^2 \gamma = 1$.

Equivalent positions: They are the positions (of atoms) obtained by successive application of symmetry operations of a point group.

Generating elements: It is possible to generate all elements of a group by starting from a certain set of elements (as their power or products) are known as generating elements.

Holosymmetry: A unit cell in its primitive form characteristically displays its maximum symmetry which is referred as holosymmetry.

Order of the group: The number of non equivalent symmetry operations forming a group is called the order of the group.

Point group: It is a group of symmetry operations which when applied to a crystal lattice, it leaves at least one lattice point unmoved.

Subgroup and Super group: A set of symmetry operations is said to be a subgroup of a bigger group (Super group) if the set itself form a group and satisfy the group conditions.

REVIEW QUESTIONS AND PROBLEMS

1. What do you understand by point group? Give the number of possible point groups for three dimensional lattices.

2. Derive the point groups based on the combined operation of rotation followed by mirror reflection perpendicular to the rotation axis. Show them diagrammatically.

$$\textbf{Ans. } m, \bar{1}, \frac{3}{m}, \bar{6}, \bar{4}, \text{ and } \bar{3}.$$

3. Derive the point groups based on axial combinations.

$$\textbf{Ans. } 222, 32, 422, 622, 23 \text{ and } 432.$$

4. Derive the point groups based on the combined operation of rotation followed by mirror reflection parallel to the rotation axis. Show them diagrammatically.

$$\textbf{Ans. } 1m, 2mm, 3m, 4mm, \text{ and } 6mm.$$

5. Derive the points based on the combined operation of rotation and mirror reflection taking place concurrently. Show them diagrammatically.

$$\textbf{Ans. } m, \frac{2}{m}, \frac{3}{m} = \bar{6}, \frac{4}{m} \text{ and } \frac{6}{m}$$

6. Derive the point groups based on the combined operation of mirror reflection perpendicular to the axial combinations taking place concurrently. Show them diagrammatically.

$$\textbf{Ans. } mmm,\ \bar{6}\,m2,\ \frac{4}{m}mm,\ \frac{6}{m}mm,\ m3,\ m3m.$$

7. Derive the point groups based on the combined operation of mirror reflection parallel to the axial combinations taking place concurrently.

$$\textbf{Ans. } \bar{4}\,2m,\ \bar{3}\,2m,\ \bar{3}\,m,\ \text{and}\ \bar{4}\,3m.$$

8. Discuss the procedure and derive all 32 point groups following the rotoinversion channel.

9. Explain the basic feature of Schoenflies and International notations. Discuss the advantages of each notation.

10. What do you understand by linear orthogonal transformation. Obtain the general transformation matrix for proper rotations.

11. Explain Euler's construction. Derive monoaxial and polyaxial point groups.

12. What do you understand by a cyclic group? Prove that a cyclic group is abelian. What about the converse of it? Prepare the character table for the point groups 3 and 6.

13. Obtain the transformation matrices of five proper rotations in terms of both orthogonal and crystallographic axes.

14. Obtain the transformation matrices of inversion and mirror placed perpendicular to the z-axis.

15. Show that the matrix multiplication of any two mutually perpendicular 2-fold axes produces the third 2-fold axis.

16. Show that the matrix multiplication of a 2-fold axis and a mirror placed perpendicular to it produces an inversion center.

17. Show that the matrix multiplication of a 2-fold axis and a mirror placed parallel to it produces another mirror perpendicular to the first mirror.

18. Show that the matrix multiplication of a 4-fold axis and a mirror placed parallel to it produces another mirror 45° away from the first mirror.

19. Show that the matrix multiplication of a 6-fold axis and a mirror placed parallel to it produces another mirror 30° away from the first mirror.

20. Show that the matrix multiplication of a 3-fold axis and a 2-fold axis perpendicular to it produces another 2-fold axis parallel to the third axis.

APPENDIX 1

Table 1   Crystallographic direction and Symmetry elements

Direction	Symmetry elements
[000]	1
[100]	$2, 2\,(H), \bar{2}\ (\equiv m), 4, \bar{4}$
[010]	$2, 2(H), \bar{2}, 4, \bar{4}$
[001]	$2, \bar{2}, 3, \bar{3}, 4, \bar{4}, 6, \bar{6}$
<110>	$2, \bar{2}$
< 111 >	$3, \bar{3}$
<120>	$2, \bar{2}$

APPENDIX 2

Table 2   Symmetry element, directions and matrices

Symmetry element	Directions	Matrices
1	[000]	$\begin{pmatrix} 1 & 0 & 0 \\ 0 & 1 & 0 \\ 0 & 0 & 1 \end{pmatrix}$
2	[100], [010], [001]	$\begin{pmatrix} 1 & 0 & 0 \\ 0 & -1 & 0 \\ 0 & 0 & -1 \end{pmatrix}, \begin{pmatrix} -1 & 0 & 0 \\ 0 & 1 & 0 \\ 0 & 0 & -1 \end{pmatrix}, \begin{pmatrix} -1 & 0 & 0 \\ 0 & -1 & 0 \\ 0 & 0 & 1 \end{pmatrix}$
	$H[100], H[010]$	$\begin{pmatrix} 1 & -1 & 0 \\ 0 & -1 & 0 \\ 0 & 0 & -1 \end{pmatrix}, \begin{pmatrix} -1 & 0 & 0 \\ -1 & 1 & 0 \\ 0 & 0 & -1 \end{pmatrix}$
	[110], [101], [011]	$\begin{pmatrix} 0 & 1 & 0 \\ 1 & 0 & 0 \\ 0 & 0 & -1 \end{pmatrix}, \begin{pmatrix} 0 & 0 & 1 \\ 0 & -1 & 0 \\ 1 & 0 & 0 \end{pmatrix}, \begin{pmatrix} -1 & 0 & 0 \\ 0 & 0 & 1 \\ 0 & 1 & 0 \end{pmatrix}$
	$[1\bar{1}0], [\bar{1}01], [01\bar{1}]$	$\begin{pmatrix} 0 & -1 & 0 \\ -1 & 0 & 0 \\ 0 & 0 & -1 \end{pmatrix}, \begin{pmatrix} 0 & 0 & -1 \\ 0 & -1 & 0 \\ -1 & 0 & 0 \end{pmatrix}, \begin{pmatrix} -1 & 0 & 0 \\ 0 & 0 & -1 \\ 0 & -1 & 0 \end{pmatrix}$
	$H[210], H[120]$	$\begin{pmatrix} 1 & 0 & 0 \\ 1 & -1 & 0 \\ 0 & 0 & -1 \end{pmatrix}, \begin{pmatrix} -1 & 1 & 0 \\ 0 & 1 & 0 \\ 0 & 0 & -1 \end{pmatrix}$

Symmetry element	Directions	Matrices
3	[001], [111]	$\begin{pmatrix} 0 & -1 & 0 \\ 1 & -1 & 0 \\ 0 & 0 & 1 \end{pmatrix}, \begin{pmatrix} 0 & 0 & 1 \\ 1 & 0 & 0 \\ 0 & 1 & 0 \end{pmatrix}$
	$[\bar{1}11]\,[1\bar{1}1]\,[11\bar{1}]$	$\begin{pmatrix} 0 & -1 & 0 \\ 0 & 0 & 1 \\ -1 & 0 & 0 \end{pmatrix}, \begin{pmatrix} 0 & -1 & 0 \\ 0 & 0 & -1 \\ 1 & 0 & 0 \end{pmatrix}, \begin{pmatrix} 0 & 1 & 0 \\ 0 & 0 & -1 \\ -1 & 0 & 0 \end{pmatrix}$
3^2	[001], [111]	$\begin{pmatrix} -1 & 1 & 0 \\ -1 & 0 & 0 \\ 0 & 0 & 1 \end{pmatrix}, \begin{pmatrix} 0 & 1 & 0 \\ 0 & 0 & 1 \\ 1 & 0 & 0 \end{pmatrix}$
	$[\bar{1}11]\,[1\bar{1}1]\,[11\bar{1}]$	$\begin{pmatrix} 0 & 0 & -1 \\ -1 & 0 & 0 \\ 0 & 1 & 0 \end{pmatrix}, \begin{pmatrix} 0 & 0 & 1 \\ -1 & 0 & 0 \\ 0 & -1 & 0 \end{pmatrix}, \begin{pmatrix} 0 & 0 & -1 \\ 1 & 0 & 0 \\ 0 & -1 & 0 \end{pmatrix}$
4	[100], [010], [001]	$\begin{pmatrix} 1 & 0 & 0 \\ 0 & 0 & 1 \\ 0 & -1 & 0 \end{pmatrix}, \begin{pmatrix} 0 & 0 & -1 \\ 0 & 1 & 0 \\ 1 & 0 & 0 \end{pmatrix}, \begin{pmatrix} 0 & -1 & 0 \\ 1 & 0 & 0 \\ 0 & 0 & 1 \end{pmatrix}$
4^3	[100], [010], [001]	$\begin{pmatrix} 1 & 0 & 0 \\ 0 & 0 & -1 \\ 0 & 1 & 0 \end{pmatrix}, \begin{pmatrix} 0 & 0 & 1 \\ 0 & 1 & 0 \\ -1 & 0 & 0 \end{pmatrix}, \begin{pmatrix} 0 & 1 & 0 \\ -1 & 0 & 0 \\ 0 & 0 & 1 \end{pmatrix}$
6	[001]	$\begin{pmatrix} 1 & -1 & 0 \\ 1 & 0 & 0 \\ 0 & 0 & 1 \end{pmatrix}$
6^5	[001]	$\begin{pmatrix} 0 & 1 & 0 \\ -1 & 1 & 0 \\ 0 & 0 & 1 \end{pmatrix}$

*1. Matrix of an inverse symmetry element can be obtained by simply changing the sign of the digits of the starting matrix.

2. $4^2 \equiv 2$, $6^2 \equiv 3$, $6^3 \equiv 2$ and $6^4 \equiv 3^2$.

CHAPTER 6

DERIVATION OF SPACE GROUPS

6.1 INTRODUCTION

In the preceding chapters, we studied about macroscopic symmetries and the derivation of point groups in detail. However, in this chapter we are going to study about the microscopic symmetries related to internal arrangement of atoms in crystals and some other important aspects which are helpful in understanding the concept of space groups as well as their derivation. We shall discuss a qualitative way to derive the space groups.

6.2 MICROSCOPIC SYMMETRY ELEMENTS

Microscopic symmetry elements are those symmetry elements which have no influence on the external shape of freely formed crystals, they are connected with the array of atoms within the crystal and involving a translation operation. They cannot be perceived by the naked eye and are detectable only by diffraction (of X-rays, electron or neutrons) methods. The microscopic symmetry elements comprise five different glide planes and eleven different screw axes.

6.3 COMBINATION OF MICROSCOPIC SYMMETRY OPERATIONS

Inclusion of translation operation gives rise to the microscopic symmetries in crystals. According to its combination with reflection or proper rotation, there exist two microscopic symmetry elements. They are:

1. Glide Planes
2. Screw axes

Glide Planes

A glide plane is defined as a plane across which a mirror plane is combined with a non primitive translation parallel to the reflecting plane (the order in which these two operations are carried out is unimportant), transforms an array of atoms to self coincidence. Three different types of glide plane exist in crystals. They are:

(i) Axial Glide

(ii) Diagonal Glide

(iii) Diamond Glide

Let us discuss them briefly in turn.

Axial Glide

A glide is called an axial glide if the translation is carried along one of the edges or axes of the unit cell. In this case, the magnitude of the translation vector T is one half of the unit cell translation parallel to the mirror (reflection) plane. An axial glide is termed as an $a-$, $b-$ or $c-$ glide according to the axis along which the translation is carried out. For example, a b-glide operation is shown in Fig. 6.1. This produces an enantiomorph of the original object as indicated by a comma in the Figure. An axial glide in no case be perpendicular to the glide direction.

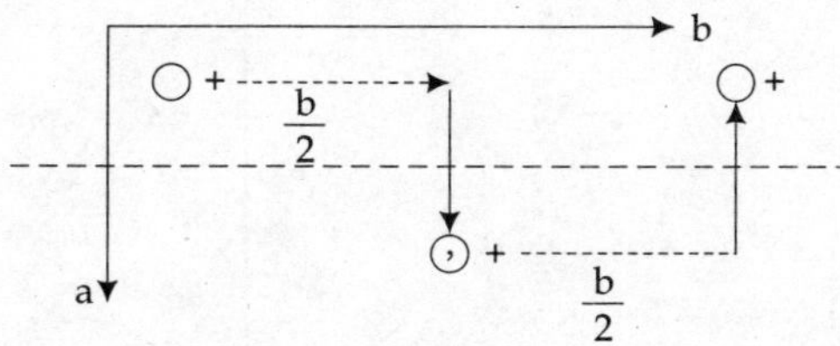

Fig. 6.1 A b-glide operation

Diagonal Glide

A glide is called a diagonal glide or usually the n-glide if the translation is carried in general along one of the face diagonals of the unit cell. In this case, the magnitude of the translation vector T is one half of the face diagonal of the unit cell. However, in case of tetragonal, rhombohedral and cubic crystal systems the n-glide can also occur along the body diagonal, where the magnitude of the translation vector is half the body diagonal. In such glides, it is possible for the glide direction to have a component perpendicular to the reflection plane. For example, an n-glide operation with a reflection across the plane perpendicular to c is shown in Fig. 6.2.

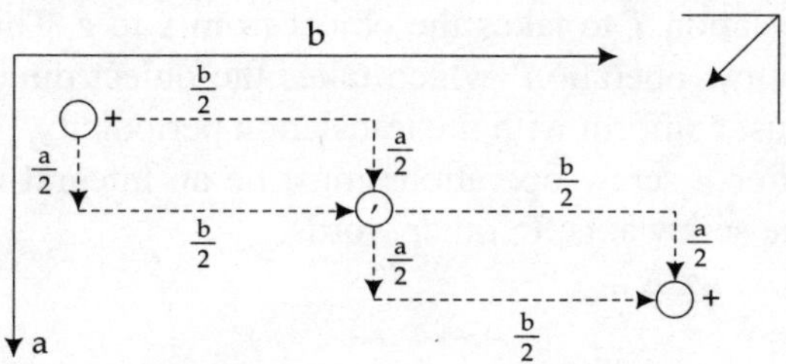

Fig. 6.2 An n-glide operation

Diamond Glide

Like a diagonal glide, the diamond glide or *d*-glide also occurs along the face diagonal in general and body diagonal in particular (in tetragonal and cubic). However, the magnitude of translation vector *T* in this case is one fourth of the face (or body) diagonals of the unit cell. In fact, *d*-glide occurs in face-and body-centered unit cells only. All symmetry planes alongwith other details are provided in Table 6.1.

Table 6.1 Symbols for Symmetry Planes

Type of glide	Symbol	Translation component *T*	Designation	
			|| *to the plane of projection*	⊥ *to the plane of projection*
Axial glide	*a*	$\dfrac{a}{2}$	⌐↓	....
Axial glide	*b*	$\dfrac{b}{2}$	↑⌐	....
Axial glide	*c*	$\dfrac{c}{2}$	—	....
Diagonal glide	*n*	$\dfrac{a}{2}+\dfrac{b}{2},\ \dfrac{b}{2}+\dfrac{c}{2}$ or $\dfrac{c}{2}+\dfrac{a}{2}$	↖	....
Diamond glide	*d*	$\dfrac{a}{4}+\dfrac{b}{4},\ \dfrac{b}{4}+\dfrac{c}{4}$ or $\dfrac{c}{4}+\dfrac{a}{4}$	↗�	 ..▸..

Screw Axes

A screw axis is defined as an axis about which a proper rotation $\left(\theta=\dfrac{2\pi}{n}\right)$ is combined with a non primitive translation parallel to the axis of rotation, transforms an array of atoms to self coincidence. Such a motion is shown in Fig. 6.3, where a rotation of an angle θ (takes the object from *x* to *y*) is combined with a translation *T* to takes the object from *y* to *z*. This is equivalent to a single hybrid operation called 'screw operation' which takes the object directly from *x* to *z*. Since, in crystals every operation must conform with the translation periodicity, this means that the cumulative translation distance after *n* screw operations must be an integral multiple (say *m*) of lattice translation (say *a*) along the screw axis. In other words,

$$nT = ma$$

or
$$T = \frac{ma}{n} \tag{1}$$

where both *m* and *n* are integers with $m < n$. We know that, *n* can take only five values 1, 2, 3, 4 and 6, where each digit represents a proper rotation. Further, $m = 0$ and $m = n$ both correspond to pure rotations. For $m = 1, 2, 3 \ldots (n-1)$ there exist only 11 unique screw axes. They are $2_1, 3_1, 3_2, 4_1, 4_2,$

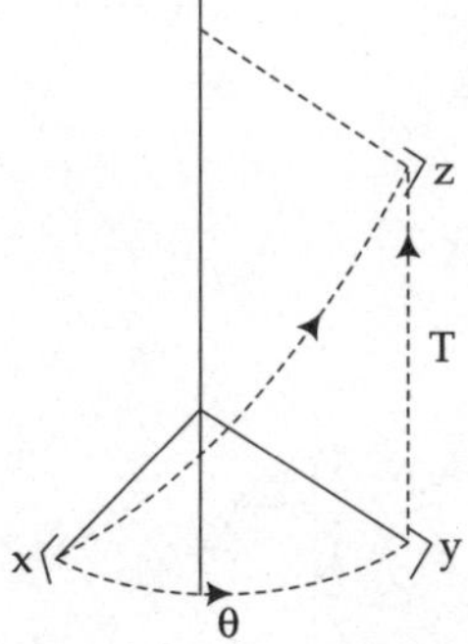

Fig. 6.3 Showing rotation and translation

Table 6.2 Symmetry axes

Symmetry	Symbol	Designation	
		|| *to plane of projection*	⊥ *to plane of projection*
Center	$\bar{1}$	○	○
2-fold axis	2	←——→	●
3-fold axis	3	—	▲
4-fold axis	4	—	■
6- fold axis	6	—	⬡
2-fold screw axis	2_1	←——→	⚡
3-fold screw axis	3_1	—	▲
3-fold screw axis	3_2	—	▲
4-fold screw axis	4_1	—	⬛
4-fold screw axis	4_2	—	⬛
4-fold screw axis	4_3	—	⬛
6-fold screw axis	6_1	—	✿
6-fold screw axis	6_2	—	✿
6-fold screw axis	6_3	—	✿
6-fold screw axis	6_4	—	✿
6-fold screw axis	6_5	—	✿

4_3, 6_1, 6_2, 6_3, 6_4 and 6_5 and symbolically represented as n_m. They are illustrated in Fig. 6.4. It can be seen in the Figure that the pair of screw axes 3_1 and 3_2 are enantiomorphic. That is, they are identical except for the sense of screw motion. Similarly, the pair of screw axes $4_1 - 4_3$, $6_1 - 6_5$ and $6_2 - 6_4$ are also enantomorphic. All symmetry axes along with other details are provided in Table 6.2.

6.4 GENERAL EQUIVALENT POSITIONS AND SPECIAL POSITIONS

In section 5.8, we observed the generation of a set of equivalent positions by successive application of non-equivalent symmetry operations of a point group. However, the non-equivalent symmetry operations of a space group can produce (i) a set of symmetry related points known as the general equivalent positions, and (ii) other sets of symmetry related point known as special positions.

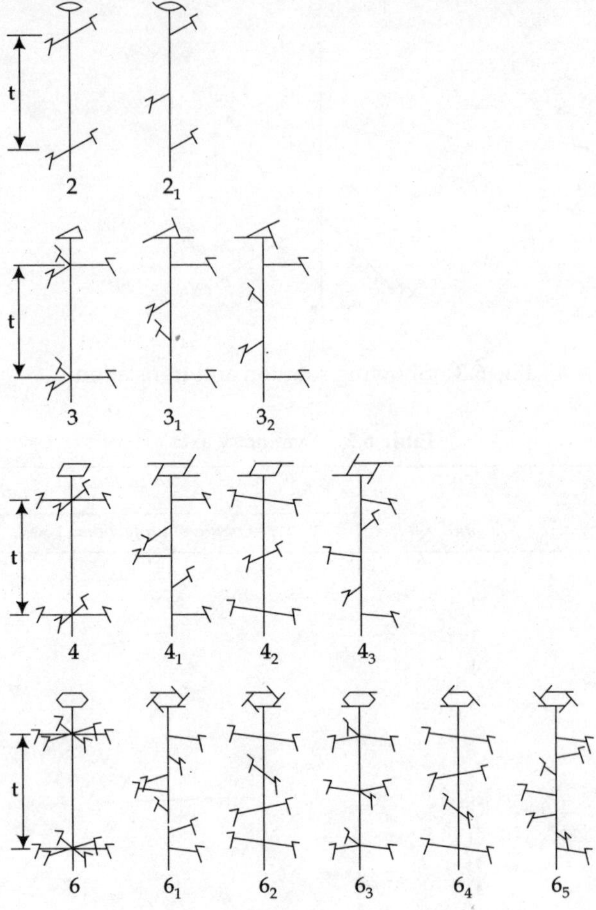

Fig. 6.4 Screw axes

When the representative point of a set of symmetry related point is specified by unspecialized coordinates $x\,y\,z$, the equivalent positions are called the general positions of the space group. The number of general equivalent positions is maximum for a given space group and is equal to the number of symmetry operations and hence it also represents the order of that space group.

When the representative point is found to lie on a symmetry element of the space group, its coordinates $x\,y\,z$ are said to be specialized and the corresponding equivalent positions are called special positions. In such cases, the space groups may either lack a translation component or in some cases despite having a translation component, certain pairs (or sets) of points are found to coalesce. Consequently, the total number of equivalent positions are reduced to become a submultiple of the 'order' of the general positions. There are usually several different set of special positions in a space group, their orders may be the same or different.

6.5   SYSTEMATIC ABSENCES

Systematic absences of some reflections are often observed in the intensity data obtained by diffraction of *X*-rays, electrons and neutrons. They are due to (i) Lattice centering and (ii) presence of microscopic symmetries like screw axes and glide planes. Let us discuss them one by one.

Systematic Absences Due to Lattice Centering

To start with let us take the case of α-Fe which exhibits bcc structure (Fig. 6.5). In this case, (100) reflection is found to be absent. It is because the effective number of corner and body centered atoms are equal and the body center atoms lie midway between the (100) planes. The planes corresponding to corner and body centered atoms diffract exactly 180° out of phase relative to each other and cancel completely. On the other hand, (200) reflections are found to be strong because there are no atoms lying between (200) planes to cause any destructive interference. It is easy to show by similar arguments that (110) reflections are observed whereas (111) reflections are systematically, absent. Hence, it can be said that for a body centered cell, the reflections for which $(h + k + l)$ is odd are absent.

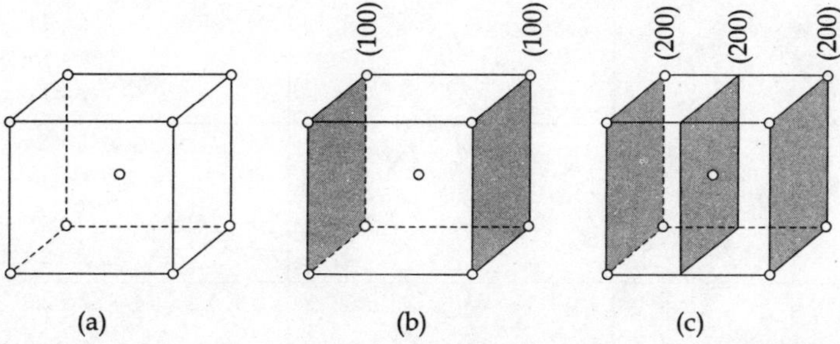

Fig. 6.5   (a) Body centered cubic of α-Fe, (b) (100) planes, (c) (200) planes

Similarly, for each non-primitive (centered) lattice, a simple characteristic formula for systematic absences can be obtained. They are provided in Table 6.3.

Systematic Absences Due to Microscopic Symmetries

Systematic absences caused by the presence of microscopic symmetry elements like screw axes and glide planes are rather complicated and hence their description below is kept simple and brief.

This kind of absences arise when a crystal containing screw axis or glide planes is held in certain orientation, the dimensions of one or more of the unit cell edges appear to be reduced (often by half). For example, a 2-fold screw axis (parallel to *x*) 2_1, represents a two-fold rotation plus a translation of half the cell. When only (*h*00) planes of the unit cell are considered, the cell repeat appears to be half. It is because the (*h*00) planes are perpendicular to *x*, any change in the rotational orientation about *x* does not affect diffraction from the (*h*00) planes. In this situation, it is not possible to differentiate between a simple translation of *a*/2 and a combination of translation (of *a*/2) and rotation (of 180°) about *x*. Thus for a 2_1 screw axis, (*h*00) reflections with $h = 2n + 1$ are systematically absent. In a similar manner, the conditions for systematic absences for other screw axes can obtained. They are presented in Table 6.3.

Table 6.3 Systematic absences

Symmetry elements		Set of reflections	Conditions for reflections to be observed
Lattice	P	hkl	none
	I		$h + k + l = 2n$
	C		$h + k = 2n$
	A		$k + l = 2n$
	B		$h + l = 2n$
	F		$h + k = 2n$
			$k + l = 2n$
			$h + l = 2n$
	R_{obv}		$-h + k + l = 2n$
	R_{rev}		$h - k + l = 3n$
Glide-plane ‖(001)	a	$hk0$	$h = 2n$
	b		$k = 2n$
	n		$h + k = 2n$
	d		$h + k = 4n$
Glide-plane ‖(100)	b	$0kl$	$k = 2n$
	c		$l = 2n$
	n		$k + l = 2n$
	d		$k + l = 4n$
Glide-plane ‖(010)	a	$h0l$	$h = 2n$
	c		$l = 2n$
	n		$h + l = 2n$
	d		$h + l = 4n$
Glide-plane ‖(010)	c	hkl	$l = 2n$
	b		$h = 2n$
	n		$h + l = 2n$
	d		$2h + l = 4n$
Screw-axis ‖c	$2_1, 4_2, 6_3$	$00l$	$l = 2n$
	$3_1, 3_2, 6_2, 6_4$		$l = 3n$
	$4_1, 4_3$		$l = 4n$
	$6_1, 6_5$		$l = 6n$
Screw-axis ‖a	$2_1, 4_2$	$h00$	$h = 2n$
	$4_1, 4_3$		$h = 4n$
Screw-axis ‖b	$2_1, 4_2$	$0k0$	$k = 2n$
	$4_1, 4_3$		$k = 4n$
Screw-axis ‖[110]	2_1 hho		$h = 2n$

In case of a crystal containing glide planes, the two parts of the unit cell are related by a reflection across a plane plus translation. For example, a c-glide may be considered as reflection across a plane perpendicular to b plus a translation (of $c/2$) parallel to c. Accordingly, the ($h0l$) reflections with l odd are systematically absent. Similarly, the conditions for other glide plane can be obtained. They are presented in Table 6.3.

6.6 SPACE GROUPS

A space group in crystallographic sense is defined as the set of geometrical operations which takes a three dimensional periodic object (a crystal) into itself. The set of operations that makes up the space group must form a group in mathematical sense, and also must include the primitive lattice translations as well as other symmetry operations. In short, a space group of symmetry specifies the full microscopic symmetry of a crystal just as a point group of symmetry specifies the full macroscopic symmetry of a crystal.

According to the International notation, all space group symbols are of the form $Apqr$, where A represents the lattice type (P, C, I or F) and p, q and r represent the primary, secondary and tertiary symmetry directions, respectively. They are provided in Table 6.4, for various crystal systems. The crystal system is not stated explicitly in the space group symbol but can be easily deduced from the associated symmetry elements.

Table 6.4 Direction associated with Apqr for space groups

Crystal system	Primary position	Secondary position	Tertiary position
Triclinic	–	–	–
Monoclinic	[010]	–	–
Orthorhombic	[100]	[010]	[001]
Trigonal	[001]	$\langle 100 \rangle^a$	$\langle 210 \rangle^a$
Tetragonal	[001]	$\langle 100 \rangle^a$	$\langle 110 \rangle^a$
Hexagonal	[001]	$\langle 100 \rangle^a$	$\langle 210 \rangle^a$
Cubic	$\langle 100 \rangle$	$\langle 111 \rangle$	$\langle 110 \rangle^a$

[a] These directions are not symmetry axes in all classes within the system.

In general, several space groups correspond to a single point group. The conversion from space group to point group is done by dropping the lattice symbol and changing the screw axes to ordinary rotation axes and glide planes to mirror planes. For example, the point group $p2_1/c$ can be easily converted into the point group $2/m$ by dropping the letter p and changing $2_1 \rightarrow 2$ and $c \rightarrow m$, respectively.

6.7 CLASSIFICATION OF SPACE GROUPS

Based on the nature of combination of symmetry elements (macroscopic or microscopic) with the 14 Bravais Lattices, the space groups can be classified into two categories. They are termed as Symmorphic and non-Symmorphic space groups.

The Symmorphic Space Groups

A symmorphic space group is the one which may be entirely specified by symmetry operations all acting at a common point only. Thus in order to obtain the symmorphic plane groups or space groups we simply combine the macroscopic symmetry elements (ten 2-dimensional and thirty two 3-dimensional point group) with 5 plane lattices and 14 Bravais Lattices, respectively. That is, each point group can be combined with all primitive (P) and non-primitive (I, F and/or C) Lattices of all crystal systems. Such combinations are found to generate 13 plane groups (two-dimensional space

Table 6.5 The 73 symmorphic space groups

Crystal System	Bravais Lattice	Space groups
Triclinic	P	$P1, P\bar{1}$
Monoclinic	P	$P2, Pm, P2/m$
	B or C	$B2, Bm\ B2/m$ (1st setting)
Orthorhombic	P	$P222, Pmm2, Pmmm$
	C, A or B	$C222, Cmm2, Amm2^*, Cmmm$
	I	$I222, Imm2, Immm$
	F	$F222, Fmm2, Fmmm$
Tetragonal	P	$P4, P\bar{4}, P4/m, P422, P4mm, P\bar{4}2m, P\bar{4}m2, P4/mmm$
	I	$I4, I\bar{4}, I4/m, I422, I4m, I4mm, I\bar{4}m2^*, I4/mmm$
Trigonal	P	$P3, P\bar{3}, P312, P321^*, P3m1, P31m^*, P\bar{3}1m, P\bar{3}m1^*$
Rhombohedral	R	$R_3, R\bar{3}, R32, R3m, R\bar{3}m$
Hexagonal	P	$P6, P\bar{6}, P6/m, P622, P6mm, P\bar{6}m2, P\bar{6}2m^*, P6/mmm$
Cubic	P	$P23, Pm3, P432, P\bar{4}3m, Pm3m$
	I	$I23, Im3, I432, I\bar{4}3m, Im3m$
	F	$F23, Fm3, F432, F\bar{4}3m, Fm3m$

*Seven extra space groups derived by taking in to account the orientation of the point group elements w.r.t the Bravais lattice.

groups) and 73 space groups (three-dimensional), respectively. The three-dimensional symmorphic space groups are listed in Table 6.5.

The Non-Symmorphic Space Groups

The non-symmorphic space groups involve the existence of microscopic symmetry elements such as screw axes and glide planes. Taking into account the combined operations of glide lines with 5 plane lattices, screw axes and glide planes with 14 Bravais lattices, additional 4 non-symmorphic plane groups (two-dimensional space group) and 157 non-symmorphic (three-dimensional) space groups can be obtained. Thus a total of 17 plane groups and 230 space groups can be obtained.

6.8 DERIVATION OF SPACE GROUPS

In section 3.9, we discussed the derivation of 17 two-dimensional space groups (or plane groups) by considering consistent combinations of ten point groups and the glide line symmetries with the five lattices. In a similar manner, one can obtain 230 three-dimensional space groups when consistent combinations of all 32 point groups with screw axes and glide planes are considered. This is a formidable task. This is also quite inappropriate to give a full treatment in an introductory text like this. Therefore, in the following we shall only attempt to provide a description within the triclinic and monoclinic systems in some what details while other systems will be delt in brief. However, the interested readers may consult the International tables for X-ray crystallography, Vol I for a comprehensive and detailed descriptions of this topic.

Triclinic Crystal System

The space groups of triclinic crystal system is simplest to determine. Since, we know that there is only one Bravias lattice, the primitive lattice P and only two point groups, 1 and $\bar{1}$ possible in triclinic system. The corresponding symmorphic space groups are $P1$ and $P\bar{1}$ as listed in Table 6.5. It is to be mentioned that other operations such as reflection, rotations of order 2 or more are not compatible with the triclinic system and hence no screw or glide operation is possible. This implies that the non-symmorphic triclinic space groups will be absent.

Monoclinic Crystal System

In the monoclinic crystal system there are two unique space lattices, P and B (*I*-setting) or C centering (*II*-setting) and three point groups, 2, m and $2/m$. Since the international union of crystallography has adopted the second setting for morphological and crystallographic studies, therefore, we shall also adopt the second setting. With this choice of axes, the symmetry planes in point groups m and $2/m$ are normal to y axis. The corresponding space groups will have a, c, and n glide planes. The consistent combinations involving the macroscopic symmetry operations (2, m and $2/m$) and microscopic symmetry operations (screw axes and glide planes) give rise to the following possible space groups under three crystal classes as shown in Table 6.6.

Table 6.6 All possible Monoclinic Space Groups

Point Group	Space Groups
2	$P2$, $P2_1$, $C2$, $C2_1$
m	Pm, Pa, Pc, Pn, Cm, Ca, Cc, Cn
$2/m$	$P2/m$, $P2/a$, $P2/c$, $P2/n$, $P2_1/m$, $P2_1a$, $P2_1/c$, $P2_1/n$, $C2/m$, $C2/a$, $C2/c$, $C2/n$, $C2_1/m$, $C2_1/a$, $C2_1/c$, $C2_1/n$

According to section 2.6, in monoclinic crystal system with second setting, we have $C = A = F = I$. This condition makes many of the above listed space groups equivalent. For example $C2 = C2_1$, $Pc = Pa = Pn$, $Cc = Ca = Cn$ etc. Thus, we are left with only thirteen independent monoclinic space groups as listed in Table 6.7.

Orthorhombic Crystal System

In this crystal system, there are three point group 222, $mm2 \equiv mm$, mmm and four Bravais lattices, P, F, I and C, where $C = A = B$. In addition 2_1 screw axis and all glide planes (axial, diagonal and diamond) are possible. Taking into account the above symmetry operations and eliminating the equivalent cases, there are 59 independent possible space groups for orthorhombic system as listed in Table 6.7. Out of these 9 belong to the point group 222, 22 to $mm2$ and 28 to mmm, respectively.

Tetragonal Crystal System

In this crystal system there are seven point group, 4, $\bar{4}$, $4/m$ 422, $4mm$, $\bar{4}2m$ and $4/mmm$ and two Bravais lattice P and I, where $I = C = F$. Combining the seven point groups with the two Bravais lattices, and also taking into account the non symmorphic operations one finds that there are 68 tetragonal space groups as listed in Table 6.7. This is the largest number of space groups associated with any one crystal system.

Table 6.7 The 230 three-dimensional space groups arranged according to crystal systems and point groups.

Crystal system	Point group	Space groups
Triclinic	1	$P1$
	$\bar{1}$	$P\bar{1}$
Monoclinic	2	$P2$, $P2_1$, $C2$
	M	Pm, Pc, Cm, Cc
	$2/m$	$P2/m$, $P2_1/m$, $C2/m$, $P2/c$, $P2_1/c$, $C2/c$
Orthorhombic	222	$P222$, $P222_1$, $P2_12_12$, $P2_12_12_1$, $C222_1$, $C222$, $F222$, $I222$, $I2_12_12_1$
	$mm2$	$Pmm2$, $Pmc2_1$, $Pcc2$, $Pma2_1$, $Pca2_1$, $Pnc2_1$, $Pmn2_1$, $Pba2$, $Pna2_1$, $Pnn2$, $Cmm2$, $Cmc2_1$, $Ccc2$, $Amm2$, $Abm2$, $Ama2$, $Aba2$, $Fmm2$, $Fdd2$, $Imm2$, $Iba2$, $Ima2$
	mmm	$Pmmm$, $Pnnn$, $Pccm$, $Pban$, $Pmmn$, $Pcca$, $pbam$, $Pccn$, $Pbcm$, $Pnnm$, $Pmmn$, $Pbcn$, $Pbca$, $Pnma$, $Cmcm$, $Cmca$, $Cmmm$, $Cccm$, $Cmma$, $Ccca$, $Fmmm$, $Fddd$, $Immm$, $Ibam$, $Ibca$, $Imma$
Tetragonal	4	$P4$, $P4_1$, $P4_2$, $P4_3$, $I4$, $I4_1$
	$\bar{4}$	$P\bar{4}$, $I\bar{4}$
	$4/m$	$P4/m$, $P4_2/m$, $P4/n$, $P4_2/n$, $4/m$, $I4_1/a$
	422	$P422$, $P42_12$, $P4_122$, $P4_12_12$, $P4_222$, $P4_22_12$, $P4_322$, $P4_32_12$, $I422$, $I4_122$.
	$4mm$	$P4mm$, $P4bm$, $P4_2cm$, $P4_2nm$, $P4cc$, $P4_2mc$, $P4_2bc$, $I4mm$, $I4cm$, $I4_1md$, $I4_1cd$
	$\bar{4}m2$	$P\bar{4}2m$, $P\bar{4}2c$, $P42_1m$, $P\bar{4}2_1c$, $P\bar{4}m2$, $P\bar{4}c2$, $P\bar{4}b2$, $P\bar{4}n2$, $I\bar{4}m2$, $I\bar{4}c2$, $I\bar{4}2m$, $I\bar{4}2d$
	$4/mmm$	$P4/mmm$, $P4/mcc$, $P4/nbm$, $P4/nnc$, $P4/mbm$, $P4/mnc$, $P4/nmm$, $P4/ncc$, $P4_2/mmc$, $P4_2/mcm$, $P4_2/nbc$, $P4_2/nnm$, $P4_2/mbc$, $P4_2mnm$, $P4_2/nmc$, $P4_2/ncm$, $I4/mm$, $I4/mm$, $I4/mcm$, $I4_1/amd$, $I4_1/acd$
Trigonal	3	$P3$, $P3_1$, $P3_2$, $R3$
	$\bar{3}$	$P3$, $R3$
	32	$P312$, $P321$, $P3112$, $P3_121$, $P3_212$, $P3_221$, $R32$
	$3m$	$P3m1$, $P31m$, $P3c1$, $P31c$, $R3m$, $R3c$
	$\bar{3}m$	$P31m$, $P31c$, $P\bar{3}m1$, $P\bar{3}c1$, $R\bar{3}m$, $R\bar{3}c$
Hexagonal	6	$P6$, $P6_1$, $P6_5$, $P6_3$, $P6_2$, $P6_4$,
	$\bar{6}$	$P\bar{6}$
	$6/m$	$P6/m$, $P6_3/m$
	622	$P622$, $P6_122$, $P6_522$, $P6_222$, $P6_422$, $P6_322$
	$6mm$	$P6mm$, $P6cc$, $P6_3cm$, $P6_3mc$
	$\bar{6}m2$	$P\bar{6}m2$, $P\bar{6}c2$, $P\bar{6}2m$, $P\bar{6}2c$
	$6/mmm$	$P6/mmm$, $P6/mc$, $P6_3/mcm$, $P6_3/mmc$
Cubic	23	$P23$, $F23$, $I23$, $P2_13$, $I2_13$
	$m\bar{3}$	$Pm\bar{3}$, $Pn\bar{3}$, $Fm\bar{3}$, $Fd\bar{3}$, $Im\bar{3}$, $Pa\bar{3}$, $Ia\bar{3}$
	432	$P432$, $P4_232$, $F432$, $F4_132$, $I432$, $P4_332$, $P4_132$, $I4_132$
	$\bar{4}3m$	$P\bar{4}3m$, $F\bar{4}3m$, $I\bar{4}3m$, $P\bar{4}3n$, $F\bar{4}3c$, $I\bar{4}3d$
	$m\bar{3}m$	$Pm\bar{3}m$, $Pn\bar{3}n$, $Pm\bar{3}n$, $Pn\bar{3}m$, $Fm\bar{3}m$, $Fm\bar{3}c$, $Fd\bar{3}m$, $Fd\bar{3}c$, $Im\bar{3}m$, $Ia\bar{3}d$

Trigonal Crystal System

In this crystal system there are five point groups 3, $\bar{3}$, 32, 3*m* and $\bar{3}$ *m* and two Bravais lattices *P* and *R*. Combining the five point groups with the two Bravais lattices, and also taking into account the non-symmorphic operations, one finds that there are 25 trigonal space groups as listed in Table 6.7.

Hexagonal Crystal System

In this crystal system there are seven point groups 6, $\bar{6}$, 6/*m*, 622, 6*mm*, $\bar{6}$*m*2 and 6/*mmm* and only one Bravais lattice, *P*. These point groups are similar to the point groups in tetragonal system and hence there is also a similarity in the derivation of space groups. Thus combining the seven point groups with the primitive hexagonal lattice and also taking into account the non-symmorphic operations, one finds 27 hexagonal space groups as listed in Table 6.7.

Cubic Crystal System

In this crystal system there are five point groups 23, *m*3, 432, $\bar{4}$3*m* and *m*3*m* and three Bravais lattices *P*, *I* and *F*. When these point groups alongwith the non-symmorphic operations are combined with the three Bravais lattices, 36 cubic space groups are obtained as listed in Table 6.7.

6.9 SUMMARY

1. Microscopic symmetry elements have no influence on the external shape of freely formed crystals, rather they are connected with the array of atoms within the crystal, involving a translational operation and are detectable only by diffraction of *X*-rays, electrons or neutrons.
2. There are broadly two microscopic symmetry elements. They are: glide planes and screw axes. Glide planes include axial glide, diagonal glide and diamond glide. Similarly, there exists eleven unique screw axes.
3. Non equivalent symmetry operations of a space group can produce sets of symmetry related point known as general equivalent positions and special positions.
4. A space group of symmetry specifies the full microscopic symmetry of a crystal just as a point group of symmetry specifies the full macroscopic symmetry of a crystal.
5. Systematic absences of reflection in the *X*-ray diffraction data are due to lattice centering and the presence of glide planes and screw axes.
6. In all there are four types of lattice centering possible in three dimensional lattices. They are: base centering, body centering, face centering and rhombohedral centering.
7. The space groups can be classified as symmorphic and non-symmorphic.

6.10 DEFINITIONS

Axial glide: It is a glide in which the translation is carried along one of the edges or axes of the unit cell.

Diagonal glide (or *n*-glide): It is a glide in which the translation is carried along one of the face diagonals of the unit cell.

Diamond glide:　It is a glide in which the translation is carried along the face diagonal in general and body diagonal in particular (in tetragonal and cubic).

General equivalent positions:　They are the symmetry related set of points (positions) obtained by the application of non-equivalent symmetry operations.

Glide plane:　It is a combination of a non-primitive translation and a mirror plane parallel to it.

Lattice centering:　Different types of centering that are possible in the primitive cells of various crystal systems without destroing their original shapes.

Microscopic symmetry elements:　They are connected with the array of atoms within the crystal and involving a translation operation. They are glide planes and screw axes.

Screw axis:　It is a combination of a non-primitive translation and proper rotation about it.

Space group:　A space group in crystallographic sense is a set of geometrical operations which takes a three-dimensional periodic object (a crystal) into itself.

Symmorphic space group:　A symmorphic space group is the one which may be entirely specified by symmetry operations acting at a common point.

REVIEW QUESTIONS AND PROBLEMS

1. What do you understand by microscopic symmetry elements? Explain glide planes and screw axes and their types.

2. What do you understand by general equivalent positions and special positions of a space groups? Prepare the equivalent positions Table of space group P_4.

3. What do you understand by space group? Differentiate between symmorphic and non-symmorphic space groups. What are their number in two and three dimensions?

4. What do you understand by systematic absences? Explain the systematic absences due to lattice centering.

5. Derive the conditions of systematic absences due to body centering and face centering.

6. Derive the condition of systematic absence due to rhombohedral centering.

7. Describe systematic absences due to the presence of microscopic symmetrics like glide planes and screw axes, taking at least one example from each category.

8. Mention some of the aspects which are helpful in deriving space groups. Briefly explain them.

9. Derive the general position for the space groups $C2/c$ and $C2_1/c$ and show that they refer to the same space group, differing only in the choice of the origin.

10. Using space group diagram compare the space groups $Cmm2$ and $Pmmm$.

11. Derive the monoclinic space groups belonging to the crystal class m. Compare your result from the Table 6.7.

12. Derive the orthorhombic space groups belonging to the crystal class 222. Compare your result from the Table 6.7.

CHAPTER 7

CRYSTAL PLANES, DIRECTIONS AND PROJECTIONS

7.1 CRYSTAL PLANES AND ZONES

A crystal is understood to be made up of a large number of different set of equidistant parallel planes passing through the lattice points as shown in Fig. 7.1. For a given crystal the interplanar spacing of each set of parallel planes is fixed. A given crystal plane can be represented by a set of three integers known as Miller indices in an orthogonal system. In order to determine the indices, let us consider a plane nearest to the origin. The intercept of the plane in question in terms of unit cell parameters a, b, c, along the axes x, y, z (Fig. 7.2) respectively are a/h, b/k and c/l. The reciprocals of these intercepts after removing the fractions will give the required indices "the Miller indices" of the plane (hkl). Thus Miller indices of a crystal plane can be obtained by adopting the following procedure:

   (i)   Determine the intercepts of the plane along x, y and z axes in terms of lattice parameters.
   (ii) Divide these intercepts by the appropriate unit translations.
   (iii) Note their reciprocals.
   (iv) If fractions result, multiply each of them by the smallest common divisor.
   (v)   Put the resulting integers in parenthesis (hkl) to get the required Miller indices of that and all parallel planes.

When a crystal plane is parallel to an axis, its intercept on that axis is infinity, the reciprocal of which is zero, gives the Miller index of that crystal plane. For example, the forward face of a cube which intersects the positive side of the x-axis only and is parallel to y and z-axes, has the Miller indices (100). The front face, which intersects the y-axis only and is parallel to other two axes has the Miller indices (010). Similarly, the top face has the Miller indices (001). These cube faces are shown in Fig. 7.3.

A zone may be defined as a set of crystal planes (or faces) which when intersect each other produce parallel edges. For example, six faces of a pencil (hexagon) intersect each other along one direction (the direction of pencil lead), constitute a zone (Fig. 7.4). There is neither any restriction in the number of faces in a zone nor they need be crystallographically equivalent. For example, the

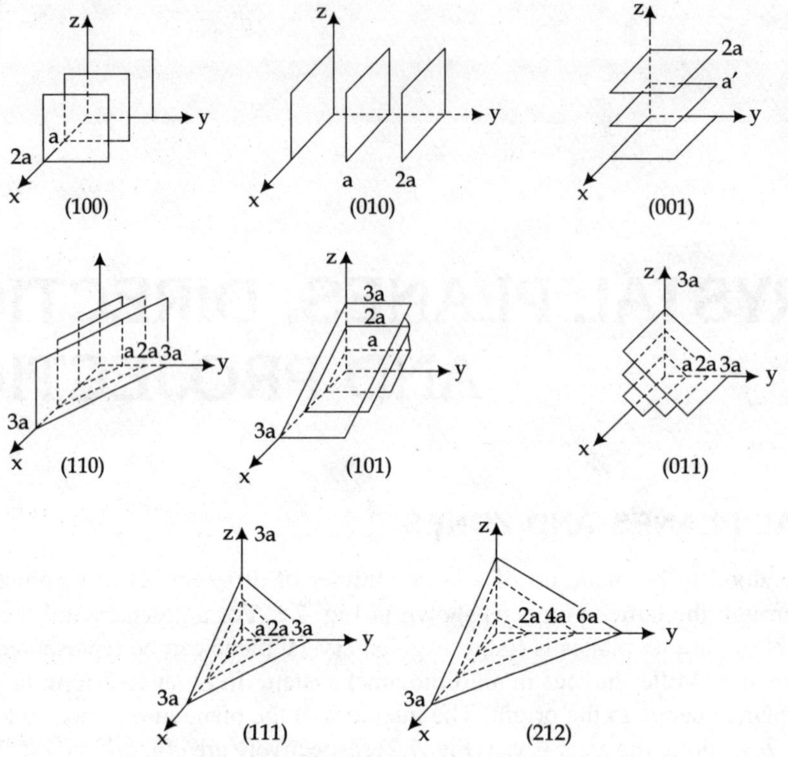

Fig. 7.1 Indices of principal planes in a cubic crystal

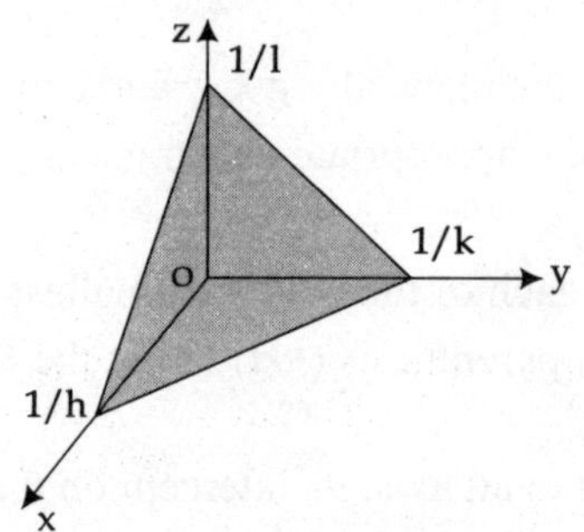

Fig. 7.2 Intercepts along x, y and z axes

edges of the pencil maybe shaved flat to give a 12-sided pencil, i.e. an additional six faces in the zone. On the other hand, a plane in a crystal can belong to a limited number of zones, 2-4.

7.2 CRYSTAL DIRECTIONS AND ZONE AXES

The crystal direction whose Miller indices is to be determined must pass through the origin of the unit cell. In order to know the indices of the given direction, the first step is to mark the point of

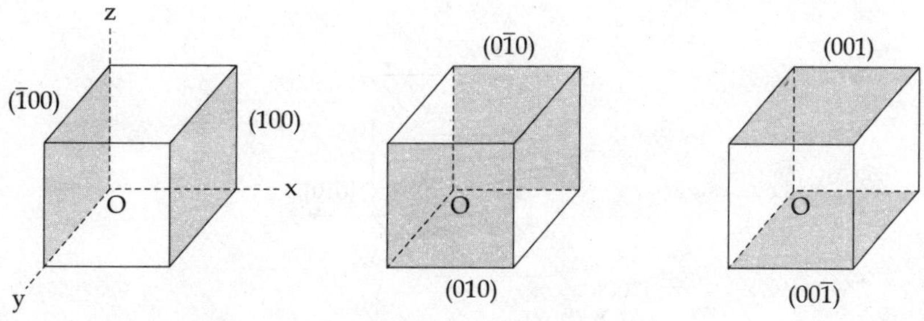

Fig. 7.3 Indices of six faces of a cube

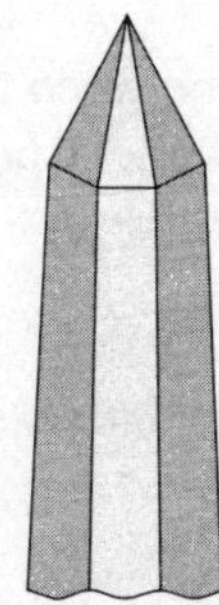

Fig. 7.4 Parallel edges in a pencil

intersection of the line at which it is leaving the unit cell. Note down the coordinates of this point in terms of fractions of unit cell parameters *a*, *b*, *c*. Remove the fractions and express them as the ratios of whole numbers. Put the digits in a set of square brackets, i.e. [*hkl*] to get the Miller indices of the required direction. Thus, Miller indices of direction can be obtained by adopting the following procedure.

   (i) Note down the coordinates of the lattice site nearest to the origin in the given direction.

  (ii) Divide the coordinates by appropriate unit translations.

 (iii) If, fractions result, multiply each of them by smallest common divisor.

 (iv) Put the resulting integers in a set of square brackets, i.e. [*hkl*] to get the required indices of that and all parallel directions.

Accordingly, the Miller indices of the lines along the three crystallographic axes *x*, *y*, *z* in an orthogonal system will be [100], [010] and [001], respectively as shown in Fig. 7.5.

The zone axis may be defined as the common direction of the intersection of the face in a zone. In a pencil, the direction of the lead is the zone axis corresponding to the six faces. In fact all crystal directions are zone axes, so the terms "crystal direction" and "zone axis" are synonymous. Thus a zone axis like a crystal direction, is represented as [*uvw*], defined as a vector *R* (measured from the crystal origin) according to the equation

$$R = ua + vb + wc \qquad\qquad (1)$$

where *a*, *b*, *c*, are translation vectors.

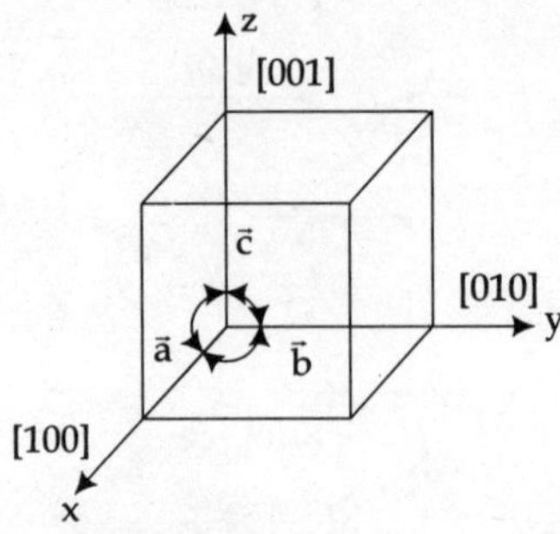

Fig. 7.5   Three crystallographic axes in an orthogonal system

In order to know the actual relationships between the zone axis and the planes in the zone, we provide here a few mathematical equations without proof. However, they will be proved later in section 7.6.

1. The Zone Law

The indices of a zone axis $[uvw]$ and a crystal plane (hkl) in the zone must obey the algebric relation

$$hu + kv + lw = 0 \tag{2}$$

2. Zone Axis at the Intersection of two Planes

The indices of the zone axis $[uvw]$ of two intersection planes $(h_1\, k_1\, l_1)$ and $(h_2\, k_2\, l_2)$ in a zone can be determined according to the following relations

$$u = \begin{vmatrix} k_1 & l_1 \\ k_2 & l_2 \end{vmatrix} \qquad v = \begin{vmatrix} l_1 & h_1 \\ l_2 & h_2 \end{vmatrix} \qquad w = \begin{vmatrix} h_1 & k_1 \\ h_2 & k_2 \end{vmatrix} \tag{3}$$

where the determinant $\begin{vmatrix} k_1 & l_1 \\ k_2 & l_2 \end{vmatrix} = (k_1 l_2 - k_2 l_1)$ etc.

These relationships can be easily remembered by using the following memogram

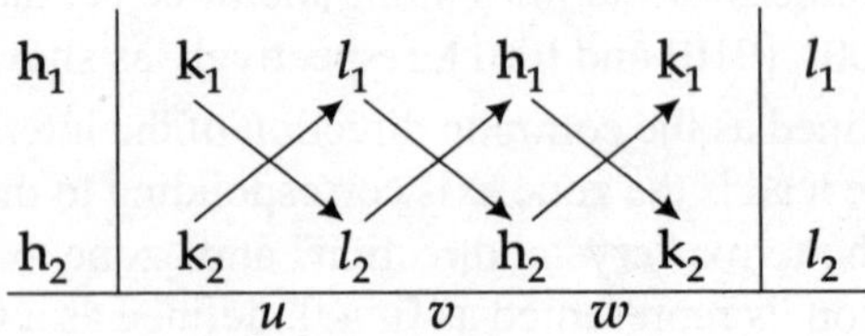

3. Plane Parallel to Two Directions

The indices of a plane (hkl) belonging to two zones with the zone axes $[u_1\, v_1\, w_1]$ and $[u_2\, v_2\, w_2]$ can be determined according to the following relations

$$h = \begin{vmatrix} v_1 & w_1 \\ v_2 & w_2 \end{vmatrix}$$

$$k = \begin{vmatrix} w_1 & u_1 \\ w_2 & u_2 \end{vmatrix} \qquad\qquad (4)$$

$$l = \begin{vmatrix} u_1 & v_1 \\ u_2 & v_2 \end{vmatrix}$$

As above, these relationships can also be easily remembered from the following memogram

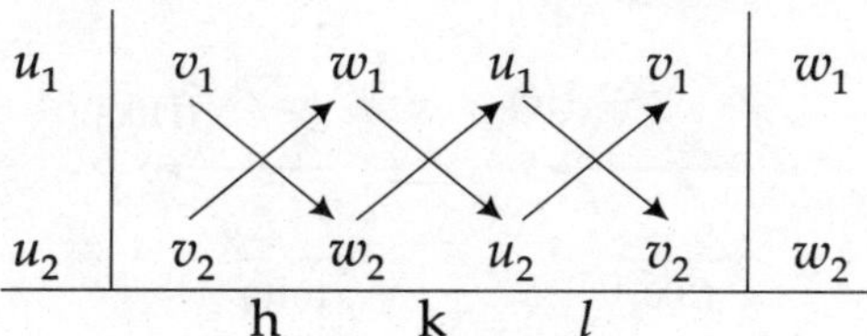

4.   The Addition Rule

The Miller indices of a plane (hkl) sandwitched between two planes of indices ($h_1k_1l_1$) and ($h_2k_2l_2$) in a zone can be determined according to the following relations

$$h = mh_1 \pm nh_2$$
$$k = mk_1 \pm nk_2 \qquad\qquad (5)$$
$$l = ml_1 \pm nl_2$$

where m and n are also whole numbers

7.3   MILLER-BRAVAIS INDICES

In order to represent a plane or a direction in a hexagonal crystal system, four indices are required. They are known as Miller-Bravais indices and are obtained exactly in a similar way as the Miller indices in an orthogonal system. Miller-Bravais indices are symbolically represented as ($hkil$). Out of the four indices, three indices correspond to three coplanar axes a_1 a_2 a_3 in the basal plane of the hexagon and the fourth index corresponds to the fourth axis (c-axis) perpendicular to the hexagonal prisms as shown in Fig. 7.6. Since the three coplanar axes a_1, a_2 and a_3 are related to one another by a rotation of 120° about the c-axis, so the corresponding three indices h, k and i are also related to one another. From a simple geometrical consideration, the relationship among the three indices is found to be

$$i = -(h + k)$$

or
$$h + k + i = 0 \qquad\qquad (6)$$

This implies that the third index is simply equal to the negative of the sum of initial two indices in the notation. The knowledge of the third index provides us a simple means of conversion from Miller indices to Miller-Bravais indices and vice-versa.

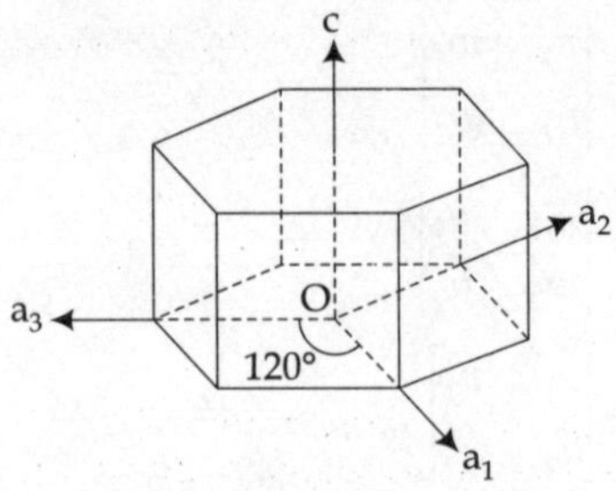

Fig. 7.6 Miller-Bravais axes

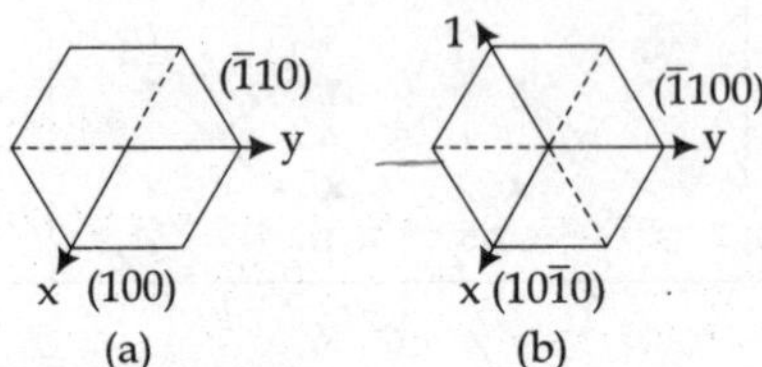

Fig. 7.7 Prismatic planes in: (a) Miller notation and (b) Miller-Bravais notation

Advantage of this conversion is fully appreciated when prismatic faces of the hexagonal unit cell are being indexed (Fig. 7.7). In the Miller index system of notation, some prismatic faces such as (100) and ($\bar{1}$10) appears crystallographically different. On the other hand, in Miller-Bravais System of notation, the above faces are represented by (10$\bar{1}$0) and ($\bar{1}$100). They are crystallographically equivalent.

Like planes, Miller-Bravais indices for directions are obtained in a similar way as Miller indices for orthogonal system and are symbolically represented as [$hkil$]. However unlike planes, the conversion of Miller indices to Miller-Bravais indices or vice-versa for directions is not simple. For a given crystal direction, the three Miller indices [HKL] and the four Miller-Bravais indices [$hkil$] are related through the following identities

$$H = h - i \qquad\qquad h = \frac{(2H - K)}{3}$$

$$K = k - i \qquad \Rightarrow \qquad k = \frac{(2K - H)}{3}$$

$$L = l \qquad\qquad i = -(h + k) = -\frac{1}{3}(H + K)$$

$$l = L$$

7.4 TRANSFORMATION OF INDICES

Different choice of axes (i.e. different unit cells) are frequently encountered in trigonal crystals (Sec. 2.5) to represent them either on rhombohedral axes or on hexagonal axes. Further, the study of solid state properties of crystals reveals that it is more convenient to choose an orthorhombic unit cell for the usual hexagonal unit cell. Similarly, it is more convenient to choose a primitive rhombohedral unit cell for the usual face centered cubic unit cell. Thus the choice of the alternative unit cells may be based on the convenience in representing the cells. However, sometimes it is advantageous also when some aspects of the atomic array of larger centered unit cell is compared with that of smaller primitive unit cell.

Whatever may be the reason, the actual transformation from one set of axes to the other is carried out quite simply by making use of the vector algebra. It is customary to take one of the unit cells as the first unit cell (correspondingly the first set of axes) and the alternative (or the transformed) unit cell as the second unit cell (correspondingly the second set of axes). Thus the second set of axes a_2, b_2 and c_2 can be defined in terms of the first set of axes a_1, b_1 and c_1 by the following simultaneous equations:

$$a_2 = m_{11}a_1 + m_{12}b_1 + m_{13}c_1$$
$$b_2 = m_{21}a_1 + m_{22}b + m_{23}c_1 \tag{7}$$
$$c_2 = m_{31}a_1 + m_{32}b_1 + m_{33}c_1$$

where m_{ij} $(i, j = 1, 2, 3)$ are the coefficients (components) of the vectors of the second set a_2, b_2 and c_2 when referred to the axes of the first set a_1, b_1 and c_1. The above equations can be written in the matrix form as

$$\begin{pmatrix} a_2 \\ b_2 \\ c_2 \end{pmatrix} = \begin{pmatrix} m_{11} & m_{12} & m_{13} \\ m_{21} & m_{22} & m_{23} \\ m_{31} & m_{32} & m_{33} \end{pmatrix} \begin{pmatrix} a_1 \\ b_1 \\ c_1 \end{pmatrix} \tag{8}$$

These equations obtained for the transformation of axes or the unit cell are also valid for indices of crystal planes. The change of axes causes a change in the unit cell volume. The two volumes are related through the expression

$$a_2 \times b_2 \cdot c_2 = \begin{vmatrix} m_{11} & m_{12} & m_{13} \\ m_{21} & m_{22} & m_{23} \\ m_{31} & m_{32} & m_{33} \end{vmatrix} a_1 \times b_1 \cdot c_1 \tag{9}$$

where the matrix of eq. 8 has been changed into determinant in eq. 9 and is known as modules of transformation. The determinant is an integer when the transformation is from smaller unit cell to larger unit cell, or a simple fraction when the transformation is otherwise.

Transformation of Indices of Crystal Planes (Unit Cell)

Let us consider a few practical cases for illustration

(i) Rhombohedral and FCC

The axial relationships between the translation vectors of the primitive rhombohedral and the *fcc* unit cells are shown in Fig. 7.8. Let us consider a crystal plane which is referred to as (*hkl*) in rhombohedral system of axes a_1, b_1, c_1 and (*HKL*) in *fcc* system of a_2, b_2, c_2.

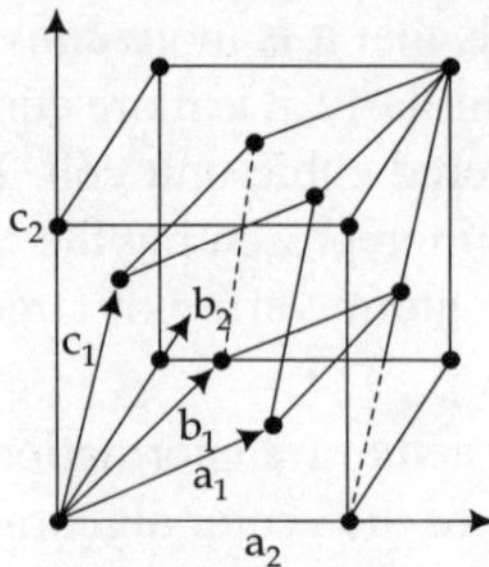

Fig. 7.8 Rhombohedral and cubic axes

Now making use of eq. 7, we can write the following transformation equation

$$H = 1h + 1k - 1l$$
$$K = 1h - 1k + 1l \tag{10}$$
$$L = -1h + 1k + 1l$$

The matrix form of these equations is

$$\begin{pmatrix} H \\ K \\ L \end{pmatrix} = \begin{pmatrix} 1 & 1 & -1 \\ 1 & -1 & 1 \\ -1 & 1 & 1 \end{pmatrix} \begin{pmatrix} h \\ k \\ l \end{pmatrix} \tag{11}$$

Consequently, the relationship between the volume of the two unit cells can be calculated from eq. 9 as

$$V_{fcc} = 4V_{rh} \tag{12}$$

The reverse relationships can be found by determining the inverse of the matrix. For this example, the inverse of the matrix is given by

$$\begin{pmatrix} 1 & 1 & -1 \\ 1 & -1 & 1 \\ -1 & 1 & 1 \end{pmatrix}^{-1} = \frac{1}{2} \begin{pmatrix} 1 & 1 & 0 \\ 1 & 0 & 1 \\ 0 & 1 & 1 \end{pmatrix}$$

and hence the reverse relationships are

$$\begin{pmatrix} h \\ k \\ l \end{pmatrix} = \frac{1}{2} \begin{pmatrix} 1 & 1 & 0 \\ 1 & 0 & 1 \\ 0 & 1 & 1 \end{pmatrix} \begin{pmatrix} H \\ K \\ L \end{pmatrix} \tag{13}$$

From equations 9 and 10, we can determine the indices of a plane in one system if the indices of other system are known.

(ii) Hexagonal and Orthorhombic

The axial relationships between the translation vectors of hexagonal and the orthorhombic unit cells are shown in Fig. 7.9. On a similar consideration as above we can express the orthorhombic indices (*HKL*) in terms of the hexagonal (*hkl*) through the following transformation equations

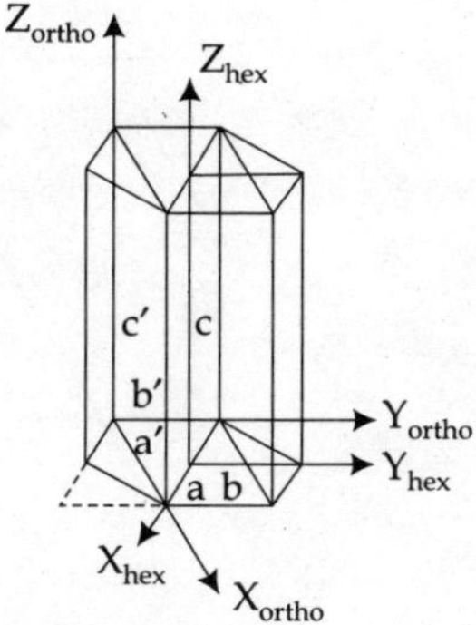

Fig. 7.9   Hexagonal and orthorhombic axes

$$H = 2h + 1k + 0l$$
$$K = 0h + 1k + 0l \tag{14}$$
$$L = 0h + 0k + 1l$$

The matrix of the equation is

$$\begin{pmatrix} H \\ K \\ L \end{pmatrix} = \begin{pmatrix} 2 & 1 & 0 \\ 0 & 1 & 0 \\ 0 & 0 & 1 \end{pmatrix} \begin{pmatrix} h \\ k \\ l \end{pmatrix} \tag{15}$$

Consequently, the relationship between the volume of the two unit cells is found as

$$V_{\text{orth}} = 2V_{\text{hex}} \tag{16}$$

After finding the inverse of the above matrix, the reverse relationships can be written as

$$\begin{pmatrix} h \\ k \\ l \end{pmatrix} = \frac{1}{2}\begin{pmatrix} 1 & -1 & 0 \\ 0 & 2 & 0 \\ 0 & 0 & 2 \end{pmatrix} \begin{pmatrix} H \\ K \\ L \end{pmatrix} \tag{17}$$

(iii) Rhombohedral and Hexagonal

Let the primitive rhombohedral cell with obverse setting is described by the unit vectors a_R, b_R, c_R such that $a_R = b_R = c_R$, $\alpha_R = \beta_R = \gamma_R$ and the 3-fold axis is along the direction $a_R + b_R + c_R$. Similarly, let the hexagonal unit cell is described by the unit vectors a_H, b_H, c_H such that $a_H = b_H$ and $c_H = a_R + b_R + c_R$.

The axial relationships of the two unit cells are shown in Fig. 7.10. From figure, we can write three sets of possible unit translations

$$a_H = a_R - b_R, \qquad b_H = b_R - c_R, \qquad c_H = a_R + b_R + c_R$$

or $\qquad a_H = b_R - c_R, \qquad b_H = c_R - a_R, \qquad c_H = a_R + b_R + c_R$

or $\qquad a_H = c_R - a_R, \qquad b_H = a_R - b_R, \qquad c_H = a_R + b_R + c_R$

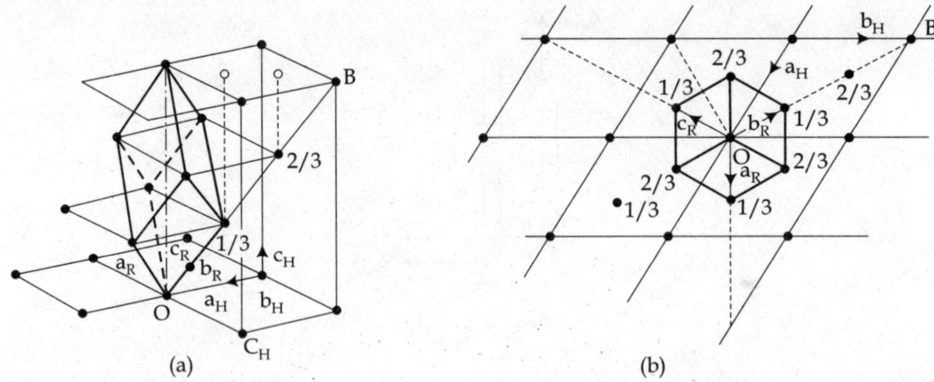

Fig. 7.10   Rhombohedral lattice. The basis of the rhombohedral cell is labelled a_R, b_R, c_R, the basis of the hexagonal centred cell is labelled a_h, b_h, c_h (numerical fractions are calculated in terms of the c_h axis) (a) Obverse setting. (b) the same figure as in (a) projected along c_h.

Now, let us suppose that a crystal plane is represented by the rhombohedral indices (*hkl*) and the hexagonal indices (*HKIL*). Making use of one of the above relationships (say first set), we can express the hexagonal indices in terms of the rhombohedral indices through the following transformation eqs.

$$H = 1h - 1k + 0l$$
$$K = 0h + 1k - 1l$$
$$L = 1h + 1k + 1l \tag{18}$$
$$I = -(H + K)$$

The matrix form of this equation is

$$\begin{pmatrix} H \\ K \\ L \end{pmatrix} = \begin{pmatrix} 1 & -1 & 0 \\ 0 & 1 & -1 \\ 1 & 1 & 1 \end{pmatrix} \begin{pmatrix} h \\ k \\ l \end{pmatrix} \tag{19}$$

The volume of the two cells are found to be related through

$$V_{hex} = 3V_{Rh} \tag{20}$$

Now, combining the hexagonal indices algebrically, we obtain

$$(-H + K + L) = 3k_{obv} \tag{21}$$

where *k* is an integer.

This is an important consequence which tells us that when the planes in a rhombohedral lattice (observe setting) are referred to hexagonal axes, only certain combinations of the hexagonal indices given by eq. 21 are allowed. Since such restrictions do not exist in primitive hexagonal, it becomes possible to distinguish the two cases by examining the list of indices. This has a real importance in the study of crystals by *X*-ray diffraction methods.

Now, we can write the reverse relationships after finding the inverse of the matrix in eq. 19 as

$$\begin{pmatrix} h \\ k \\ l \end{pmatrix} = \frac{1}{3} \begin{pmatrix} 2 & 1 & 1 \\ -1 & 1 & 1 \\ -1 & -2 & 1 \end{pmatrix} \begin{pmatrix} H \\ K \\ L \end{pmatrix} \tag{22}$$

Transformation of Indices of Direction (Zone Axes)

Unlike the transformation of indices of planes, the transformation of indices of direction is not identical to the transformation of axes. Let us find out the general equation for transformation of direction first and then we shall take a few practical examples. Let $[u_1 v_1 w_1]$ and $[u_2 v_2 w_2]$ are the indices of the same direction (a vector r) in the two sets of axes, respectively. The vector r in the lattice can be written in terms of the components of the two sets of the unit cell vectors, i.e.,

$$r = u_1 a_1 + v_1 b_1 + w_1 c_1 = u_2 a_2 + v_2 b_2 + w_2 c_2 \tag{23}$$

Substituting the values of a_2, b_2 and c_2 from eq. 7 in to eq. 23, we hence

$$r = u_1 a_1 + v_1 b_1 + w_1 c_1 = u_2(m_1 a_1 + m_{12} b_1 + m_{13} c_1) + v_2(m_{21} a_1 + m_{22} b_1 + m_{23} c_1)$$
$$+ w_2(m_{31} a_1 + m_{32} b_1 + m_{33} c_1)$$

Comparing the coefficients of $u_1 v_1$ and w_1, we have

$$\begin{aligned} u_1 &= m_{11} u_2 + m_{21} v_2 + m_{31} w_2 \\ v_1 &= m_{12} u_2 + m_{22} v_2 + m_{32} w_2 \\ w_1 &= m_{13} u_2 + m_{23} v_2 + m_{33} w_2 \end{aligned} \tag{24}$$

The matrix form of these equation is

$$\begin{pmatrix} u_1 \\ v_1 \\ w_1 \end{pmatrix} = \begin{pmatrix} m_{11} & m_{21} & m_{31} \\ m_{12} & m_{22} & m_{32} \\ m_{13} & m_{23} & m_{33} \end{pmatrix} \begin{pmatrix} u_2 \\ v_2 \\ w_2 \end{pmatrix} \tag{25}$$

Comparing this with eq. 8, the indices of the two sets as well as the matrix elements (rows and columns) are found to be interchanged. It is interesting to note that eq. 25 can also be used to transform the reciprocal lattice vectors and the coordinates of positions in the unit cell from one set of axes to another and vice-versa.

Hexagonal (4-index System) and Trigonal (3-index System)

Let $[u_1 v_1 i w_1]$ be the direction referred to the hexagonal axes (4-index system) and $[u_2 v_2 w_2]$ represent the same direction referred to the trigonal axes (3-index system), respectively. The axial relationships between the two systems is illustrated in Fig. 7.11. Thus the indices of direction in 3-index system can be written in terms of the indices of direction in the 4-index system by the following equations

$$\begin{aligned} u_2 &= 2u_1 + 1v_1 + 0w_1 = u_1 - i \\ v_2 &= 1u_1 + 2v_1 + 0w_1 = v_1 - i \\ w_2 &= 0u_1 + 0v_1 + 1w_1 = w_1 \end{aligned} \tag{26}$$

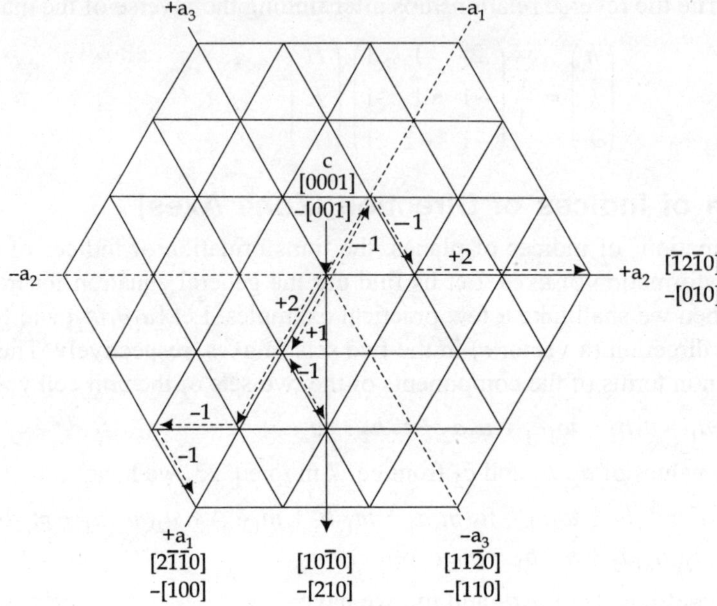

Fig. 7.11 Indices of directions in the hexagonal system with both three and four-digit indices

The matrix from these of equation is

$$\begin{pmatrix} u_2 \\ v_2 \\ w_2 \end{pmatrix} = \begin{pmatrix} 2 & 1 & 0 \\ 1 & 2 & 0 \\ 0 & 0 & 1 \end{pmatrix} \begin{pmatrix} u_1 \\ v_1 \\ w_1 \end{pmatrix} \qquad (27)$$

After determining the inverse of the matrix, the reverse relationships can be written as

$$\begin{pmatrix} u_1 \\ v_1 \\ w_1 \end{pmatrix} = \frac{1}{3} \begin{pmatrix} 2 & -1 & 0 \\ -1 & 2 & 0 \\ 0 & 0 & 3 \end{pmatrix} \begin{pmatrix} u_2 \\ v_2 \\ w_2 \end{pmatrix}$$

where the fourth index $i = -(u_1 + v_1)$.

7.5 CRYSTAL PROJECTIONS

In general crystal projections can be divided into two different categories.

1. Projection of atoms/ions in the unit cell
2. Projection of crystal faces (Miller Planes)

(i) Projection of Atoms/ions in the Unit Cell

When we deal with smaller unit cells or simpler crystal structures (for example the primitive Bravais lattices), it is easy to visualize the positions of the atoms/ions in the unit cell even without being plane projected. However, in longer unit cells or more complex crystal structures, it is not so easy to visualize them from simple diagrams or models because some of the atoms/ions may be hidden behind others and cannot be seen, hence the crystal projection is needed. Under this category, a plane projection of the crystal is taken perpendicular to the z-axis.

The first step is to define the origin and the three crystallographic axes of the given unit cell. Back left hand corner of the unit cell is chosen as the origin (in a cubic unit cell all the eight corners are equivalent and any one of them can be chosen as origin), the z-axis to the upwards, the y-axis to the right and the x-axis to the forward (out of the page).

The position of atoms/ions (or their coordinates) in the unit cell are specified as the fractions of the cell parameters along x, y and z directions. In this representation, the coordinates of a *bcc* unit cell are (000) and $\left(\dfrac{1}{2}\dfrac{1}{2}\dfrac{1}{2}\right)$. They specify the positions of atoms/ions at the origin (corner) and the body center of the unit cell. This actually means that there are two atoms/ions in a *bcc* unit cell, one at the origin and the other at the body center. Similarly, the coordinates of four atoms/ions in an *fcc* unit cell are (000), $\left(\dfrac{1}{2}\dfrac{1}{2}0\right)$, $\left(\dfrac{1}{2}0\dfrac{1}{2}\right)$, $\left(0\dfrac{1}{2}\dfrac{1}{2}\right)$, i.e., one atom/ions is at the origin and other three are at three face centers of the unit cell.

They are diagrammatically shown in Fig. 7.12. It is to be noted that only the z-coordinates are indicated in these diagrams; the x and y coordinates are clear from the projection and hence need not be written. However, the z-coordinates for corner atoms/ions need not be indicated because all the eight corners in a unit cell represent equivalent positions, as mentioned above. More complex structures can also be represented by this type of crystal projections. This type of crystal projections also help us to understand the similarities and differences between structures which is normally difficult to understand them otherwise. For example, Fig. 7.13(a) and 7.13(b) show the same crystal structure (perovskite, $CaTiO_3$). They look different because the origin of the cells have been chosen to coincide with different atoms/ions.

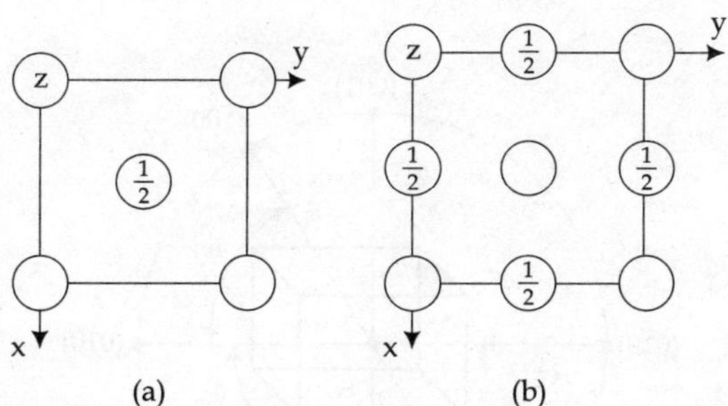

Fig. 7.12   Plane projection of *bcc* and *fcc* structures

(ii) Projection of Crystal Faces (Miller Planes)

We know that a crystal is a three dimensional object. Further, geometric measurements reveal that the crystal of the same substance can have different habits depending upon the number and size of

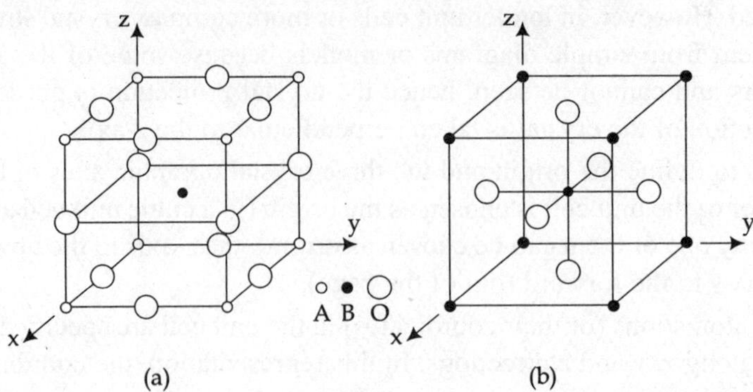

Fig. 7.13   Alternative unit cells of the perovskite ABO$_3$ structure.

the crystal faces, but the angles between their corresponding faces remain constant. For a better understanding of the crystallographic details, crystal projection are extremely useful. This problem is similar to the problem of a geographer who represents the spherical earth on to a globe and a plane map. In a similar manner, crystallographers like to represent the crystallographic information like angular relationships, symmetry, etc. on a spherical surface using plane projection and on a plane surface using plane projections.

Spherical Projection

This is the most simplest and convenient way of crystal projection. In this case, crystal is kept at the center of the sphere and normals are drawn from the point of intersection (coinciding with the center of the sphere) of the crystal axes to all faces of the crystal. They are extended until intersect the surface of the sphere. These normals correctly represent the orientation of various crystal planes being projected. The resulting collection of points (dots) on the surface of the sphere are called poles of the faces and constitute a spherical projection. The spherical projection of {100} planes of a simple cubic crystal is shown in Fig. 7.14.

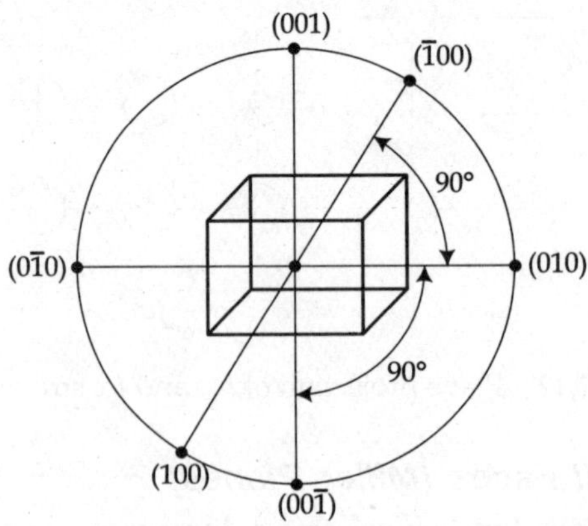

Fig. 7.14   The spherical projection {100} planes of a simple cubic crystal

The spherical projection has the following advantages.

(i) Mutually intersecting crystal planes are represented by poles (points) on a single spherical surface.

(ii) Angular relationships between the crystal planes are correctly represented by the poles, as illustrated in Fig. 7.14.

(iii) The zone axis of any two planes lie 90° away from their corresponding poles.

However, the most important disadvantage of the spherical projection is the same as that of a globe. In order to make it more convenient and meaningful, the following projections are used by the crystallographers to transfer a spherical drawing into a plane drawing.

Gnomonic Projection

Gnomonic projection is the simpler form of the plane projection derived from the spherical projection. In this case, the plane of the projection is tangent to the north pole of the sphere and the view point is placed at the center of the sphere. The straight lines connecting the center with the face poles (spherical poles) are extended until they intersect the tangential plane on which the gnomonic poles of the crystal are obtained. The distance of a gnomonic pole from the point of contact N (taken as the center of the projection) is given by

$$D_g = r \tan \alpha \tag{28}$$

where r is the radius of the sphere and α is the angle between the gnomonic pole and the center of projection. Fig. 7.15 illustrates the relation between the spherical projection and the corresponding gnomonic projection.

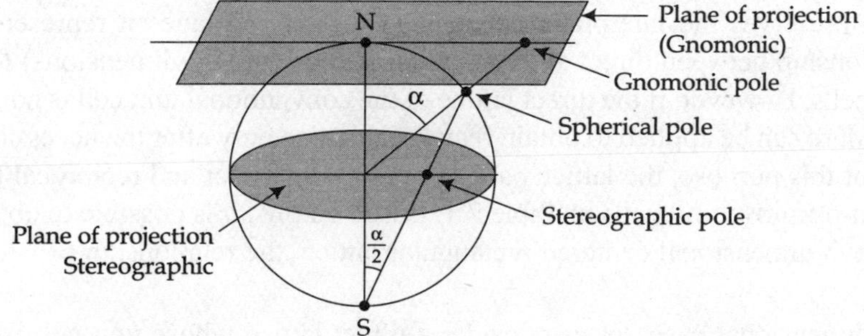

Fig. 7.15   Gnomonic and stereographic plane of projections and poles.

Stereographic Projection

The stereographic projection is another plane projection derived from spherical projection. In this case, the plane of the projection is the equatorial plane passing through the center of the sphere. The view point is placed on one of the poles (north or south) of the sphere. The locations of the stereographic poles are the point of intersections with the equational planes of the lines connecting the spherical poles with the view point (say) at south pole S. The center of the sphere is the center of the stereographic projection. The distance of an stereographic pole from the centre of the projection is given by

$$D_s = r \tan \frac{\alpha}{2} \qquad (29)$$

where r and α have the same meaning as in gnomonic case. Fig. 7.15 illustrate the relation between spherical projection and corresponding stereographic projection.

7.6 THE RECIPROCAL LATTICE

In the above discussion of crystal projections, viz. gnomonic, stereographic etc. we observed that they display only the orientation of the planes with respect to the crystallographic axes and not the interplanar springs. However, in the following we will see that the reciprocal lattice is a kind of projection which displays both.

In fact, to each direct lattice (one, two or three dimensional) there exists a corresponding reciprocal lattice, which has the same symmetry and can therefore be referred to the same type of coordinate axes. The reciprocal lattice to a given direct (primitive) lattice can be obtained according to the following procedure:

 (i) From a common origin draw normals to all sets of crystal planes

 (ii) Set the length of each normal equal to or 2π times the reciprocal of interplanar spacing d_{hkl}

 (iii) Mark a point at the end of each normal called the reciprocal lattice point.

A collection of points obtained in this way from various crystal planes form a lattice array and is known as reciprocal lattice. The coordinates of each reciprocal lattice point is denoted by the symbol hkl (without brackets) represents (hkl) plane of the direct lattice. In fact, the reciprocal lattice points preserve all the important characteristics of the planes they represent. The direction from the origin preserve the orientation of the plane, and the distance of the reciprocal lattice point from the origin preserves the interplanar spacing of the set of planes it represents. Figure 7.16 shows the relationship between direct and reciprocal lattice (in two dimensions) for some simple primitive unite cells. However, if the direct lattice of the conventional unit cell is non-primitive then the above procedure can be applied to obtain reciprocal lattice only after the necessary modification is made in it. For this purpose, the lattice parameters of both direct and reciprocal lattices are provided for all non-primitive unit cells in Table 7.1. Based on this, it is possible to obtain the reciprocal lattice of a two dimensional centered rectangular lattice, the relationship of which is shown in Fig. 7.17.

In a three dimensional case, let us consider a direct lattice whose unit cell is defined by the vectors a, b and c (Fig. 7.18). Since the unit cells are stacked tightly in a periodic manner to construct a space lattice, the vector to any unit cell corner can be written as

$$R = ua + vb + wc \qquad (30)$$

where u, v, w are integers and are the coordinates of the point at the end of the vector R. We now define a reciprocal lattice that is related to the space lattice in a particular way. For this purpose, let us erect from the same origin the reciprocal axes x^* normal to $b - c$ plane, y^* normal to $c - a$ plane and z^* normal to $a - b$ plane, respectively. Construct a reciprocal unit cell defined by the vectors a^*, b^* and c^* and define a reciprocal lattice vector similar to equation 30 as

$$R^* = ha^* + kb^* + lc^* \qquad ...(31)$$

where h, R and l are integers.

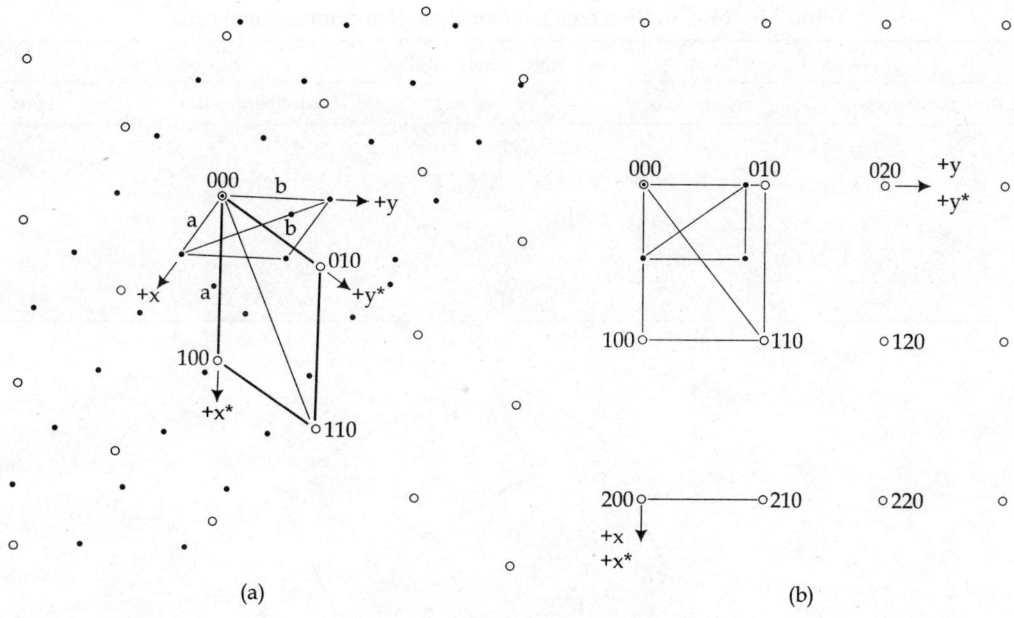

Fig. 7.16   (a) Oblique direct lattice and the reciprocal lattice (b) Primitive rectangular direct lattice and the reciprocal lattice.

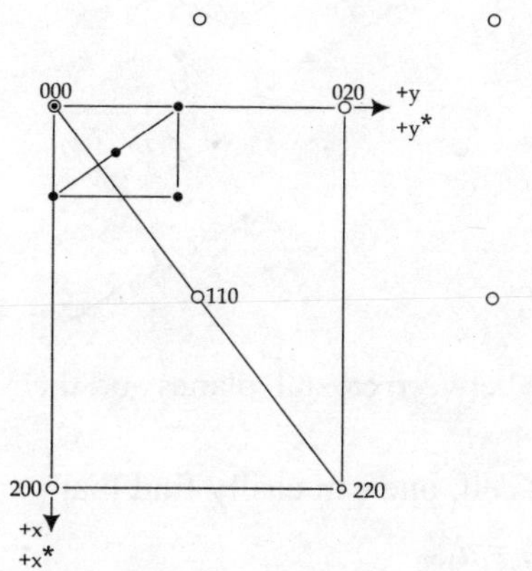

Fig. 7.17   Centred rectangular direct lattice and the reciprocal lattice.

Since we know that the scalar product (dot product) of two vectors is the product of their magnitudes times the cosine of the angle between them, it follows that

$$a^* \cdot b = a^* \cdot c = b^* \cdot a = b^* \cdot c = c^* \cdot a = c^* \cdot b = 0 \qquad (32)$$

Further, let us choose the length of a^* according to the procedure mentioned above such that

$a^* \cdot a = K$, where $K = 1, \lambda$ or 2π depending upon the situation. Here let $a^* \cdot a = 2\pi$

$$\Rightarrow \quad a^* a \cos \delta = 2\pi \quad \text{or} \quad a^* = \frac{2\pi}{a \cos \delta}, \text{ where } \delta \text{ is the angle between } a \text{ and } a^*.$$

Table 7.1 Modification required to include Non-primitive unit cells

Direct lattice		Lattice type	Reciprocal lattice	
Lattice parameters	Volume		Lattice parameters	Volume
		P and R	a^*, b^*, c^*	V^*
		A	$a^*, 2b^*, 2c^*$	$4V^*$
a, b, c	V	B	$2a^*, b^*, 2c^*$	$4V^*$
		C	$2a^*, 2b^*, c^*$	$4V^*$
		F	$2a^*, 2b^* \, 2c^*$	$8V^*$
		I	$2a^*, 2b^*, 2c^*$	$8V^*$

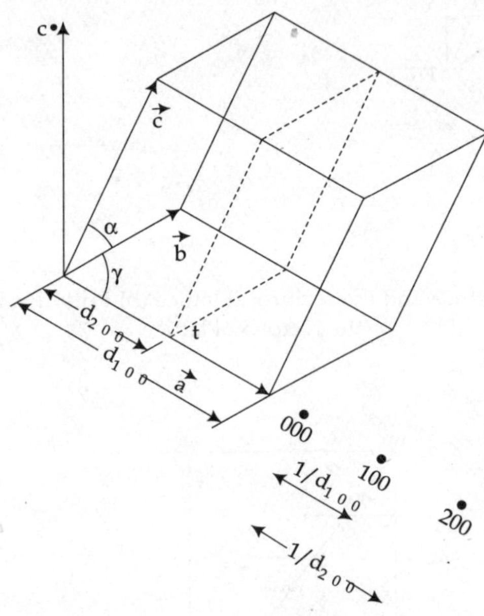

Fig. 7.18 Relationship between crystal planes and their reciprocal lattice points.

From a general triclinic unit cell, one can easily find that

$$a \cos \delta = d_{100}$$

Also, from vector algebra we know that the area of bc-plane, $A = bc \sin \alpha = b \times c$ then the volume of the unit cell is given by

$$V = \text{Area of bc-plane} \times \text{height} = A \times d_{100} = a \cdot b \times c$$

So that

$$a^* = \frac{2\pi}{d_{100}} = \frac{2\pi A}{V} = 2\pi \frac{b \times c}{a \cdot b \times c}$$

Similarly,

$$b^* = \frac{2\pi}{d_{010}} = 2\pi \frac{c \times a}{a \cdot b \times c}$$

and
$$c^* = \frac{2\pi}{d_{001}} = 2\pi \frac{a \times b}{a \cdot b \times c}$$

The exact relationships between the direct and the reciprocal lattice parameters and interaxial angles of primitive unit cells of various crystal systems are given in Table 7.2.

Table 7.2   Relationship between direct and reciprocal lattice parameters and interaxial angles

Triclinic

$$a^* = \frac{Kbc \sin \alpha}{V}; \quad b^* = \frac{Kca \sin \beta}{V}; \quad c^* = \frac{Kab \sin \gamma}{V}$$

where $\quad V = abc \{1 + 2 \cos \alpha \cos \beta \cos \gamma - \cos^2 \alpha - \cos^2 \beta - \cos^2 \gamma\}^{1/2}$

$$= 2\, abc \{\sin s . \sin (s - \alpha) . \sin (s - \beta) . \sin (s - \gamma)\}^{1/2}; \quad V^* = \frac{1}{V}$$

$$2s = \alpha + \beta + \gamma$$

$$\cos \alpha^* = \frac{\cos \beta \cos \gamma - \cos \alpha}{\sin \beta \sin \gamma}; \quad \cos \beta^* = \frac{\cos \gamma \cos \alpha - \cos \beta}{\sin \gamma \sin \alpha}; \quad \cos \gamma^* = \frac{\cos \alpha \cos \beta - \cos \gamma}{\sin \alpha \sin \beta}$$

Monoclinic

1st setting $\quad a^* = \dfrac{K}{a \sin \gamma}; \quad b^* = \dfrac{K}{b \sin \gamma}; \quad c^* = \dfrac{K}{c}; \quad \alpha^* = \beta^* = 90°; \quad \gamma^* = 180° - \gamma$

2nd setting $\quad a^* = \dfrac{K}{a \sin \beta}; \quad b^* = \dfrac{K}{b}; \quad c^* = \dfrac{K}{c \sin \beta}; \quad \alpha^* = \gamma^* = 90°; \quad \beta^* = 180° - \beta$

Orthorhombic	*Tetragonal*	*Cubic*
$a^* = \dfrac{K}{a}; b^* = \dfrac{K}{b}; c^* = \dfrac{K}{c}$	$a^* = b^* = \dfrac{K}{a}; c^* = \dfrac{K}{c}$	$a^* = b^* = c^* = \dfrac{K}{a}$
$\alpha^* = \beta^* = \gamma^* = 90°$	$\alpha^* = \beta^* = \gamma^* = 90°$	$\alpha^* = \beta^* = \gamma^* = 90°$

Hexagonal

$$a^* = b^* = \frac{2K}{a\sqrt{3}}; \quad c^* = \frac{K}{c}$$

$$\alpha^* = \beta^* = 90°; \gamma^* = 60°$$

Rhombohedral

$$a^* = b^* = c^* = \frac{K.a^2 \sin \alpha}{V} \quad \text{where} \quad V = a^3[1 - 3 \cos^2 \alpha + 2 \cos^3 \alpha]^{1/2}$$

$$\cos \alpha^* = \cos \beta^* = \cos \gamma^* = \frac{\cos^2 \alpha - \cos \alpha}{\sin^2 \alpha} = -\frac{\cos \alpha}{(1 + \cos \alpha)}$$

Some Geometrical Relationships

Some geometrical relationships, stated earlier without proof, can be proved easily by reciprocal lattice concept. Elementary knowledge of vector algebra will also be used.

1. The Zone Law

If a plane (hkl) lies in a zone $[uvw]$, then d^*_{hkl} is perpendicular to R_{uvw}, i.e.

$$d^*_{hkl} \cdot R_{uvw} = 0$$

or $\qquad (ha^* + kb^* + lc^*) \cdot (ua + vb + wc) = 0$

$\Rightarrow \qquad hu + kv + lw = 0$

where $a^* \cdot a = b^* \cdot b = c^* \cdot c = 1$

and $a^* \cdot b = a^* \cdot c = 0$, etc.

2. Zone Axis at the Intersection of Two Planes

The zone axis R_{uvw} lying at the intersection of the planes $(h_1 k_1 l_1)$ and $(h_2 k_2 l_2)$ is perpendicular to both their normals $d^*_{h_1 k_1 l_1}$ and $d^*_{h_2 k_2 l_2}$. Therefore, from simple geometry the zone axis can be expressed as

$$R_{uvw} = \frac{d^*_{h_1 k_1 l_1} \times d^*_{h_2 k_2 l_2}}{V^*} \tag{33}$$

where V^* is the volume of the reciprocal unit cell. substituting the values of various quantities, we have

$$ua + vb + wc = \frac{(h_1 a^* + k_1 b^* + l_1 c^*) \times (h_2 a^* + k_2 b^* + l_2 c^*)}{V^*}$$

$$= \frac{1}{V^*} [h_1 k_2 (a^* \times b^*) + h_1 l_2 (a^* \times c^*) + k_1 h_2 (b^* \times a^*) + k_1 l_2 (b^* \times c^*)$$

$$+ l_1 h_2 (c^* \times a^*) + l_1 k_2 (c^* \times b^*)]$$

$$= \frac{1}{V^*} [(a^* \times b^*)(h_1 k_2 - k_1 h_2) + (b^* \times c^*)(k_1 l_2 - l_1 k_2)$$

$$+ (c^* \times a^*)(l_1 h_2 - l_2 h_1)]$$

using the identies $(b^* \times a^*) = -(a^* \times b^*)$, etc.

again using the identities $a = \dfrac{b^* \times c^*}{V^*}$, etc. we can write $ua + vb + wc = c\,(h_1 k_2 - k_1 h_2) + a(k_1 l_2 - l_1 k_2) + b(l_1 h_2 - l_2 h_1)$.

Now, comparing the coefficients of a, b and c, we have

$$u = (k_1 l_2 - l_1 k_2), \quad v = (l_1 h_2 - l_2 h_1), \quad w = (h_1 k_2 - k_1 h_2).$$

3. A Plane (hkl) Containing Two Directions $[u_1 v_1 w_1]$ and $[u_2 v_2 w_2]$

The plane (hkl) is defined by its normal d^*_{hkl} is perpendicular to both directions $\gamma_{u_1 v_1 w_1}$ and $\gamma_{u_2 v_2 w_2}$. Therefore, d^*_{hkl} can be expressed as

$$d^*_{hkl} = \frac{\gamma_{u_1 v_1 w_1} \times \gamma_{u_2 v_2 w_2}}{V} \tag{34}$$

Substituting the values of various quantities, we have

$$ha^* + kb^* + lc^* = \frac{(u_1 a + v_1 b + w_1 c) \times (u_2 a + v_2 b + w_2 c)}{V}$$

$$= \frac{1}{V} [u_1 v_2 (a \times b) + u_1 w_2 (a \times c) + v_1 u_2 (b \times a) + v_1 w_2 (b \times c) + w_1 u_2 (c \times a)$$

$$+ w_1 v_2 (c \times b)]$$

$$= \frac{1}{V} [(a \times b)(u_1 v_2 - v_1 u_2) + (b \times c)(v_1 w_2 - w_1 v_2) + (c \times a)(w_1 u_2 - u_1 w_2)]$$

Making use of the identities $a^* = \dfrac{b \times c}{V}$, etc. and then comparing the coefficients of a^*, b^* and c^*, we have

$$h = u_1 w_2 - w_1 v_2, \quad k = w_1 u_2 - u_1 w_2, \quad l = (u_1 v_2 - v_1 u_2)$$

4. d-spacing between the planes (*hkl*) for orthogonal system

For orthogonal system, interplanar spacing can be defined as

$$\frac{1}{d_{hkl}} = |d^*_{hkl}| = (d^*_{hkl} \cdot d^*_{hkl})^{1/2}$$

$$= [(ha^* + kb^* + lc^*) \cdot (ha^* + kb^* + lc^*)]^{1/2}$$

$$= (h^2 a^{*2} + k^2 b^{*2} + l^2 c^{*2})^{1/2}$$

using identities $a^* \cdot b^* = 0$, etc.

Further, using $a^* = \dfrac{1}{a}$, etc., we have

$$\Rightarrow \qquad \frac{1}{d_{hkl}} = \left[\frac{h^2}{a^2} + \frac{k^2}{b^2} + \frac{l^2}{c^2} \right]^{1/2} \tag{35}$$

This is a general formula and is applicable to the primitive lattice of orthorhombic, tetragonal and cubic systems.

Special Cases:

Case I: Tetragonal system: $a = b \neq c$

Applying this condition, eq. 35 will reduces to

$$\frac{1}{d_{hkl}} = \left[\frac{h^2 + k^2}{a^2} + \frac{l^2}{c^2} \right]^{1/2} \tag{36}$$

Case II: Cubic system: $a = b = c$

Applying this condition, eq. 35 will reduce to

$$\frac{1}{d_{hkl}} = \left(\frac{h^2 + k^2 + l^2}{a^2} \right)^{1/2}$$

or
$$d_{hkl} = \frac{a}{\left(h^2 + k^2 + l^2 \right)^{1/2}} \tag{37}$$

7.7 SUMMARY

1. A zone in crystals includes all those planes (or faces) which when intersect, produce parallel edges. Further, in a zone there is neither any limitation in the number of planes nor they need be crystallographically equivalent. On the other hand, a plane in a crystal can belong to a limited number of zones, 2 – 4.

2. A zone axis in crystals is the common direction of the intersecting crystal planes in a zone. All crystal directions are also called zone axes, they are synonymous.

3. The Miller indices is a three index system of notation based on three orthogonal axes, while the Miller-Bravais indices is a four index system of notation based on four axes in the hexagonal system.

4. The study of transformation of indices (of planes or directions) provides many important information such as simplicity in representation of a unit cell or the study of solid state properties, the relationship between two sets of parameters etc.

5. Crystal projections are useful in many ways. They are helpful in (i) two dimensional representation of complex structures between two different but similar looking structures, (ii) understanding the similarities and differences between two different but similar looking structures, (iii) representing the crystallographic information like angular relationships, symmetries etc. on spherical or plane surfaces.

6. The reciprocal lattice is a special kind of projection which displays the orientation of the crystal planes w.r.t the crystallographic axes as well as the interplanar spacings.

7.8 DEFINITIONS

Crystal plane: A crystal is supposed to be made up of a large number of different set of equidistant parallel planes passing through the lattice points. Each crystal plane can be represented by a set of three or four integers.

Crystal direction: Any direction which is passing through the origin is a crystal direction. Each crystal direction can be represented by a set of three or four integers.

Gnomic Projection: When the lines drawn from a point within the crystal and normal to various crystal planes are extended to meet a plane to gives rise gnomonic poles.

Miller-Bravais indices (*hkil*):   It is a four index system of notation based on four axes in the hexagonal system.

Miller indices (*hkl*):   Index relating a plane to the reference axes of a crystal. Reciprocal of axial intercepts, cleared of fractions and of common multiples.

Reciprocal lattice: A theoretical lattice associated with a crystal lattice. The reciprocal lattice points are obtained by drawing normal to each crystal plane from a common origin and lying at a distance equal to the reciprocal of interlayer spacing.

Spherical Projection: When the lines drawn from a point within the crystal and normal to various crystal planes are extended to intersect the surface of the sphere to produce corresponding poles.

Steriographic Projection: When the lines drawn from one of the poles (north or south) of a sphere are passing through the equatorial plane and the spherical poles intersect a plane, producing the corresponding stereographic projection.

REVIEW QUESTIONS AND PROBLEMS

1. Sketch the following planes and directions in cubic unit cells: (123), [123], (112), [112], ($\bar{1}$10), [$\bar{1}$10]. Show that each of the above planes contains the [11$\bar{1}$] direction.

2. Draw ($\bar{1}$11), (1$\bar{1}$2) and ($\bar{2}$10) in cubic and orthorhombic unit cells. Also draw [1$\bar{1}$1], [1$\bar{2}$1] and [0$\bar{1}$2] directions in cubic and tetragonal unit cells.

3. Sketch the family of crystallographically equivalent face diagonal and body diagonal planes in a cubic unit cell. Determine the number of planes in each family.

4. Draw the family of crystallographically equivalent directions along face diagonals and body diagonals in a cubic unit cell. Determine the number of directions in each family.

5. What are the Miller indices of the line of intersection (i.e. the zone axis) of (111) and (11$\bar{1}$) planes in a cubic crystal and sketch them.   **Ans.** [$\bar{1}$10].

6. Which of the planes (101), (112), (321), (1$\bar{1}$1) lie parallel to the [1$\bar{1}$$\bar{1}$] direction?   **Ans.** (101), (321).

7. Which of the directions [1$\bar{1}$0], [432], [$\bar{2}$10], [23$\bar{1}$] lie parallel to the plane (115)?   **Ans.** [1$\bar{1}$0], [23$\bar{1}$].

8. Find the planes which lie parallel to the directions [131], [0$\bar{1}$1] and [102], [$\bar{1}$11].   **Ans.** ($\bar{4}$11) or (4$\bar{1}$$\bar{1}$), ($\bar{2}$$\bar{3}$1) or (23$\bar{1}$).

9. Find the directions which lie parallel to the planes (21$\bar{3}$), (110) and (210), (1$\bar{1}$1).   **Ans.** [3$\bar{3}$1] or [$\bar{3}$3$\bar{1}$], [1$\bar{2}$3] or [$\bar{1}$23].

10. Draw the directions (zone axes) [001], [010], [210], [110] in a hexagonal unit cell and determine their Miller Bravais indices.   **Ans.** [0001], [$\bar{1}$2$\bar{1}$0], [10$\bar{1}$0], [11$\bar{2}$0].

11. Obtain the matrix form of the equation for transformation of axes or planes. Determine the Miller indices in rhombohedral representation when the planes have the Miller indices (100), (110) and (111) in *fcc* representation.   **Ans.** (11$\bar{1}$), (200), (111)

12. Determine the Miller indices in orthorhombic representation when the planes have the indices (100), (110) and (111) in hexagonal representation.   **Ans.** (200), (310), (311).

13. Find the indices of directions in rhombohedral representation when the same directions in hexagonal system are [10$\bar{1}$0], [$\bar{1}$2$\bar{1}$0] and [11$\bar{2}$0].   **Ans.** [210], [030] = [010], [330] = [110].

14. Prove that the reciprocal lattice of *bcc* lattice is an *fcc* lattice.

15. Prove that the reciprocal lattice of *fcc* lattice is a *bcc* lattice.

CHAPTER 8

EXPERIMENT AND THEORY OF CRYSTAL GROWTH

8.1 INTRODUCTION

The study of crystal growth techniques is an important aspect in the field of crystallography and Materials science. Crystals of a number of elements and compounds are found in nature. They can also be grown in the laboratory using various growth techniques. However, in general crystals are found to grow during the course of phase transformations, i.e., when a substance changes from one physical state to another. It could be:

   (i) a change from the vapour phase to the solid-crystallization occurs by sublimation.

   (ii) a change from the liquid phase to the solid-crystallization occurs from a solution or a melt.

   (iii) a change from one solid phase to another-the phenomenon often called recrystallization.

However, the experimental data in the field show that three processes of crystal growth mentioned above are not equally prevalent; the second process is more common than the other two. Accordingly, we shall study the more widely used methods under the solution or the melt growth.

8.2 METHODS OF CRYSTAL GROWTH

As stated above, we shall confine our discussion to the methods only under the solution or the melt growth. Accordingly, the methods to be discussed are:

1. Solution Growth
    - (a) Aqueous solution method
    - (b) Flux method
    - (c) Hydrothermal method
2. Melt Growth
    - (a) Bridgman – Stockbarger method
    - (b) Czochralski method
    - (c) Zone – refining method
    - (d) Floating zone method

Before we discuss any method of crystal growth, it is necessary to check whether the particular method follows a general criteria or not. The general criteria for assessing a method are:

1. The universality, i.e., the number of materials to which the method may be applied.
2. The size and quality of the grown crystals.
3. The requirement of chemicals and apparatus.
4. The requirement of experience and time.
5. Industrial application and cost of per acceptable crystal.

Solution and melt growth in general satisfy all the above said criteria and are widely used.

8.3 SOLUTION GROWTH

The expression "solution" is most commonly used to describe the homogeneous mixture of two or more substances (they may be gases, liquid or solid). It is usually convenient to think of the bulk liquid as solvent—the dissolving substance as the solute and the resultant mixture as solution.

The solute material is crushed and made in powder form so that it quickly gets dissolved in the solvent. Stirring or vigorous shaking helps to promote the rapid mixing of the two components.

The solubility of a substance depends on the temperature of the solvent at that instant of time. It can be increased by raising the temperature of the solvent. For most substances, the solubility versus temperature curve will look like as shown in Fig. 8.1. However, the solubility of different solute materials will be different for a given solvent.

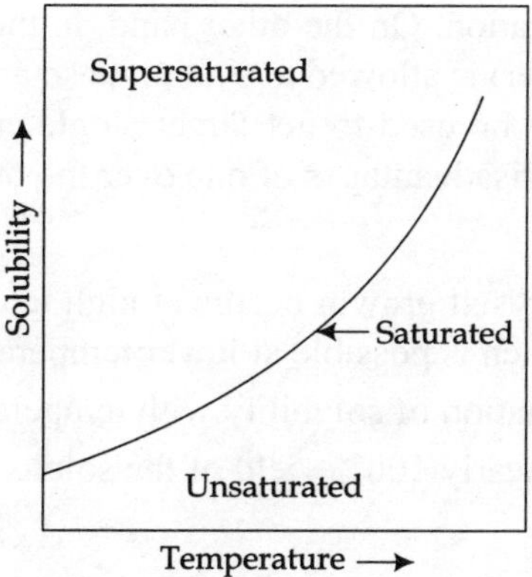

Fig. 8.1 Solubility vs. temperature curve

A solution that is in equilibrium with the solid phase (solute) is said to be saturated with respect to that solid. If it has more dissolved solid than that represented by saturated condition, it is known as supersaturated solution. This is achieved when the temperature of the solution is raised slightly above the normal dissolution temperature. This can also be achieved by allowing the solution to evaporate without disturbance in a dust free atmosphere. The rate of evaporation (as in summer) will increase the rate of supersaturation. In fact, this is an essential feature of all crystallization.

The supersaturation of a solution can be found if the concentration (C) of the solution is measured at a given temperature and the corresponding equilibrium concentration (C^*) is known. The two concentrations provide us the following:

Concentration driving force, $\Delta C = C - C^*$

Supersaturation ratio, $S = \dfrac{C}{C^*}$

and

Relative supersaturation, $\sigma = \dfrac{\Delta C}{C^*} = S - 1$

The concentration of a given solution can be determined directly by chemical analysis or indirectly by measuring some of the properties of the system that is sensitive to concentration. They include density, viscosity, refracture index and electrical conductivity.

(a) Aqueous Solution Method

Crystal growth from aqueous solution is probably the oldest and easiest method. This method has vast industrial applications. One can grow a large number of crystals ranging from table salt (or sugar) to copper sulphate on an industrial scale. Once a homogeneous solution is prepared, the crystals can be grown by two different methods: (i) slow cooling method and (ii) slow evaporation method

In the first case, the temperature of the system is lowered slowly ($\sim 1°$/hr) so as to reduce the solubility leading to the crystallization. On the other hand, in the second case, the temperature is held constant but the solvent (water) is allowed to evaporate so as to reduce the solubility to obtain crystallization. Seed crystals may be used to get larger single crystals. The two methods can be compared to see the advantages/ disadvantages of one over the other. They are:

Advantages:

1. In evaporation method, crystal growth occurs at high temperature. This avoids the formation of any compound which is possible at lower temperature.
2. Crystals having small variation of solubility with temperature can be grown.
3. If carried to completion, nearly 100% yield of the solute phase is obtained.

Disadvantages:

1. Nucleation at the melt surface may be excessive and thus intergrown crystals may result.
2. There is less control over the growth process than in slow cooling, which typically produced better quality crystals.

(b) Flux Method

Flux growth is the term which is most commonly used to describe the growth of crystals from molten solvents at high temperatures. Single crystals of some important materials like Al_2O_3 (ruby gem), $BaTiO_3$ (ferroelectric materials) and YIG (Yttrium Iron Garnet) are grown by this method on large scale.

One of the important aspects of this method is the choice of the suitable flux for a given solute material. In general, this is difficult because of lack of experimental data on (i) phase diagram, vapour pressure and viscosity at normal temperature and (ii) density, coefficient of thermal expansion, thermal and electrical conductivity and heat capacity at higher temperatures. However, with some efforts, an optimum solvent can be obtained. It is to be mentioned here that the flux material normally chosen should have at least one ion common with the solute material. For example, BaO/ Ba_2O_3 is used as a flux for YIG ($Ba_2Zn_2Fe_{12}O_{12}$).

The flux should have the following characteristic features:

1. High solubility for solute and very low solubility for the crucible material in general.

2. In order to arrive at a compromise between the solubility and the volatility, mixed solvents are used. An eutectic composition is the most suitable choice (when two or more substances are mixed, each component has the property of lowering the freezing point of the other, the minimum freezing point attainable for all possible mixtures is termed as eutectic mixture).

3. Viscosity should be in the range 1 to 10 centipoise.

4. Melting point should be low.

5. Volatility should be low at the highest temperature.

Flux growth involves high temperature furnace, temperature controller, temperature measuring device and crucibles. An schematic diagram of the experimental set up is shown in Fig. 8.2. For high temperatures, a resistance heating or an induction heating system may be used. For resistance heating, normally Kanthal, Al or Si–Carbides and MoS_2 (molybdenum disulphide) heating elements are used. Thermal insulation is achieved by fire bricks. ZrO_2 phaque is used to separate the crucible from heating element. In general, platinum crucibles are used as container for the growth from melt. However, crucibles of quartz glass, graphite and other ceremic materials may also be used. The whole system is enclosed in a high temperature resistant metal shell. An adequate ventilation is provided because many of the flux materials such as volatile oxides and fluorides are toxic.

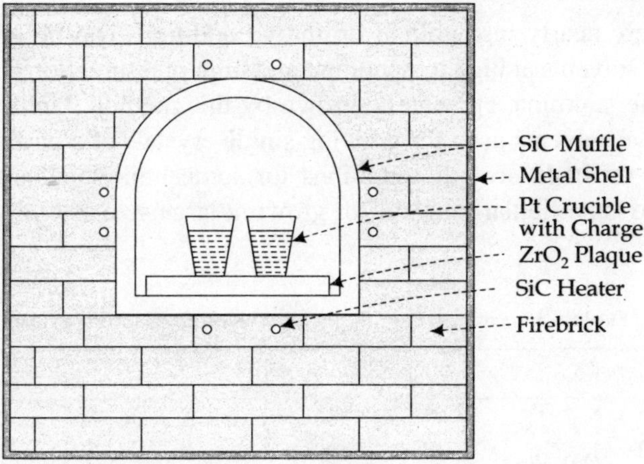

Fig. 8.2   Schematic of small scale flux growth furnace

Solute and flux materials are mixed and kept at a temperature well above the melting point for about 12–24 hours to ensure complete dissolution. The molten solvent is then left for growth using one of the following methods:

(i) Slow cooling method

(ii) Slow evaporation method

(iii) Temperature gradient method

First two methods are the same as discussed in aqueous solution method. On comparison, we observe some advantages/disadvantages of evaporation method over slow cooling method as pointed out earlier.

In order to grow crystals by temperature gradient method we can use either travelling heater type or travelling solvent zone type heating arrangement. In either case, the temperature difference across the melt is 5–50°C to vary the supersaturation and consequently the rate of crystallization.

Advantages of Flux Growth:

1. The crystals are relatively free from thermal strain.

2. The container problem for refractory materials with very high melting point is avoided.

3. Materials normally decomposes below m.p. can be flux grown much below that temperature.

4. Materials volatile at m.p. can be flux grown much below that temperature.

5. Dislocation density are low as compared to other methods ($\sim 10^2/cm^2$)

Disadvantages:

1. Relatively small size crystals are obtained.

2. Invariably the ions of solvent phase are present in the crystal. Suitable choice of flux can minimise this.

(c) Hydrothermal Method

The materials which are nearly insoluble at ordinary conditions can be grown by hydrothermal method using aqueous solvents at high temperature and high pressure. Some of the important materials like quartz, calcite, alumina, etc. can be grown by this method. Unlike the flux method, the composition of effective solvents used for growing single crystals of a particular solute material is fairly standard in this method. Growth conditions for some important materials are provided in Table 8.1. This method is particularly useful for growing large size crystals or crystals with faster rates.

Table 8.1 Growth conditions for some important materials

Material	*Solvent*	*Temperature (°C)*	*Fill (%)*	*Rate (miles/day)*
SiO_2	1 M NaOH	380	82	80
Al_2O_3	1 M K_2CO_3	490	80	10
ZnO	5 M KOH	350	80	10
$Y_3Fe_5O_{12}$	20 M KOH	350	88	5
$YbFeO_3$	20 M KOH	350	88	5
ZnS	5 M NaOH	350	85	2

Hydrothermal growth involves an autoclave, heating system, temperature controller and temperature measuring devices. Since the growth is carried at high temperature (~500°C) and high pressure (~3 k bars) using aqueous solvents, an autoclave becomes most important part of the system. Another important part is a furnace capable of producing, controlling and measuring temperature. A pressure measuring device is also used. An ideal autoclave should have the following characteristic features:

1. Inertness to acids, bases and oxidizing agents.
2. Ease in assembling and dismantling.
3. Sufficient length for producing large internal temperature differences.
4. Sufficient size for the growth of relatively large size crystals.
5. Moderate cost.
6. Convenient in use, safe in operation and simple to fabricate.

Different types of autoclaves are in use. They are provided in Table 8.2. Depending upon the pressure-temperature range, they can be grouped as low, medium or high pressure autoclaves. The choice of a suitable autoclave depends on the working pressure-temperature range and the nature of aqueous media used. Glass vessels are used for low pressure work. This has the advantage that one can directly observe the growth process through it. Steel vessels are normally used for medium pressure work.

Table 8.2   Different autoclaves with their working conditions

Autoclave	*I.D/Wall Thickness*	*Pressure–Temperature Range*
Pyrax tube	5 mm/2 mm	6 bar at 250°C
Quartz tube	5 mm/2 mm	6 bar at 300°C
More type	2.5 cm	400 bar at 400°C
Modified Bridgman-type	2.5 cm	3700 bar at 500°C
Tuttle-Roy-type	5 mm	5000 bar at 750°C

Wall thickness of the autoclave is calculated on the basis of some specified parameters: the internal diameter D, the length L, the pressure p and the temperature T. Usually, L/D ratio is taken as 10. Further, for thick walled autoclave,

$$\frac{2(S - C)}{D} \geq 0.5$$

where S is wall thickness and C is allowed safety margin for actual construction and D is the internal diameter.

The system employed for the growth of most materials is similar to that used for quartz and is as follows. Crushed fine particles of α-quartz (called nutrient) is placed in the bottom part of the vessel, called the dissolution region and single crystal seed plates of α–quartz are suspended in the upper part of the vessel, called the growth region as shown in Fig. 8.3. The vessel is filled to some predetermined fraction of its volume, typically 0.80 (often referred to as 80% fill) with a basic

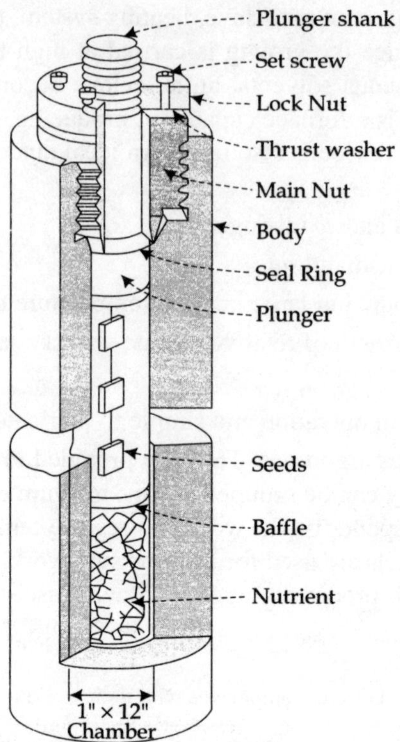

Fig. 8.3   Modified Bridgman hydrothermal autoclave

solution such as 0.50 M NaOH. The fill percent is referred to the volume after deducting the volume of the starting material. For example, if the volume of the vessel is 100 ml and the volume of α–quartz is 20 ml, then the free volume is 80 ml. Therefore, 80% of 80 ml is 64 ml, is called 80% fill.

The vessel is then placed in furnace designed to heat the lower (dissolution) region isothermally hotter than the upper (growth) region, which is also maintained isothermal. A perforated metal disc called "baffle" is placed within the vessel separating the dissolution and growth regions to help in localizing the temperature differential.

As the temperature is raised (the pressure also increases), the liquid level rises. The saturated solution in the lower region slowly rises by convection in relatively cooler growth region where the solubility decreases and the solution becomes supersaturated. This helps in growing the seeds placed in this region and the unspent solution returns to the dissolution region. This process continues till the whole charge is recrystallized. The solvent simply acts as a transport agent for the solid phase, so that large amount of substance can be recrystallized to obtain large crystals. Typical operating conditions for quartz are: Dissolution temperature = 400°C, growth temperature = 360°C, degree of fill = 80%, base solution – 1 M NaOH, pressure ~2 K bar, normal growth rate ~ 1 mm/day.

8.4   MELT GROWTH

The expression "melt" is understood to be a congruent melt of a given material (element or compound) above its melting paint. The materials to be grown by melt should have following characteristic features:

(i) low m.p.

(ii) melt congruently and without decomposition.

(iii) no solid state transformation below its m.p.

Once a congruent melt is obtained, a single crystal can be growth by either Bridgman–stockbarger method or Czochralski method. Further, purification of materials as well as their growth can be carried out by zone refining and floating zone methods. A brief description of each method is provided below.

(a) Bridgman–Stockbarger Method

The Bridgman–Stockbarger set up includes three important components. They are:

(i) A crucible (or a boat) of a suitable material and of suitable geometry.

(ii) A suitable furnace capable of producing the desired temperature gradient, and

(iii) Temperature controlling and measuring device

Cylindrical crucible of suitable material and of different geometry are adopted (Fig. 8.4). However, in general a cylinderical crucible having conical shape at the lower end (Fig. 8.4b) is preferred so that initially only a small volume of the melt is solidified. This favours the formation of a single nucleus which is turn favours the formation of a single crystal. Sometimes, a horizontal boat with conical shape at one end is also used (Fig. 8.4g). In this case, the furnace axis is horizontal rather than vertical. It is to be noted that the growing material must not react with the crucible material at the growth temperature which may give rise altogether different crystal material. It is also desirable that the solid does not stick to the crucible. This helps to minimize the strain on the crystal surface and also makes possible to remove the grown crystal without destroying the crucible.

Different crucible/ampoule materials are in use. The choice of a suitable crucible material depends on the working temperature range. Pyrax glass could be used upto about 600°C. Similarly,

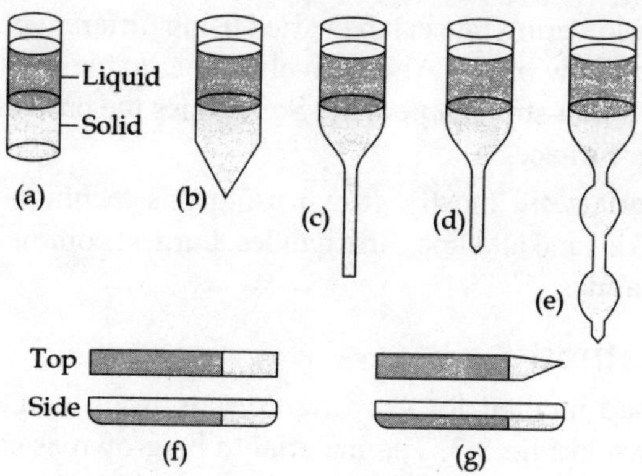

Fig. 8.4   Crucibles of different geometry

Vycor (glass) upto ~1000°C, vitreous silica ~1200°C, alumina, noble metals or graphite ~2500°C (in non oxidizing atmosphere) could be used. In case of graphite crucible, it is necessary to pass inert gas through the furnace.

Like crucible/ampoule materials, furnaces of different temperature gradients are in use. Some of the most common furnace gradients used are shown in Fig. 8.5. When the crucible/ampoule is lowered through the furnace (or the furnace is raised about the crucible). Usually the gradient of type (a) is used. This has two isothermal regions with a suitable gradient in between them. This permits annealing of the crystal after growth without introducing large thermal strains caused by excessive temperature gradients. On the other hand, when the crucible/ampoule is kept stationary and the whole furnace is cooled, a furnace gradient of type (b) is used. However, this introduces relatively large thermal strains.

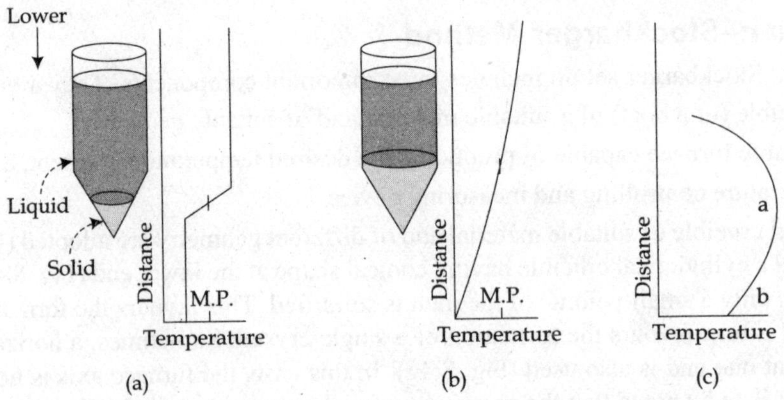

Fig. 8.5 Different types of temperature gradients

The block diagram of conventional Bridgman–Stockbarger set up is shown in Fig. 8.6. The material to be grown as single crystal is put into the crucible/ampoule, which is then evacuated, sealed and kept in the furnace. The crucible/ampoule is attached to a wire passing through a pully and a gear assembly. A counter weight is suspended at the other end of the wire to ensure a smooth motion of the crucible. The lowering rate can be varied using different gear assemblies. Usually, the lowering rate is kept about a few *mm/h*. Alternatively, the crucible may be directly connected to a slow speed motor through a non-stretchable wire. Sometimes the crucible is rotated to even out the thermal assymmetry in the furnace.

Three classes of materials are mostly grown using this technique. They are: (i) metals (ii) semiconductors and (iii) alkali and alkaline earth halides. Largest commercial application has been to alkali and alkaline earth halides.

(b)   Czochralski Method

This is another widely used method for growing crystals from melt. The block diagram of the Czochralski set up is shown in Fig. 8.7. The material to be grown as single crystal is crushed and kept in a crucible, which is then placed inside the furnace capable of producing the desired tempera-

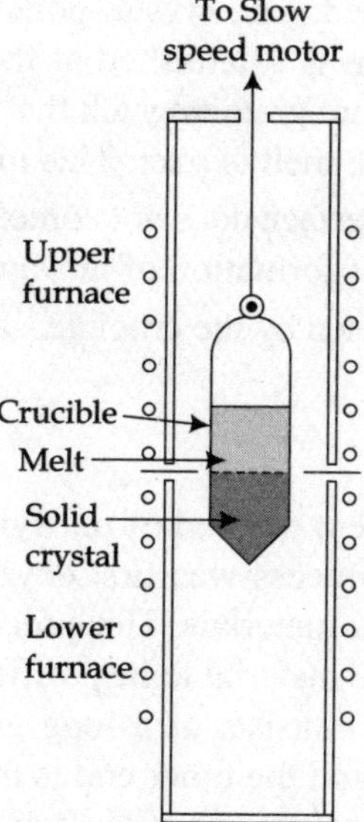

Fig. 8.6 Conventional Bridgman set up.

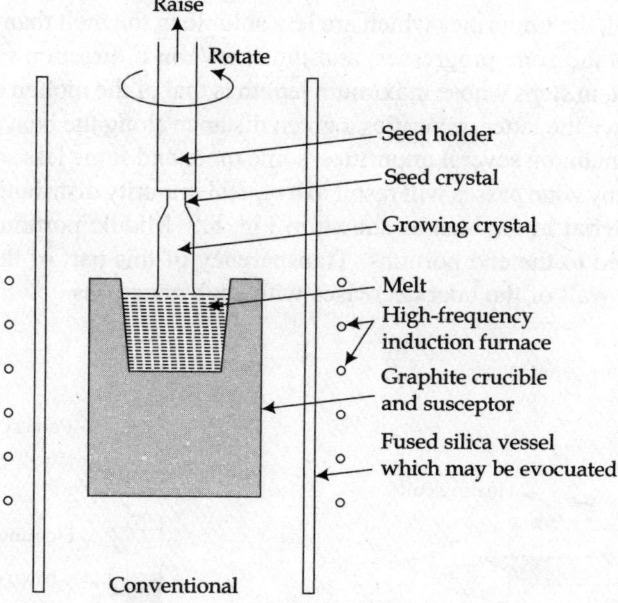

Fig. 8.7 Conventional Czochralski set up.

ture so that a congruent melt is obtained. The crystal puller holding a seed crystal is lowered to touch the melt. As soon as equilibrium is established at the seed and melt interface, the seed is slowly lifted under a suitable temperature gradient such that a solid–liquid interface is maintained. The process of growth continues till the melt is over. This method has the following advantages:

(i) In this case, the solid–liquid interface does not come in contact with the wall of the crucible which reduces the possibility of formation of unwanted nuclei.

(ii) The grown crystal is not contained by the crucible, so it is not subjected to any deformation resulting due to contact.

(c) Zone Refining Method

Zone refining is a simple and convenient method of purifying those solid materials which can be melted without decomposition. This process was first developed by W.G. Pfann in 1952 for the purification of inorganic semiconductor materials. This process involves repeated passage of a narrow molten zone along the length of the material to be purified. Fig. 8.8 shows a schematic diagram of horizontal zone refining set up. It consists of a long glass/quartz tube, one end of which is connected to the inlet of the inert gas and the other end is made narrow for the gas to escape. The material to be purified is kept in a boat also made of glass or quartz is placed inside the long tube. A liquid zone is created by melting a small amount of the material on the narrow end of the boat. It is then made to traverse slowly throughout the boat containing the charge. Impurities which are more soluble in the melt than the host material are carried along the direction of movement of the molten zone. On the other hand, the impurities which are less soluble in the melt than the host material will tend to solidify first as the zone progresses, and thus move in a direction opposite to that of the molten zone movement, in steps whose maximum length is that of the molten zone. More passes are therefore needed to move the latter impurities a given distance along the boat than for the former. If the starting material containing several impurities, some more and some less soluble in the melt than in the host material, many zone passes will result in a typical impurity distribution along the length of the zone – refined material in the boat as shown in Fig. 8.8. Middle portion of the ingot is much more purer as compared to the end portions. Transparency of this part of the ingot increases and the stickiness with the wall of the boat decreases with each zone pass.

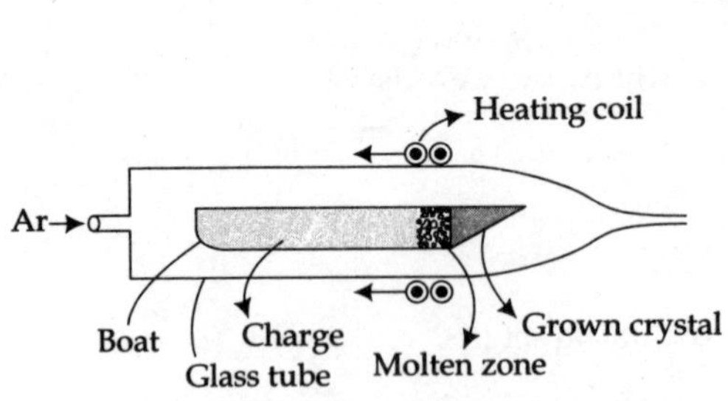

Fig. 8.8   Schematic zone refining set up.

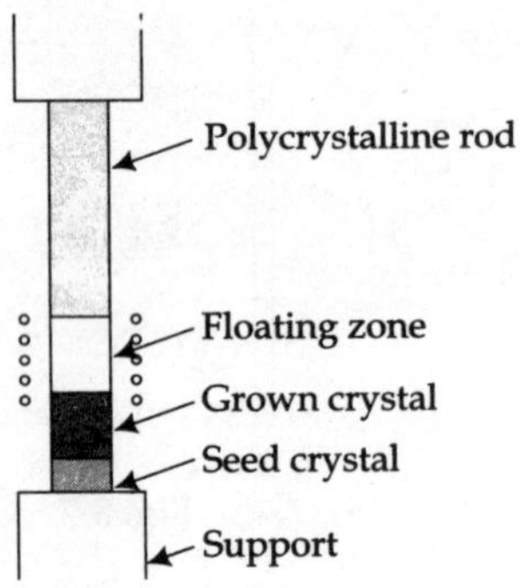

Fig. 8.9   The float zone method

(d) Float Zone Method

This technique was first developed by Keek and Golay in 1953 for the purification of silicon rod initially grown by some other methods such as Bridgman – Stockbarger method or Czochralski method. In this case, a cylinderical (polycrystalline) rod to be purified is held vertically as shown in Fig. 8.9. A small section of the rod is heated to melting, where the surface – tension holds the molten zone. A slow movement of the heater or of the rod along the axis leads to a progressive melting on one side of the zone and a corresponding solidification on the other side. If the latter is a single crystal seed, the whole rod can be converted into a single crystal. This method has yet another advantage that this does not require any crucible or boat and hence reactivity is no longer any problem. Silicon rod is purified by this method.

8.5 NUCLEATION

Nucleation is the most fundamental requirement for crystal growth. In a supersaturated (or super-cooled) solution, when a few atoms or molecules join together and form a cluster, a reduction in the free energy takes place. The cluster consisting of a group of atoms or molecules is normally termed as 'embryo'. An embryo may continue to grow to form a crystal provided it acquires a critical size (known as critical nucleus) or disintegrate completely. Thus the birth of a critical nucleus is an important event in crystal growth. The following four stages are involved to form a stable nucleus.

  (i) The development of supersaturated stage. This can be achieved by chemical reaction, change in pressure, change in temperature or any other physical or chemical conditions.

 (ii) The generation of embryo. This can be achieved either by homogeneous process in which the atoms or molecules build the embryo themselves (spontaneous process) in the interior of the parent system or by heterogeneous process where the atoms or molecules are collected around an impurity or other imperfections.

(iii) Birth of the critical nucleus. This can be achieved when the embryo continues to grow from an unstable or metastable state to a state of critical size.

 (iv) The relaxation stage. This is the stage where the texture of the newly formed nucleus alters. That is the nucleus starts taking a shape of crystal lattice.

Nucleation may occur spontaneously or it may be induced artificially. They are referred to as homogeneous and hetrogeneous nucleation, respectively. The interaction between the monomers and the nucleus is direct in homogeneous case while direct and indirect both in heterogeneous case (Fig. 8.10). Together they are referred to as primary nucleation. On the other hand, if the nuclei are generated in the vicinity of crystallites present in a supersaturated system, is referred to as the secondary nucleation.

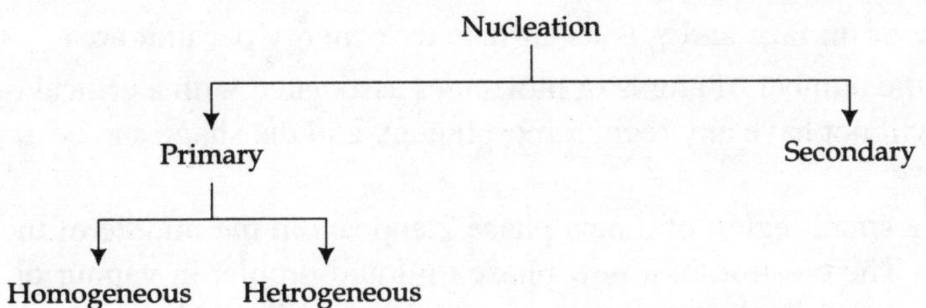

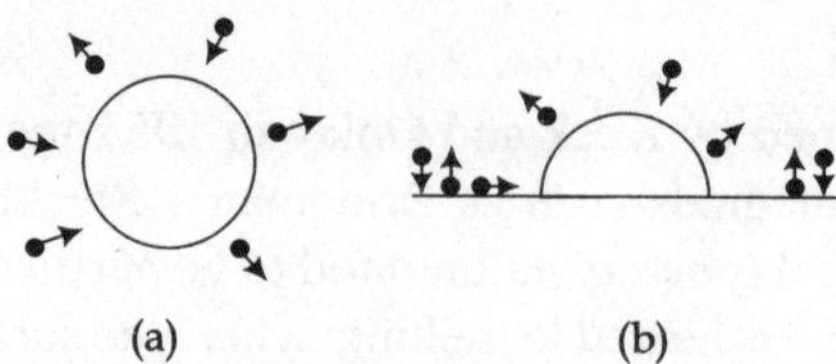

Fig. 8.10 Interaction between monomers and nucleus (a) Homogeneous nucleation (b) Heterogeneous nucleation

Heterogeneous nucleation can often be induced by external influences like agitation, mechanical shock, friction, pressure, electric and magnetic fields, spark discharge, ultra violet, X-rays, γ-rays, sonic, ultrasonic irradiations and so on.

The formation of crystal nuclei is a difficult and complex process. In doing so, the atoms or molecules not only coagulate against the tendency of dissolution but also have to orient themselves into a fixed lattice. The formation of such crystal nuclei seems to be improbable from classical theory where the atoms or molecules are in random motion and collide with each other. However, a number of atoms or molecules are understood to come together as a result of statistical fluctuation and form an ordinary cluster, known as embryo. This embryo may redissolve unless it reaches a critical (minimum) size, called the critical nucleus. This does not necessarily mean that a macroscopic crystal will grow, since it will depend on the availability of sufficient solute material in the system.

8.6 ENERGY OF FORMATION OF A NUCLEUS

In the following, we shall determine the energy of formation, of nucleus under both homogenous and heterogeneous nucleations.

(i) Homogeneous Nucleation

We observe in nature that an isolated droplet of a liquid (or fluid) is always spherical in shape. It is because of the fact that for a given volume, a spherical shape has minimum surface area and thus its surface free energy will be minimum. Gibbs considered that the growth of an embryo or a crystal as a special case of this principle. Accordingly, for a given volume, the total surface free energy of a crystal in equilibrium with its surrounding at constant temperature and pressure would be minimum. Therefore, when the volume free energy per unit volume is taken to be constant, then

$$\sum_i^n a_i\gamma_i = \text{minimum} \tag{1}$$

where a_i is the area of ith face and γ_i is its surface free energy per unit area.

Further, when the number of atoms or molecules associated with a critical nucleus is only few tens in number, it will not have any regular morphology and the shape can be approximately taken as spherical.

Now, suppose a small region of a new phase 2 appears in the middle of the initial phase 1 as shown in Fig. 8.11. The creation of a new phase (a liquid droplet in vapour or a solid particle in liquid) demands the expenditure of certain amount of energy. From thermodynamics, a decrease in

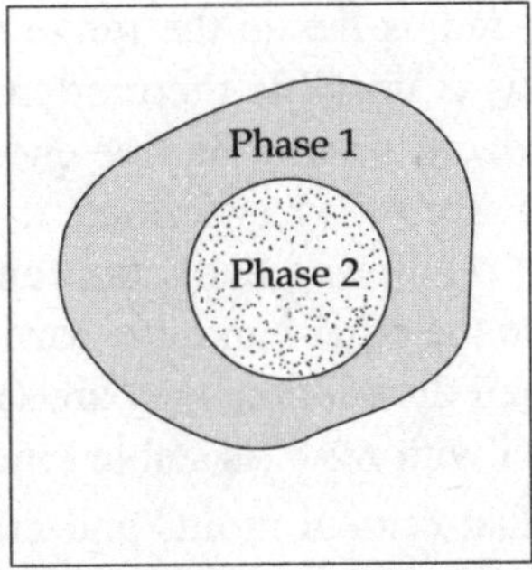

Fig. 8.11 Development of a new phase

the volume free energy is expected for the stability of the new phase. However, the new phase is bounded by a surface, which has a positive surface free energy per unit area associated with it. Thus the total change in the free energy can be written in terms of volume free energy and the surface free according to the relation.

$$\Delta G = \Delta G_S + \Delta G_V \tag{2}$$

where ΔG_S and ΔG_V are excess surface free energy and excess volume free energy, respectively. Here, ΔG_S is $+ve$ and is proportional to the surface area of the new phase (i.e., nucleus) and ΔG_V is $-ve$ and is proportional to the volume of the nucleus. Assuming the new (second) phase to be spherical, eq. 2 can be written as

$$\Delta G = 4\pi\, r^2\gamma + \frac{4}{3}\,\pi r^3 \Delta g_V \tag{3}$$

where γ is the surface free energy per unit area, r is the radius of the nucleus and Δg_V is the free energy per unit volume relative to the parent phase, respectively. The variation in quantities ΔG, ΔG_S and ΔG_V with respect to r is represented in Fig. 8.12. Since the surface energy term increases with r^2 and the volume energy term decreases with r^3, the net change in the free energy ΔG initially increases with r attains a maximum value at $r = r^*$ and then decreases with r. The size of the nucleus corresponding to which the change in the free energy is maximum is known as critical nucleus. For $r > r^*$, ΔG decreases with increasing r, at $r = r_S$, $\Delta G = 0$ and for $r > r_S$, ΔG is $-ve$.

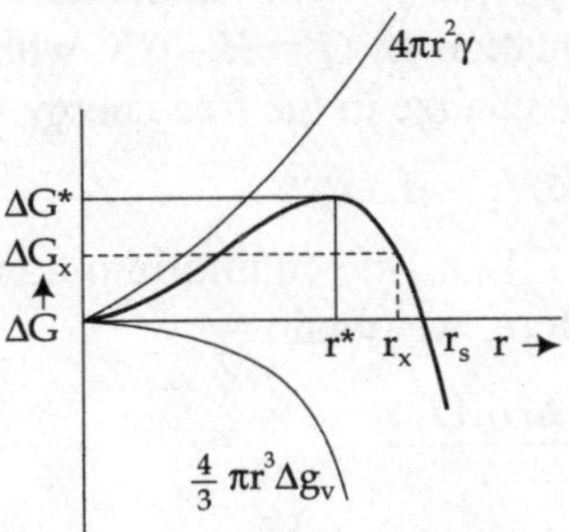

Fig. 8.12 Free energy change for nucleation

Let us consider a nucleus whose radius lies in the range $r^* < r_x < r_S$ and corresponding free energy is $+\Delta G_x$. While such a nucleus is unstable thermodynamically, it is stable kinetically because if we start dissolving, r would decrease and the free energy would begin to rise towards the critical value ΔG^*. Once a nucleus of size $r > r^*$ is formed, it is kinetically stable and continue to grow. Since the particles with $r < r^*$ require no activation energy for their dissolution, they are unstable and are termed as embryo. On the other hand, the particles in the range $r^* < r_x < r_S$ require an activation energy, $\Delta G^* - \Delta G_x$ for their dissolution, they are said to be in a metastable state and are termed as nuclei. Further, all the nuclei with $r > r_S$ is stable kinetically as well thermodynamically.

Now from Fig. 8.12, it is clear that critical radius and critical energy of the nucleus can be obtained by maximizing the eq. 3. This gives us

$$\frac{d\Delta G}{dr} = 0 \quad \text{at} \quad r = r^*. \text{ Therefore,}$$

$$\frac{d\Delta G}{dr} = 8\pi r\gamma + 4\pi r^2 \Delta g_V = 0$$

Solving for $r = r^*$, we obtain

$$r^* = -\frac{2\gamma}{\Delta g_V} \tag{4}$$

Substituting the value of r^* in eq. 3, we can obtain the free energy change associated with the formation of the critical nucleus, i.e.

$$\Delta G^* = \frac{16\pi\gamma^3}{3(\Delta g_V)^2} \tag{5}$$

Equation 5 can further be written as

$$\Delta G^* = \frac{4}{3}\gamma\, r^{*2}$$

$$= \frac{1}{3}O^*\gamma \tag{6}$$

where O^* is the surface area of the critical nucleus. Let us examine the variation of ΔG^* and r^* with temperature. A good approximation at melting point could be found if the variation of volume enthalphy (ΔH_V) and volume entropy (ΔS_V) with temperature is supposed to be very slow. We know that the Gibbs free energy equation is, $G = H - TS$, where H is enthalpy and S is entropy, respectively. At any temperature, the change in the free energy will become

$$\Delta G_V = \Delta H_V - T\Delta S_V \tag{7}$$

However, at the melting point (T_o) i.e., the equilibrium transformation temperature, $\Delta G_V = 0$. This implies that $\Delta g_V = 0$ and therefore, we obtain

$$\Delta S_V(T_o) = \frac{\Delta H_V(T_o)}{T_o} \tag{8}$$

Equation 7 can be rewritten approximately as

$$\Delta g_V(T) = \Delta H_V(T_o) - T\Delta S_V(T_o)$$

$$= \Delta H_V(T_o)\left(1 - \frac{T}{T_o}\right) \tag{9}$$

which is quite good approximation at the melting point for thermodynamic parameters like enthalpy and entropy. The surface energy is also found to be relatively insensitive to temperature, and therefore the approximate temperature variation of ΔG^* and r^* is obtained by substituting eq. 9 into eqs. 4 and 5. They are

$$r^* = \frac{-2\gamma T_o}{\Delta H_V(T_o)[T_o - T]} \tag{10}$$

and

$$\Delta G^* = \frac{16\pi\gamma^3 T_o^2}{3[\{\Delta H_V(T_o)\}(T_o - T)]^2} \tag{11}$$

They are plotted with temperature as shown in Fig. 8.13. This indicates that ΔG^* falls very rapidly with the decrease of temperature and a consequent increase in the rate of nucleation.

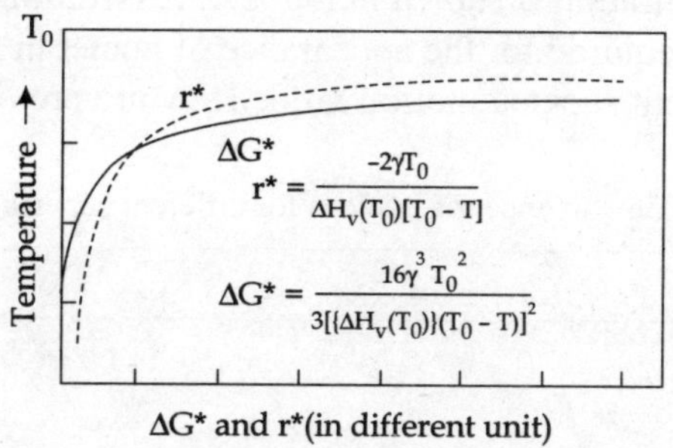

Fig. 8.13   Temperature variation of ΔG^* and r^* on the basis of eqs. 10 and 11, T_0 is the equilibrium transformation temperature.

From the rate of nucleation, we actually mean the number of critical nucleus formed per unit time per unit volume. Since the nucleation process is basically a thermally activated process, the rate of nucleation can be given by Arrhenius type equation such as

$$J = A\exp\left(\frac{-\Delta G}{kT}\right) \tag{12}$$

where A is a pre – exponancial constant, $\Delta G\,(= \Delta G^* + \Delta G_a)$ is the sum of critical free energy and activation energy (also known as overall activation energy). However, for critical nucleus, $\Delta G^* \gg \Delta G_a$.

In order to estimate the value of ΔG^* in terms of supersaturation ratio, let us make use of Gibbs–Thomson equation connecting the vapour pressure and the radius of liquid nucleus. For liquid solid growth system, Gibbs–Thomson equation in terms of supersaturation ratio is given by

$$\ln S = \frac{2\gamma V}{kTr} \qquad (13)$$

where γ is the surface free energy, V is the molecular volume and r is the radius of the solid nucleus. For a critical nucleus $r = r^*$, therefore, eqs. 4 and 13 give us

$$-\Delta g_v = \frac{2\gamma}{r^*} = \frac{kT \ln S}{V}$$

Substituting this in eq. 5, ΔG^* becomes

$$\Delta G^* = \frac{16\pi\gamma^3 V^2}{3k^2 T^2 (\ln S)^2} \qquad (14)$$

Further, substituting ΔG^* for ΔG in eq. 12, the rate of nucleation is obtained as

$$J = A \exp\left[-\frac{16\pi\gamma^3 V^2}{3k^3 T^2 (\ln S)^2}\right] \qquad (15)$$

Equation 15 shows that the rate of nucleation depends on the temperature T (Fig. 8.14), interfacial tension γ and supersaturation ratio S. A graph plotted between rate of nucleation (both theory and experiment) and supersaturation is shown in Fig. 8.15. Assuming the value of pre–exponential factor A to be 10^{25}, the time required for the appearance of nuclei in supercooled water vapour can be easily estimated for different supersaturation ratio. They are provided in Table 8.3.

Table 8.3 Time to appear a nucleus for different supersaturation ratio

Supersaturation ratio $S = C^*/C$	Time, t
1.0	—
2.0	10^{62} years
3.0	10^3 years
4.0	0.1 second
5.0	10^{-13} second

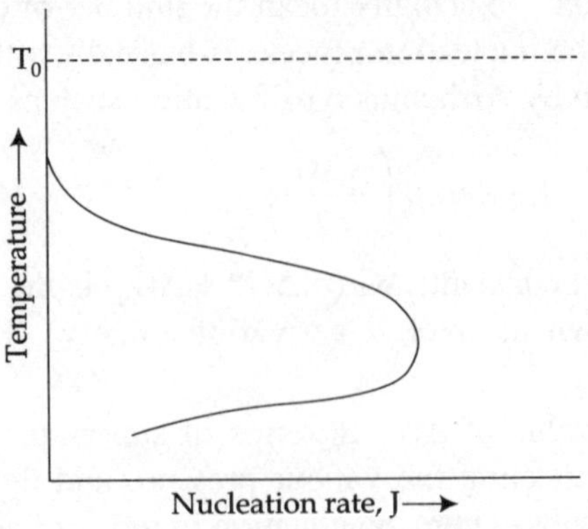

Fig. 8.14 Homogeneous nucleation rate

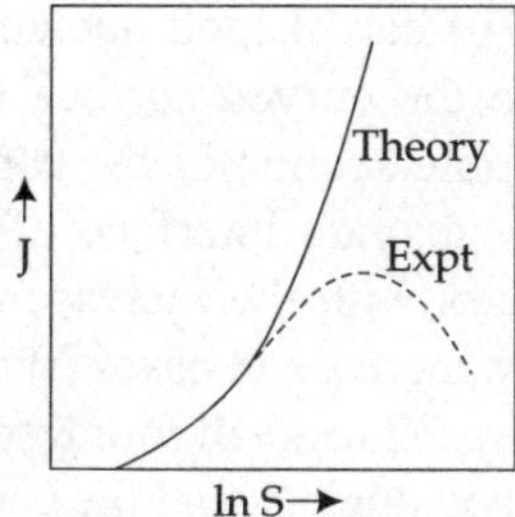

Fig. 8.15 Rate of nucleation vs. supersaturation

(ii) Hetrogeneous Nucleation

Unlike homogeneous nucleation, heterogeneous nucleation is initiated through some external influence or agency such as impurity. They act as catalyst or nucleant. They may be a solid particle suspended in the liquid, the surface of the container or a solid surface (substrate). The heterogeneous nucleation is common to all systems, namely the vapour–liquid, vapour–solid or liquid–solid transformation.

In order to estimate the energy of formation of a critical nucleus in this case, the properties of clusters are considered to be equal to that of the properties of the bulk. Further, because of mathematical complexities only simple shapes of embryos like cap or disc (cylinder) will be discussed.

Cap Shaped Nucleus

Let us consider a heterogeneous nucleation taking place on the surface of a substrate.

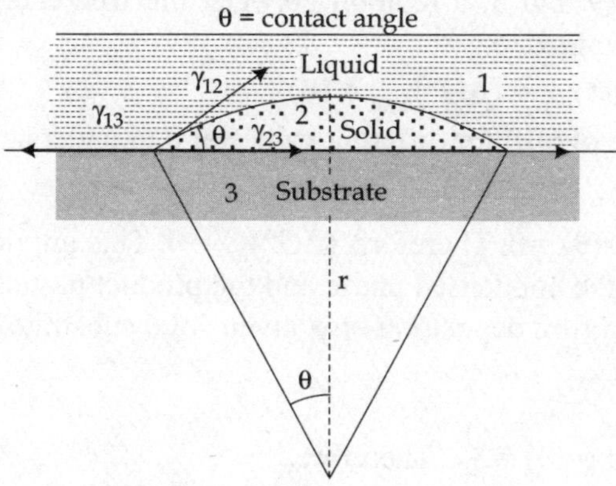

Fig. 8.16 Cap – shaped nucleus

Figure 8.16 represents the schematic diagram of cap-shaped nucleus. Let the contact (wetting) angle between the substrate and embryo be θ and r be the radius of curvature of the embryo. Let γ_{12}, γ_{13} and γ_{23} be the interfacial angles between the liquid-solid, liquid-substrate and solid-substrate, respecively. Such an embryo will be in equilibrium only if

$$\gamma_{13} = \gamma_{23} + \gamma_{12} \cos \theta \tag{16}$$

During the process of nucleation of cap-shaped nucleus, we observe that (i) A new $(1-2)$ liquid-solid interface of area equal to the curved surface is created. (ii) A new solid-substrate interface equal in area to the circle of intersection of the cap with the substrate surface is created. (iii) In this process, $a\,(1-3)$ liquid-substrate interface of the same circular area is consumed. Further, the volume of the cap increases with the increase in radius of curvature and is taken as positive. The surface areas are found to increase in cases (i) and (ii) and are taken as positive while decreases in (iii) and is taken as negative. Thus in all four terms (one volume and three surfaces) are involved in the formation of a cap-shaped nucleus and the corresponding free energy can be written as

$$\Delta G = \frac{4}{3}\pi r^3 \frac{(2-3\cos\theta+\cos^3\theta)}{4}\Delta g_V + 4\pi r^2 \frac{(1-\cos\theta)}{2}\gamma_{12} + \pi r^2 \sin^2\theta(\gamma_{23}-\gamma_{13}) \tag{17}$$

As discussed earlier, at $r = r^*$, $\dfrac{d\Delta G}{dr} = 0$. Accordingly, the radius of the critical nucleus can be obtained for all values of θ in the range $0 < \theta < 180°$ as

$$r^* = -\frac{2\gamma_{12}}{\Delta g_V} \tag{18}$$

Similarly, from eq. 17 and 18, we obtain

$$\Delta G^* = \frac{16\pi\gamma_{12}^3}{3(\Delta g_V)^2}\,\phi(\theta) \tag{19}$$

where
$$\phi(\theta) = \frac{2-3\cos\theta+\cos^3\theta}{4}$$

Now, comparing eqs. 19 and 5, a relation between the free energies of heterogeneous and homogeneous nucleus is obtained, e.g.

$$(\Delta G^*)_{het} = (\Delta G^*)_{hom}\,\phi(\theta) \tag{20}$$

Let us analyse the value of $\phi(\theta)$ for different values of θ to understand the nature of nucleation.

Case I: $\theta = 0$

This gives us $\cos\theta = 1$ and $\phi(\theta) = 0$. Therefore, $(\Delta G^*)_{het} = 0$. This implies that there is no activation barrier to the nucleation of the condensed phase and the product particles completely wet the substrate as in the case of a thin film deposition on a given solid substrate.

Case II: $\theta = 90$ and $\gamma_{13} = \gamma_{23}$

This gives us $\cos\theta = 0$ and $\phi(\theta) = \dfrac{1}{2}$. Therefore,

$$(\Delta G^*)_{het} = \frac{1}{2}(\Delta G^*)_{hom}$$

Here the product particle is hemi-spherical in shape.

Case III: $\theta = 180$

This gives us $\cos\theta = -1$ and $\phi(\theta) = 1$. Therefore,

$$(\Delta G^*)_{het} = (\Delta G^*)_{hom}$$

This implies that the substrate has no effect on the nucleus process. In fact, the product particle is just in point contact with it. They are shown in Fig. 8.17.

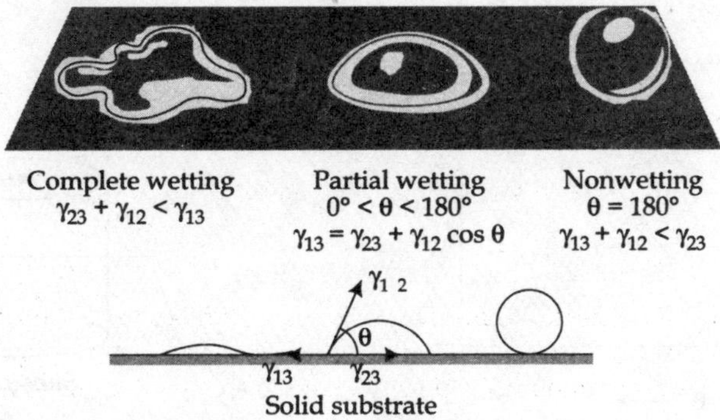

Fig. 8.17   Wetting of a substrate by a liquid.

The above discussed cases provide us some clues for selecting the hetrogeneous nucleation agent when the nucleation process is to be induced deliberately. A small value of contact angle promotes hetrogeneous nucleation. The effect of contact angle could be seen when we compare the volume of the cap-shaped nucleus of radius r and height h, V_c with a spherical nucleus of the same radius, V_s as

$$\frac{V_c}{V_s} = \frac{\dfrac{1}{3}\pi h^2 (3r - h)}{\dfrac{4}{3}\pi r^3}$$

$$= \frac{h^2}{4r^3}(3r - h)$$

$$= \frac{q^2(3-q)}{4} \tag{21}$$

where $q = \dfrac{h}{r}$. Since the ratio $\dfrac{h}{r}$ depends on the contact angle and hence the ratio $\dfrac{V_c}{V_s}$ will also depend on it. Fig. 8.18 provides the dependence of the relative volume of the embryo on the contact angle.

Disc–shaped (Cylinderical) Nucleus

Disc or cylindrical-shaped nucleus is a good possibility under a nonspherical shape. Let r be the radius and h be the height of the cylindrical nucleus as shown in Fig. 8.19. Further, assume that the interfacial energies between phases 2, 3 as γ_{23} between 1, 2 as γ_{12}, between 1, 3 as γ_{13} and the interfacial energy of the curved surface be γ_e (energy of the end faces, top and bottom). Then the free energy of formation of the nucleus is

$$\Delta G = \pi r^2 h \Delta g_V + 2\pi r h \gamma_e + \pi r^2(\gamma_{12} + \gamma_{23} - \gamma_{13}) \tag{22}$$

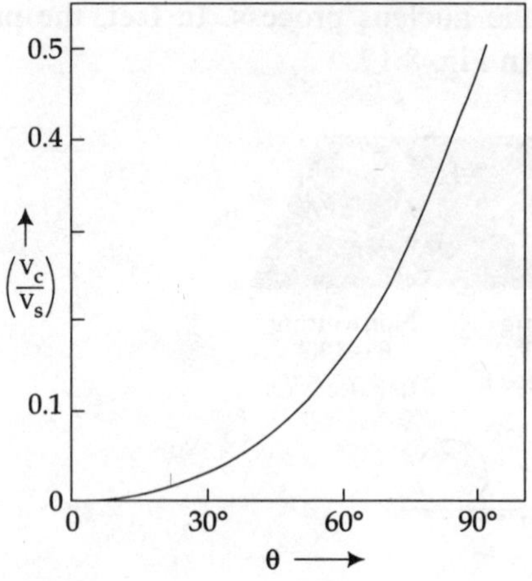

Fig. 8.18 (V_c/V_s) Vs contact angle

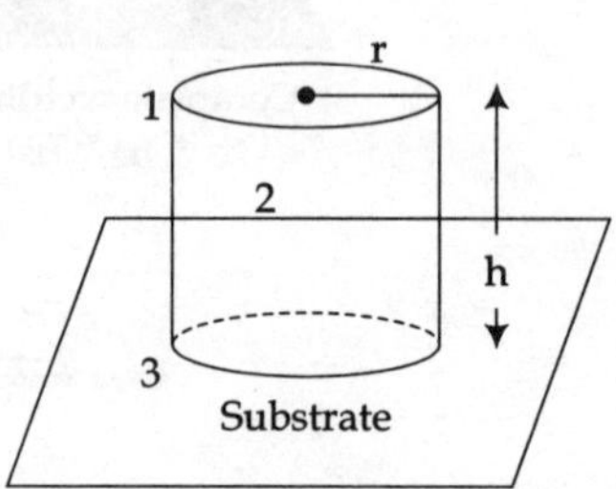

Fig. 8.19 Cylindrical–shaped nucleus

As discussed earlier, at $h = h^*$ and $r = r^*$, we have

$$\frac{d\Delta G}{dh} = 0 \text{ and } \frac{d\Delta G}{dr} = 0, \text{ respectively. Consequently,}$$

$$r^* = -\frac{2\gamma_e}{\Delta g_V} \text{ and } h^* = \frac{2(\gamma_{12} + \gamma_{23} - \gamma_{13})}{\Delta g_V} \tag{23}$$

Substituting eq. (23) in eq. (22), we obtain

$$\Delta G^* = \frac{4\pi\gamma_e^2}{\Delta g_V^2}(\gamma_{12} + \gamma_{23} - \gamma_{13}) \tag{24}$$

If $\theta = 90°$, then $\gamma_{23} = \gamma_{13}$ so that eq. 23 becomes

$$\Delta G^* = \frac{4\pi\gamma_e^2}{\Delta g_V^2}\gamma_{12} \tag{25}$$

8.7 VELOCITY OF GROWTH

In preceeding sections, we discussed the formation of critical nucleus in a supersaturated or super-cooled system under both homogeneous and heterogeneous cases. We also know that once a critical nucleus is formed, it may grow into a crystal of macroscopic size with well-developed faces. The external faces of a crystal are in general different except in cubic suggesting different growth velocities in different crystal planes or crystal directions. G. Wulff in his experiment first of all measured the thickness of the new layers deposited on different faces of the crystal and thereby established the relative growth velocities of different faces. The knowledge of the growth velocities of different faces of a crystal makes it possible to use a simple geometrical construction to predict the faces that will persist and the faces that gets eliminated during growth process.

Draw two vectors from a common point O (as the origin) normal to any two crystal faces such that their lengths are proportional to the relative growth velocities of these faces. Join the intersection point of the faces with the origin. Finally draw a line intersecting the growth velocity vectors and is normal to the line OQ as shown in Fig. 8.20. In this construction, we observe that:

(i) the Δ's OPQ and Oqp are similar (because $\angle QOP \equiv \angle qOp$ and $\angle OPQ = \angle Oqp = 90°$). Therefore,

$$\frac{OP}{Oq} = \frac{OQ}{Op} \quad \text{or} \quad Op = \frac{OQ}{OP} \cdot Oq \tag{26}$$

(ii) the Δ's ORQ and Oqr are similar, therefore,

$$\frac{OR}{Oq} = \frac{OQ}{Or} \quad \text{or} \quad Or = \frac{OQ}{OR} \cdot Oq \tag{27}$$

In eqs. 26 and 27, we observe that the lines Op and Or are respectively proportional to the reciprocals of the growth velocity vectors. These lines actually indicate about the presence or absence of a crystal face at a particular instant of time. Consequently, they are called "index vectors".

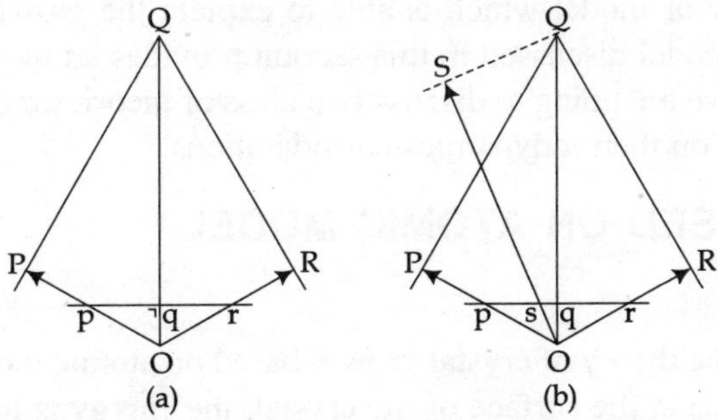

Fig. 8.20 Geometrical representation of growth – velocity

Now, let us add a virtual third face QS (Fig. 8.20b) and similarly obtain the index vector in terms of its velocity vector as

$$Os = \frac{OQ}{OS} \cdot Oq \tag{28}$$

From eq. 28, it is clear that if the index vector Os lies inside the ΔOpr then the new face does not appear. Conversely, if Os lies outside the ΔOpr, the new face will appear on the crystal. Further, if Os terminate on pr, then the new face just appears (or disappears). In each of the above three eqs. (26 – 28) it is observed that the index vector is inversely proportional to the respective growth velocity vector. This follows that the fast growing faces will get eliminated whereas the slow growing faces will persist in a crystal (Fig. 8.21a). Fig. 8.21b shows a cross section of an actual hexagonal crystal with well defined faces and the corresponding index vectors.

The relationship between the faces present on a crystal called its form or habit. The tendency of certain faces to grow more slowly and hence to predominate in crystal's habit is related in a simple

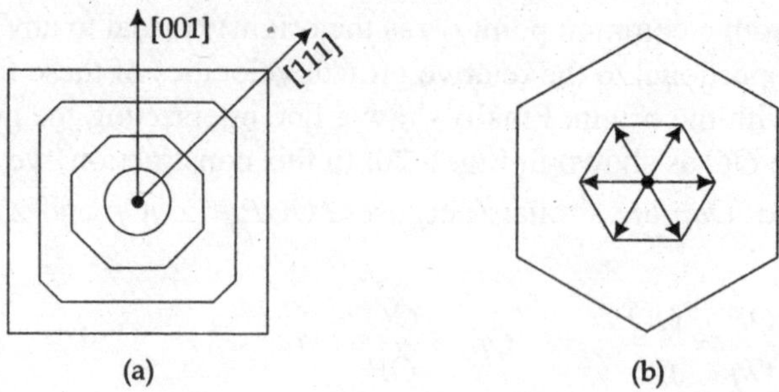

Fig. 8.21   (a) Growth and elimination of faces (b) Index vectors in a hexagonal crystal

way to its crystal structure. Certain faces (planes) in the crystal are more dense (contains more atoms per unit area) and take longer time to complete a new layer than the less dense ones. Therefore, in general it can be said that the growth velocity is inversely proportional to its atomic density.

8.8   THEORIES/MODEL OF CRYSTAL GROWTH

As such there is no theory or model which is able to explain the growth of crystals completely. However, each theory or model discussed in this section provides an over all idea of growth in its own way. Here, basically we are going to discuss two class of theories/model: one based on atomic model and the other based on thermodynamics considerations.

8.9   THEORIES BASED ON ATOMIC MODEL

1. Kossel's Theory

Kossel (1927) developed the theory of crystal growth based on atomic model. He assumed that due to unsaturated atomic bonds at the surface of the crystal, the energy is lowered when an atom (or a growth unit) attaches itself to the surface. The amount of energy released by such attachment is called the attachment energy (or lattice energy) and is different at different parts of the crystal.

For simplicity, let us consider the growth of a homopolar cubic crystal where the growth unit contains only one atom in it as shown in Fig. 8.22. In Fig. 8.22a, the growth unit is embedded inside the crystal which experiences forces from all directions. On the other hand, when a crystal is still in the process of growth (Fig. 8.22b) the growth unit will experience the forces just half than the first case. For this reason, according to Kossel, a growing crystal is called a 'half crystal'.

The attachment energy associated with a growing unit can be resolved along three mutually perpendicular directions as shown in Fig. 8.23, such that

$$\phi = \phi_1 + \phi_2 + \phi_3 \tag{29}$$

This can be understood by considering the growth of a fresh layer of the surface as shown in Fig. 8.23. When a single growth unit settles on a flat surface, it releases ϕ_3 amount of energy directed below. Another unit adjacent to the first releases $\phi_2 + \phi_3$ amount of energy where ϕ_2 component is due to the side pull. When a unit sits at a step (as in Fig. 8.23), it releases the amount of energy given by eq. 28, where ϕ_1 component is due to the pull by another side wall.

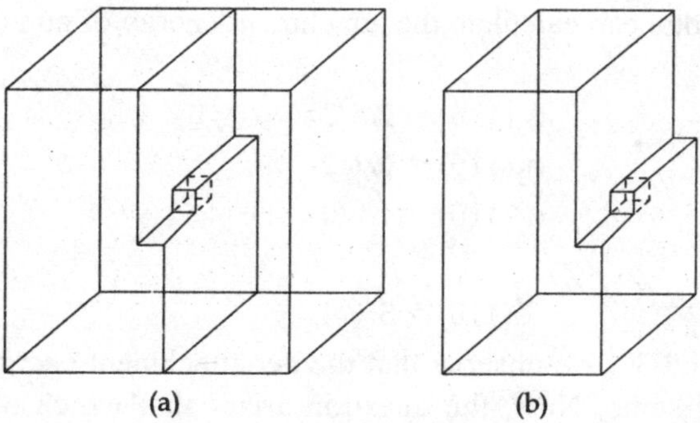

Fig. 8.22   (a) Growth unit embedded inside, (b) Growth unit lies outside

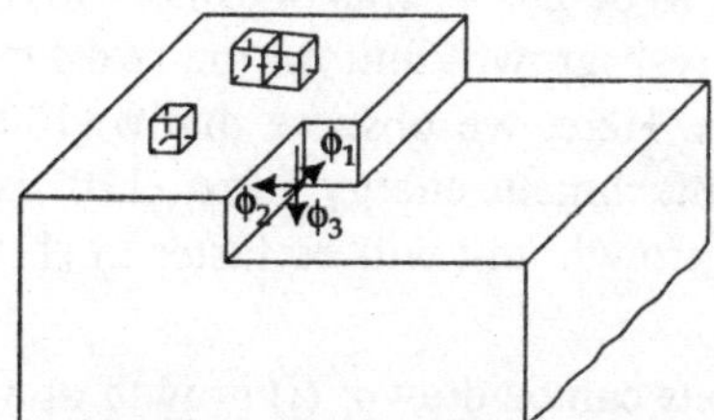

Fig. 8.23   The attachment energy released at different sites

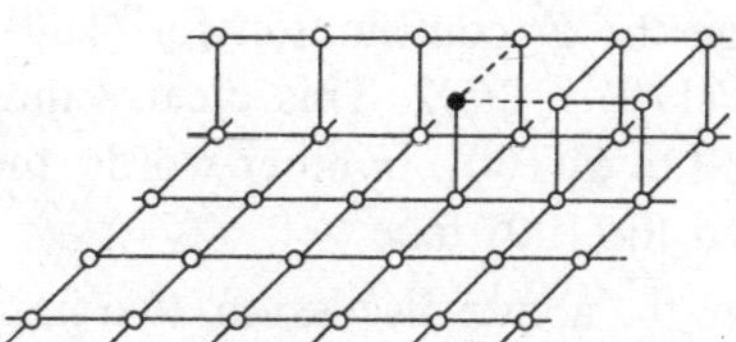

Fig. 8.24   Arrangement of atoms in a simple cubic crystal

It is now possible to compute the amount of energy released at different parts of the crystal in terms of the components of ϕ. This calculation differs and depends on the type of bonding in the crystal. Let us discuss the energy associated in the case of homopolar (covalent) simple cubic crystal. However, a similar argument could be applied to hetropolar (ionic) crystals also.

The arrangement of atoms in a simple cubic crystal is shown in Fig. 8.24 where the side of the unit cell is taken as unity. For such an arrangement there are 6 first nearest neighbour atoms lying along the cube edges at a unit distance, 12 second nearest neighbour atoms lying along the face diagonals at a distance $\sqrt{2}$ and 8 third nearest neighbour atoms lying along the body diagonals at a distance of $\sqrt{3}$. Only these 26 nearest neighbour atoms will contribute towards the attachment energy when a growth unit (or an atom) attaches itself on a growing face. Thus for an atom sitting on (100) face has only one first nearest neighbour atom at a unit distance, 4 second nearest neighbour atoms at a distance $\sqrt{2}$ and 4 third nearest neighbour atoms at a distance $\sqrt{3}$. The energy associated with ϕ_3 component is given by

$$\phi_3(100) = 1/4/4$$

Similarly, other energy components can be computed. They are:

$$\phi_2(100) = 1/2/0$$

and
$$\phi_1(100) = 1/0/0$$

This give us the attachment energy of an atom on (100) face according to eq. 29 as

$$\phi(100) = 3/6/4 \tag{30}$$

In a similar manner, one can calculate the attachment energy of an atom on (110) face. Three energy components are:

$$\phi_3(110) = 2/5/2$$
$$\phi_2(110) = 0/1/2$$
$$\phi_1(110) = 1/0/0$$

This gives us

$$\phi(110) = 3/6/4 \tag{31}$$

Comparing eqs. 30 and 31, we observe that the net attachment energy of an atom on the two faces (100) and (110) is same. Now, the question arises that which of the two faces is more favourable for crystal growth? In order to know it, let us check the case of attachment of a growth unit adjacent to a unit already existed on the two faces, i.e., to determine $\phi_2 + \phi_3$ for (100) and (110) faces. It is observed that in both cases $\phi_2 + \phi_3$ comes out to be 2/6/4. This does not provide us any clue. Therefore, now let us check the attachment of a fresh growth unit on the two faces, i.e. to determine the ϕ_3 contribution for (100) and (110) faces. Here, we observe that $\phi_3(100) = 1/4/4$ while $\phi_3(110) = 2/5/2$. This clearly indicates that the attachment energy for $\phi_3(110)$ is more as compared to $\phi_3(100)$. In other words, the attachment of growth unit will be faster in (110) face as compared to (100) face.

From the above discussion, two important conclusions can be drawn: (i) growth of a crystal is a discrete process and not a continuous one, and (ii) a new layer is more favourable to begin from the center of the face.

The second conclusion can be verified by computing the ϕ_3 component of energies at different sites of a given face. Such calculations for (100) face gives us ϕ_3 (mid face) = 1/4/4, ϕ_3(edge) = 1/3/2 and ϕ_3(corner) = 1/2/1. This indicates that the attachment energy for a given face decreases in the order.

$$\text{mid face} > \text{edge} > \text{corner}$$

Thus a new layer will always start from the center of the growing face.

2. Screw Dislocation Theory

In Kossel's theory, we discussed the growth of an ideal crystal where the atoms are arranged in a perfectly periodic manner. However, real crystals do contain some impurities or imperfections and hence the arrangement of atoms in them no longer remain perfectly periodic.

The nature of deviation depends on the nature of impurities or imperfections (point, linear planar) in them. Frank (1951 – 52) put forward a theory for the growth of crystals containing imperfections. According to him, initially at high supersaturations, crystals grow into thin plates by surface–nucleation mechanism in accordance with the theory of growth of perfect crystals. These plates then become self–stressed through non uniform distribution of impurities or thermal stresses, till the theoretical yield stress is exceeded and the plate shears or buckels. This raises terminated steps on the crystal surface. A typical growth step resulting from a screw dislocation is shown in Fig. 8.25. As discussed in the preceeding section that atoms prefer to attach themselves to such steps in order to decrease their surface free energy. Therefore, the growth proceeds by a continuous growth of same layers in the form of a spiral and continues even under low supersaturation. Several stages of development of these so called growth spirals are shown in Fig. 8.26. At a steady

Fig. 8.25 A typical growth step resulting from a screw dislocation

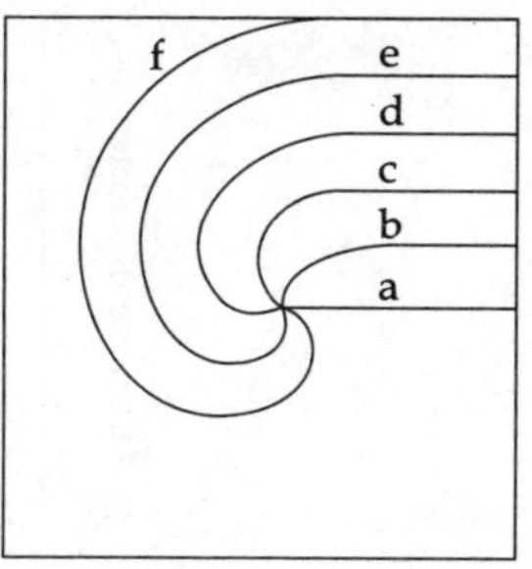

Fig. 8.26 Several stages of development of a growth spiral

state (or equilibrium) it assumes the shape of a perfect spiral. This is then continues to move around the crystal's surface.

The dislocation theory of crystal growth is supported by the direct observation of spirals on the crystal's surface of many materials like paraffin, silicon carbide or cadmium iodide, etc. The spiral step heights can be measured directly by using very high resolution electron microscope or indirectly by the use of multiple–beam interferometry.

8.10 THEORIES BASED ON THERMODYNAMIC CONSIDERATIONS

1. The Diffusion Theory

The first diffusion theory was proposed by Noyes and Whitney in 1897. They assumed that growth is a reverse process of dissolution and both are governed by diffusion of atoms or molecules from one phase to another as a result of concentration gradient. The amount of solute materials that will be deposited over the surface of the growing crystal within a supersaturated solution is given by

$$\frac{dm}{dt} = k_{m}A(C - C^*) \tag{32}$$

where dm is the mass of the solute material deposited over the crystal surface of area A in time dt, C is the actual solute concentration of C^* is solute concentration at saturation.

Later on Nernst assumed that the there exists a thin stagnant layer of thickness δ near the crystal surface and accordingly he modified eq. 32 as

$$\frac{dm}{dt} = \frac{D}{\delta}A(C - C^*) \tag{33}$$

where D is the diffusion coefficient of the solute. The value of δ depends on the relative motion between the crystal surface and the solution. Under unstirred condition δ has been found to vary from face to face in the thickness range 20 – 50 microns. However, under stirred condition δ approaches zero.

Experimental observations suggest that the dissolution process is faster than the growth process under identical conditions and hence they cannot be exactly inverse of one another. Keeping this in view, Berthoud suggested that the growth process is not merely the diffusion process alone

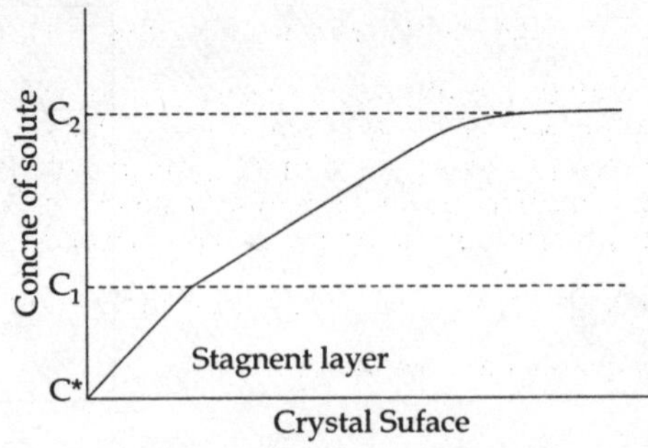

Fig. 8.27 Different concentration layers over crystal surface

but also associated with reaction at crystal's surface. Thus according to him, the concentration of the bulk solution is C_2, the stagnant layer C_1 and at saturation C^* as shown in Fig. 8.27, such that $C_2 > C_1 > C^*$. In a supersaturated solution, the value of C_2 remains the same w.r.t time during growth. However, the value of C_1 depends on the rate of reaction (or rate of deposition) k_r such that C_1 is less when the reaction rate is fast, but under no condition C_1 becomes less or equal to C^* (i.e. C_1 is always greater than C^*). In a simplest case, one can consider that the growth rate is directly proportional to the supersaturation. Therefore, eq. 32 can be written as

$$\frac{dm}{dt} = k_r A(C_1 - C^*) \tag{34}$$

Further, there exists a concentration gradient $(C_2 - C_1)$ between the bulk and the stagnant layer, the solute is transported through diffusion. Hence eq. 33 can be written as

$$\frac{dm}{dt} = \frac{DA}{\delta}(C_2 - C_1) \tag{35}$$

Substituting the value of C_1 from eq. 34 into eq. 35, we obtain

$$\frac{1}{A}\frac{dm}{dt} = \frac{D}{\delta + \dfrac{D}{k_r}}(C_2 - C^*)$$

or

$$\frac{1}{A}\frac{dm}{dt} = R(C_2 - C^*) \tag{36}$$

This equation was first of all derived by Berthoud. Here R is known as over all growth coefficient which helps in understanding the over all growth of crystals in a qualitative way. Let us consider the value of R from eq. 36, i.e.

$$R = \frac{D}{\delta + \dfrac{D}{k_r}}$$

or

$$\frac{1}{R} = \frac{1}{k_d} + \frac{1}{k_r} \tag{37}$$

where $\dfrac{D}{\delta} = k_d$, known as the mass transfer coefficient by diffusion. Let us consider two separate cases:

(i) **Case I:** When the reaction at the crystal surface is infinity, i.e., $k_r \to \infty$, then eq. 37 gives us

$$R = k_d$$

As a result eq. 36 reduces to eq. 33. This indicates that the growth is completely controlled by diffusion alone.

(ii) **Case II:** When the reaction at the crystal surface is negligible as compared to diffusion, i.e., $k_r \ll k_d$, then eq. 37 gives us

$$R = k_r$$

This implies that the growth is completely controlled by the surface reaction alone.

In deriving equations $34 - 36$, the reaction rate at the growing surface was assumed to be of first order meaning thereby that the growth habit (or form) should not change. However, it is well known that supersaturation changes the growth habit of a crystal very much. This leads to the conclusion that the mass deposited per unit area per unit time due to surface reaction is neither a linear function of $(C_1 - C^*)$ nor $(C_2 - C^*)$. Accordingly, the above mentioned equations can be rewritten as

$$\frac{1}{A}\frac{dm}{dt} = k_r (C_1 - C^*)^z \tag{34$'$}$$

$$\frac{1}{A}\frac{dm}{dt} = k_d (C_2 - C_1) \tag{35$'$}$$

and
$$\frac{1}{A}\frac{dm}{dt} = R\,(C_2 - C^*)^n \tag{36$'$}$$

where n varies from 1 to 2 for different systems. These equations represent the rate equations for surface reaction, diffusion and overall growth process, respectively.

Friedel successfully extended the diffusion theory to melt. It is well know that solidification from melt does not require the movement of material, but a temperature gradient is set up near the crystal surface. Accordingly, Friedel applied eq. 36 to represent the diffusion of heat and the temperature gradient providing the 'tempo' of reaction.

If we look back the above discussion thoroughly we observe the following important assumptions were made to develop the diffusion theory. Also, we observe a few serious objections to this theory. They are:

Important Assumptions

1. There exit three stages of concentrations C_2, C_1 and C^* in the vicinity of a growing surface such that $C_2 > C_1 > C^*$.
2. The growth process is reverse of a dissolution process.

Serious Objections

1. Experimental value of k_r for growth and dissolution are found to be different. In general, the rate of dissolution is found to be faster than the rate of growth under same condition.

2. When dissolution takes place in an unsaturated solution, the concentration of the 'solution layer' in contact with the crystal surface is expected to be less than the saturation concentration. However, Miers proved it other way round. This is quite contrary to the diffusion theory.

3. Its vagueness when compared with the theories developed from atomic models.

2. Bulk Diffusion Model

The principal assumption of this model is that the molecules in contact with the surface of a growing crystal are adsorbed quickly. As a result, a concentration gradient is produced between the bulk of the solution and the growing crystal surface. This causes the transfer of solute from the bulk to the surface of growing crystal, which in turn controls the growth process. The rate of mass transfer of solute from bulk to the surface of area A through the stagnant layer is given by

$$J = k_C A(C_B - C_B^*) \tag{38}$$

where k_C is the solute mass transfer coefficient due to the concentration driving force, C_B is the actual solution concentration and C_B^* is the equilibrium concentration. Now,

$$C_B - C_B^* = C_B^* \left(\frac{C_B - C_B^*}{C_B^*} \right) = C_B^* \frac{\Delta C_B}{C_B^*} = C_B^* \sigma$$

where σ is the supersaturation ratio. Therefore, eq. 38 can be written as

$$J = k_C A C_B^* \sigma \tag{39}$$

This gives us the amount of solute molecules adsorbing on to the crystal surface of area A per unit time. Thus the growth rate of the crystal face can be given by

$$R = \frac{JV}{A} = k_C V C_B^* \sigma \tag{40}$$

where V is the molecular volume of the solute. The value of k_C is given by a simple equation connecting Sherwood number (N_{Sh}), Reynolds number (N_{Re}) and Schmidt number (N_{Sc}) as

$$N_{Sh} = 2 + 0.6 N_{Re}^{1/2} N_{Sc}^{1/3} = k_C \frac{d}{D} \tag{41}$$

$$N_{Re} = \frac{du}{\eta} \tag{42}$$

and
$$N_{Sc} = \frac{\eta}{D} \tag{43}$$

where d is the dimension of the crystal, D is the diffusion coefficient, u is the relative motion of the crystal and solution during growth, η is the coefficient of viscosity of the liquid. Now, solving eqs. 41, 42 and 43, we get,

$$k_C = \frac{2D}{d} + \frac{0.6 D^{2/3} u^{1/2}}{d^{1/2} \eta^{1/6}} \tag{44}$$

Let us discuss two cases:

Case I: When the relative motion u is very small, then eq. 44 reduces to

$$k_C = \frac{2d}{D} \tag{45}$$

Case II: When u is large as compared to the diffusion coefficient D, then eq. 44 reduces to

$$k_C = \frac{0.6D^{2/3}u^{1/2}}{d^{1/2}\eta^{1/6}} \tag{46}$$

From eq. 38, one can predict that for a particular system of known supersaturation the growth rate R is proportional to k_C. On the other hand, from eqs. 44 – 46 we observe that k_C depends on d, D and η is temperature dimension. Since k_C depend on dimension of the crystal, eq. 44 can not be treated as universal.

3. BCF Bulk Diffusion Model

Burton, Cabrera and Frank (BCF) jointly put forward yet another model based on diffusion to explain crystal growth. They assumed that the kinks are the most favorable positions for crystal growth and hence the solute molecules directly enter into them (i.e. molecules never get adsorbed first on the surface and then enter into the kinks). In a growing crystal, each step has a number of equi-spaced kinks with an average separation distance x_o. Also, a growing surface contains a number of steps with an average separation distance y_o as shown in Fig. 8.28. They also assumed that the relative motion of the kinks and the steps (x_o and y_o are constants) is negligibly small as compared to the bulk diffusion flux. Under these conditions, the diffusion equation, for the concentration of solute near the growing surface is given by

$$\nabla^2 C_B = \frac{\partial^2 C_B}{\partial x^2} + \frac{\partial^2 C_B}{\partial y^2} + \frac{\partial^2 C_B}{\partial z^2} = 0 \tag{47}$$

where C_B is the actual solute concentration in the bulk. Equation 47 can be solved by considering three types of force fields, one each created by kinks, steps and boundary layer, respectively.

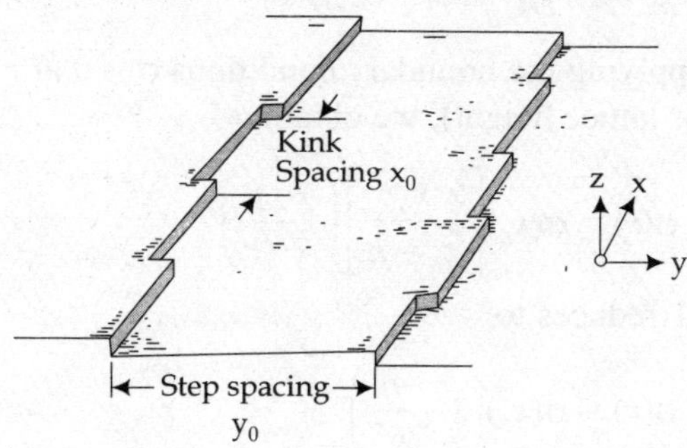

Fig. 8.28   Steps and Kinks in a growing crystal surface

A hemi-spherical force field is created by the kinks around it with radius x_o. Any solute molecule within this region will be attracted towards the kink. Hence the boundary condition is

$$C_B = C_B(r) \text{ in the range } 0 < r < x_o$$

where r is the spherical radial distance.

A semi – cylindrical force field is created by the steps with the step as the axis and radius y_o. This field acts in such a way that any molecule within this region will be attracted radically along the axis of the cylinder. Hence the boundary condition is $C_B = C_B(r)$ in the region $x_o < r < y_o$ where r is cylindrical radial distance.

Finally a plane force field is created by the concentration boundary layer of thickness δ. Any molecule within this boundary will be attracted normal to the surface. Hence the boundary condition is $C_B = C_B(z)$ in the region of $y_o \leq z < \delta$ where z is the distance from the surface and δ is the concentration boundary layer thickness. All the three force fields are schematically represented in Fig. 8.29. Also, the average kink spacing, step spacing and boundary layer thickness are assumed to be in the order $x_o < y_o < \delta$.

Now, let us consider three types of force fields one by one.

(i) Hemi-Spherical Force Field

In this case, the solute flux (mass of solute per area per unit time) diffusing towards the kink is given by

$$J_k(r) = \frac{k}{r^2} \tag{48}$$

where k is the solute mass transfer coefficient. Also, from Fick's law of diffusion $J_k(r)$ is given by

$$J_k(r) = DC_B^* \frac{d\sigma(r)}{dr} \tag{49}$$

where $\sigma(r)$ is the supersaturation depending on r. Now, from eq. 48 and 49, we obtain

$$d\sigma(r) = \frac{k}{DC_B^*} \cdot \frac{dr}{r^2} \tag{50}$$

Integrating eq. 50 and applying the boundary conditions $\sigma = 0$ ar $r = h$ and $\sigma = \sigma(x_o)$ at $r = x_o$ where h is the step height (or lattice height), we obtain

$$\sigma(r) = \sigma(x_o) \left[\frac{x_o}{x_o - h} \right] \left(1 - \frac{h}{r} \right) \tag{51}$$

Here $x_o \gg h$ then eq. 51 reduces to

$$\sigma(r) = \sigma(x_o) \left(1 - \frac{h}{r^2} \right) \tag{52}$$

Now, differentiating this w.r.t.r and substituting the value in eq. 49, we obtain

$$J_k(r) = DC_B^* \sigma(x_o) \frac{h}{r^2} \tag{53}$$

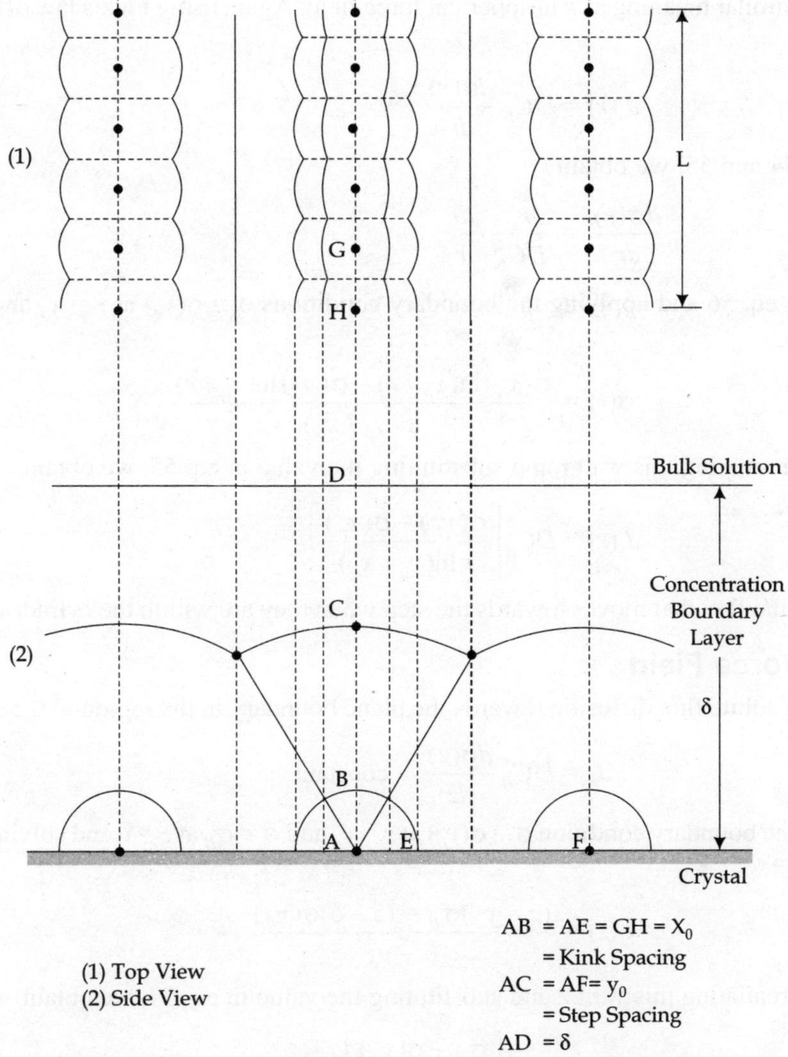

Fig. 8.29   Three force fields acting on a growing crystal surface

This gives the flux that moves towards the kink when the molecules are within the spherical force field.

(ii) Semi-Cylinderical Force Field

In this case, the solute flux diffusing towards the step can be written as

$$J_s(r) = \frac{k_1}{r} \tag{54}$$

where k_1 has a similar meaning as k in spherical force field. Again using Fick's law of diffusion $J_s(r)$ is given by

$$J_s(r) = DC_B^* \frac{d\sigma(r)}{dr} \qquad (55)$$

From eq. 54 and 55, we obtain

$$\frac{d\sigma(r)}{dr} = \frac{k_1}{DC_B^*} \cdot \frac{dr}{r} \qquad (56)$$

Integrating eq. 56 and applying the boundary conditions $\sigma = \sigma(x_o)$ at $r = x_o$ and $\sigma = \sigma(y_o)$ at $r = y_o$,

$$\sigma(r) = \frac{\sigma(x_o)\ln(y_o/r) - \sigma(y_o)\ln(x_o/r)}{\ln(y_o/x_o)} \qquad (57)$$

Now, differentiating this w.r.t.r and substituting the value in eq. 55, we obtain

$$J_s(r) = DC_B^* \left[\frac{\sigma(y_o) - \sigma(x_o)}{r\ln(y_o/x_o)} \right] \qquad (58)$$

This gives the flux that moves towards the step when they are within the cylindrical force field.

(iii) Plane Force Field

In this case, the solute flux diffusing towards the plane boundary in the region $y_o \le z < \delta$ is given as

$$J_b = DC_B^* \frac{d\sigma(z)}{dz} = \text{constant} \qquad (59)$$

Applying the boundary condition $\sigma = \sigma(y_o)$ at $z = y_o$ and $\sigma = \sigma_B$ at $z = \delta$ and solving for $\sigma(z)$, we obtain

$$\sigma(z) = \frac{(z - y_o)\sigma_B - (z - \delta)\sigma(y_o)}{\delta - y_o} \qquad (60)$$

Now, differentiating this w.r.t.z and substituting the value in eq. 59, we obtain

$$J_b = DC_B^* \left[\frac{\sigma_B - \sigma(y_o)}{\delta - y_o} \right]$$

This gives the flux that moves towards the boundary. Now matching the fluxes at x_o and y_o, we obtain

$$J_k(x_o) \simeq J_s(x_o) \text{ and } J_s(y_o) \simeq J_b(y_o) \qquad (61)$$

Substituting the actual values and simplifying for $\sigma(x_o)$, we obtain

$$\sigma(x_o) = \frac{\sigma_B}{\left[\dfrac{h(\delta - y_o)}{x_o y_o} + \dfrac{h}{x_o}\ln\left(\dfrac{y_o}{x_o}\right) + 1 \right]} \qquad (62)$$

Now, the advancing velocity of a straight step of unit length consisting of kinks equispaced with x_o in a direction perpendicular to the advancement of steps is the product of the number of molecules arriving at the kink per unit time, the number of kinks available per unit length in the step and the volume to height ratio of the growth unit, i.e.

$$V_\infty = (2\pi h^2)\, J_k(h) \left(\frac{1}{x_o}\right)\left(\frac{h^3}{h}\right)$$

or
$$V_\infty = \frac{2\pi h^4}{x_o}\, J_k(h) \tag{63}$$

Substituting the value of $J_k(r = h)$ from eq. 53 into eq. 63 we obtain

$$V_\infty = 2\pi h^3 DC_B^* \frac{\sigma(x_o)}{x_o} \tag{64}$$

Combining eqs. 62 and 64, we obtain

$$V_\infty = \frac{2\pi h^3 DC_B^* \sigma_B}{\left[\dfrac{h(\delta - y_o)}{x_o y_o} + \dfrac{h}{x_o}\ln\left(\dfrac{y_o}{x_o}\right) + 1\right]} \tag{65}$$

Now, assuming that the parallel steps are equispaced with y_o, the growth rate R will be given by

$$R = \frac{hV_\infty}{y_o} = \frac{2\pi h^4 DC_B^* \sigma_B}{x_o y_o\left[\dfrac{h(\delta - y_o)}{x_o y_o} + \dfrac{h}{x_o}\ln\left(\dfrac{y_o}{x_o}\right) + 1\right]} \tag{66}$$

In order to calculate R, *BCF* made use of the following equations, i.e.

$$y_o = 4\pi r^* \ (\text{or } 19r^*)$$
$$r^* = \gamma Vm/kTln\,(1 + \sigma_B)$$

and
$$V_r = V_\infty\left(1 - \frac{r^*}{r}\right)$$

where r^* is the critical radius of a two dimensional nucleus, V_m is the volume of a molecule, V_r is the step velocity of a curved step of radius of curvature r and γ is the interfacial energy. Further, recalling the relation $h \ll x_o < y_o < \delta$, let us discuss two different cases:

Case I: When σ_B is very small (i.e. y_o is large), then

$$R \propto \frac{\sigma_B}{y_o\ln(y_o / x_o)} \quad \text{or } R \propto \frac{\sigma_B^2}{C - \ln\sigma_B^2} \tag{67}$$

where C is constant

Here, it is to be noted that in this case R is not affected by variation in the boundary layer thickness δ.

Case II: When σ_B is large, but still small enough to employ the approximation, $\ln(1 + \sigma_B) \simeq \sigma_B$, then

$$R \propto \frac{DC_B^* \sigma_B}{\delta} \tag{68}$$

This indicates that R is linear with σ_B but inversely proportional to δ.

From the above discussion, we can conclude that *BCF* bulk diffusion model provides some interesting predictions regarding R but has very little basis.

8.11 CONCENTRATION OF KINKS AND MOBILITY OF ADSORBED MOLECULES

BCF calculated the concentration of kinks in a step of monoatomic layer in terms of interatomic distance 'a' at a given temperature T. They found that the average kink separation x_o is given by

$$x_o \sim \frac{1}{2} a e^{\omega/kT} \tag{69}$$

where ω is the energy necessary to form a kink. They estimated ω for close packed crystals with homopolar binding and found it to be of the order of $W/12$, where W is the evaporation energy. Introducing ϕ, the nearest – neighbour interaction as in Kossel's model, which is related to the evaporation energy W according to the equation.

$$\phi = \frac{1}{6} W \tag{70}$$

Equation 69 can be written as

$$x_o \sim \frac{1}{2} a e^{\phi/2kT} \tag{71}$$

For the growth of a crystal under certain conditions such that the temperature lies between 0.5 to 0.8 times the boiling point, $\phi/kT \sim 4$. Accordingly,

$$x_o \sim 4a \tag{72}$$

Thus, under above growth conditions, there appears a kink after every four atoms or so in a step. In fact, the concentration of kinks in a step depends only on the temperature and remains practically unchanged even if the vapour surrounding the crystal is supersaturated.

BCF also examined the possibility of surface. According to them the mean displacement x_s of absorbed molecules on the crystal surface is given by

$$x_s = a e^{(\Delta G_{des} - \Delta G_{diff})/2kT} \tag{73}$$

where ΔG_{diff} is the activation energy for surface diffusion, i.e., the energy needed for migration from one surface site to another. It has been estimated to be of the order of $W/20$. On the other hand, ΔG_{des} is the energy of desorption (evaporation). Considering again the nearest neighbour only for the half crystal of Kossel type, the number of nearest neighbour is 3. Hence,

$$\Delta G_{des} = 3\phi = \omega = \frac{\omega}{2}$$

and
$$x_s \sim ae^{3\phi/2kT} \tag{74}$$

Again, taking $\phi/kT \sim 4$ for similar growth conditions, we have

$$x_s \sim 4 \times 10^2 a \tag{75}$$

This indicates that x_s is large as compared to 'a' and hence the adsorbed molecule travel considerable distances before re-evaporation.

8.12 SUMMARY

1. There are several methods of (single) crystal growth based vapour-solid, liquid-solid and solid-solid phase transformations. However, the methods based on solid-solid transformation are more prevalent than the other two.

2. Under solution growth, single crystal of high melting point materials can be grown at comparatively very low temperatures using Flux method or hydrothermal method.

3. For high temperature growth, we need to have a crucible (or a boat) of a suitable material and of suitable geometry, a suitable furnace capable of producing the desired temperature gradient, suitable temperature controller and temperature measuring device.

4. Once a homogeneous solution is prepared, crystals can be grown by slow cooling method, slow evaporation method or temperature gradient method.

5. Under melt growth, single crystals of a large number of materials are generally grown by Bridgeman-Stockbarger method or an improved Czochralski method. On the other hand, zone refining and floating zone methods are used to purify the materials as well as for their growth.

6. Nucleation is the most fundamental requirement for crystal growth. Nucleation may occur spontaneously (as in homogeneous nucleation) or it may be induced artificially (as in hetrogeneous nucleation).

7. Growth process can be broadly divided into four stages. They are: (i) the development of supersaturated stage, (ii) the generation of embryo, (iii) birth of the critical nucleus, and (iv) the relaxation stage where the texture of the newly formed nucleus alters.

8. A nucleus with a radius $r < r^*$ is called embryo and is unstable. At $r = r^*$, it is called critical nucleus. A nucleus with a radius in the range $r^* < r < r_s$ is thermodynamically unstable but kinetically stable while for $r > r_s$, it is stable in both ways.

9. The growth velocity is found to be different in different directions in the crystal. Alternatively, the growth velocity of different crystal planes is found to be different, the fast growing faces gets eliminated whereas the slow growing faces persist in a crystal.

10. Prime facie, crystal growth appears to be simply the reverse of dissolution process. This is not true because for a given system dissolution process is much faster than the growth process.

11. Crystal growth is the discrete process and not a continuous process. A new layer is more favourable to begin from mid face.

12. The attachment energy is found to be different on different parts of the crystal. For a given face, the attachment energy decreases in the order mid face > edge > corner.

13. Most of the existing theories/model of crystal growth are based on either atomic model or on thermodynamics considerations. For example, Kossels theory is based atomic model whereas diffusion theories are based on thermodynamic considerations.

14. Kinks are supposed to be the most favourable positions for the crystal growth. A growing crystal contains a large number of kinks in each step and a large number of steps in each crystal face.

8.13 DEFINITIONS

Activation energy: It is the additional energy required above the average (minimum) energy for a thermally activated reaction to take place.

Attachment energy: It is the amount of energy released when an atom (or a growing unit) attaches itself to the surface of a growing crystal.

Critical free energy: The maximum change in the free energy corresponding to a particular (critical) size of the nucleus.

Critical nucleus: The size of the nucleus corresponding to which the change in the free energy is maximum.

Embryo: A cluster of atoms of less than critical size.

Equilibrium concentration: A homogeneous solution that is in equilibrium with the solid phase (solute) is saturated w.r.t that solid.

Eutectic mixture: The mixture (of two or more substances) corresponding to which the minimum freezing point is obtained for all possible mixtures.

Fill percentage: It is the percentage fractional volume of the vessel obtained after deducing the volume of the starting material.

Flux: It is the molten solvent at high temperatures.

Hetrogeneous nucleation: The process of nucleation induced by some external agency (even the surface of the container).

Homogeneous nucleation: The process of nucleation initiated spontaneously within the system.

Index vector: Reciprocal of the growth velocity vector and acts as an indicator about the presence or absence of a crystal plane at a particular instant of time.

Supersaturation: It defines the stage of a solution in which the amount of dissolved solid (solute) is more than the saturated (equilibrium) condition.

REVIEW QUESTIONS AND PROBLEMS

1. Define solute, solvent and solution. What do you understand by solubility? How it varies with temperature in general?

2. Define supersaturation and explain how it can be measured?

3. Enumerate general criteria employed to assess a method of crystal growth. Describe aqueous solution method. Differentiate between slow cooling and slow evaporation methods. Write their advantages and disadvantages.

4. What do you understand by the term 'flux'? Enumerate the essential requirements of a flux. Describe the general procedure of flux growth employing temperature gradient method. Give a few advantages and disadvantages of this method.

5. What do you understand by the term 'autoclave'? Enumerate the characteristic features of an ideal autoclave. Discuss the growth of Quartz crystal by Hydrothermal method.

6. Name the important methods employed to grow single crystals by melt. Describe Bridgman-Stockbarger method.

7. Describe Czochralski method of crystal growth by melt. Discuss the merits of this method over Bridgman-Stockbarger method.

8. Describe the zone-refining and zone-melting methods of purification and growth of crystals by melt. Discuss their merits over other methods.

9. What do you understand by the term 'nucleation'? Explain primary and secondary nucleation. Obtain the expression for the critical radius and the corresponding change in the critical free energy for a homogeneous nucleus.

10. What do you understand by hetrogeneous nucleation? Obtain the expression for the critical radius and the corresponding change in the critical free energy for a cap-shaped nucleus. Discuss various cases.

11. What do you understand by the term 'index vector'? Based on geometrical considerations, show that the fast growing faces get eliminated and slow growing faces persist.

12. What do you understand by the term 'attachment energy'? Describe Kossel's theory of crystal growth. Explain the term 'half crystal'. Show that the new layer is more favourable to start at the center of the growing face.

13. Describe diffusion theory of crystal growth. Explain the importance of stagnant layer and surface reaction in the process of growth.

14. Describe BCF bulk diffusion model of crystal growth. Determine the fluxes which are diffusing towards kinks, steps and boundary layer and obtain the overall growth rate.

15. Describe the calculations of kink concentration and mobility of adsorbed molecules made by BCF during growth. What conclusions you draw from these results?

CHAPTER 9

CRYSTAL IMPERFECTIONS

9.1 INTRODUCTION

In an ideal crystal, the arrangement of atoms is in perfectly periodic manner. That is, an ideal crystal is completely free from imperfections, which is true only for theoretical considerations. However, real crystals grown in the laboratory do contain some degree of imperfections in them. The presence of imperfections has a strong influence on structure sensitive properties of crystals, such as strength, electrical conductivity and hysteresis loss of ferromagnets. Crystal imperfections can be classified on the basis of their geometry as follows:

    (a) Point imperfection (zero-dimensional defects)

        (i) Vacancy (Schottky imperfection)

        (ii) Self-interstitial (Frenkel imperfection)

    (b) Line imperfections (One-dimensional defects)

        (i) Edge dislocation

        (ii) Screw dislocation

    (c) Surface imperfections (two-dimensional defects)

        (i) Grain boundary

        (ii) Stacking faults

    (d) Volume imperfections (three-dimensional defects)

        (i) Foreign particle inclusions

        (ii) Large voids or pores

9.2 CONCENTRATION OF POINT IMPERFECTIONS

Schottky Imperfection (Monoatomic Solid)

According to thermodynamics, the equilibrium state of a solid is the state of minimum free energy

$$F = E - TS \tag{1}$$

where E is the internal energy, S is the entropy and T is the temperature in absolute scale. The free energy expression suggests the existence of a certain amount of disorder in the lattice at all temperature $T > 0K$ (which otherwise seems to be a perfeet crystal). We shall consider only those imperfections which are present as a result of thermodynamic conditions and not otherwise, such as accidental faults during growth.

A vacancy may be inherent or may be created delibrately by transferring an interior atom from its regular site to a site on the surface of a two dimensional crystal as shown in Fig. 9.1. Let us suppose that the energy required to create such a vacancy is E_v and for n such vacancies

$$\Delta E = nE_v$$

where ΔE is the change in internal energy of the crystal as a result of the creation of vacancies.

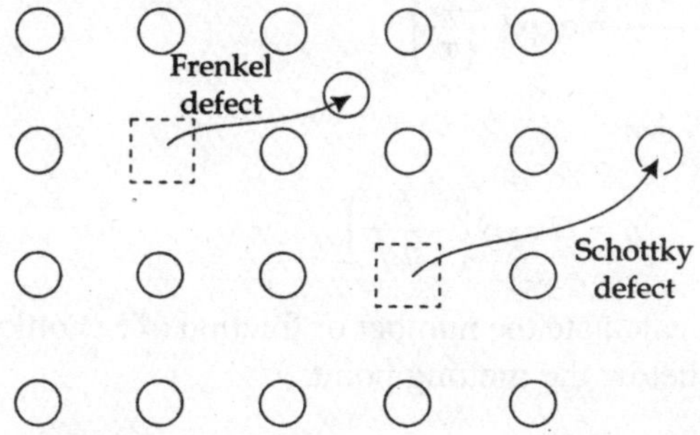

Fig. 9.1 The formation of Schottky and Frankel imperfections

As far as entropy of the crystal is concerned, for simplicity let us consider only configurational entropy. This is determined by considering the number of different ways (W) in which the atoms may be arranged over the available number of lattice sites. For example, let us consider that there are N lattice sites in the crystal and n vacancies that are produced. Then the number of ways of arranging n vacancies and $(N-n)$ atoms on N lattice site, is given by

$$W = \frac{N(N-1)(N-2)...(N-n+1)}{n!} = \frac{N!}{(N-n)!\,n!}$$

Further, the entropy change associated with the process is given by

$$S = k\ln W = k\ln\left[\frac{N!}{(N-n)!\,n!}\right]$$

where k is the Boltzmann constant.

The logarithmic fractional terms can be simplified by using Stirling's approximation,

$$\ln x! \cong x\ln x - x \text{ for } x >> 1$$

so that
$$\ln\left[\frac{N!}{(N-n)!\,n!}\right] = N\ln N - (N-n)\ln(N-n) - n\ln n$$

Now, the change in the free energy according to eq. 1 is given by

$$\Delta F = \Delta E - T\Delta S = nE_v - kT\left[N\ln N - (N-n)\ln(N-n) - n\ln n\right] \tag{2}$$

When the equilibrium is reached at a given temperature, there is no further change in the free energy of the system. Therefore, at equilibrium

$$\left(\frac{\partial \Delta F}{\partial n}\right)_T = E_v kT \ln\left(\frac{N-n}{n}\right) = 0$$

or

$$E_v = kT \ln\left(\frac{N-n}{n}\right)$$

or

$$\frac{N-n}{n} = \exp\left(\frac{E_v}{kT}\right)$$

If $n \ll N$, then

$$N - n \simeq N$$

so that

$$n = N \exp\left(-\frac{E_v}{kT}\right) \tag{3}$$

This is normally used to calculate the number or fraction of Schottky imperfections (vacancies) at a given temperature well below the melting point.

Frenkel Imperfection (Monoatomic Solid)

In this case, an atoms is being transferred from a regular site to an interstitial site (Fig. 9.1). A Frenkel imperfection thus consists of two components: a vacancy and an interstitial atoms. Let E_F be the energy required to form one Frenkel defect. Further, if there are N lattice sites in the crystal, N_i interstitial sites and n Frenkel defects are to be produced, then the number of ways of arranging $(N-n)$ identical atoms and n identical vacancies on N sites is

$$\frac{N!}{(N-n)!\,n!}$$

Also, the number of ways of arranging n identical vacancies and $(N_i - n)$ identical atoms on N_i identical sites is

$$\frac{N_i!}{(N_i-n)!\,n!}$$

so that, the total probability will become

$$W = \frac{N!}{(N-n)!\,n!} \times \frac{N_i!}{(N_i-n)!\,n!}$$

Now following a similar procedure as in the preceeding section, the expression for Frenkel imperfections can be obtained as

$$\frac{n}{(NN_i)^{1/2}} = \exp\left(-\frac{E_F}{2kT}\right) \tag{4}$$

Schottky Imperfection (Ionic solid)

Let us consider a perfect ionic crystal composed of equal number of positively and negatively charged ions, in which it is energetically favorable to form ion-pair vacancies (Fig. 9.2), such that the whole crystal remains electrically neutral even on the local scale. Further, let E_p be the energy required to create such an ion-pair (one cation and one anion) vacancy, then the number of ways of producing n ion-pair vacancies will be given by

$$W = \left(\frac{N!}{(N-n)!\,n!} \right)^2$$

so that the change in the energy according to eq.1 is given by

$$\Delta F = \Delta E - T\Delta S = nE_p - 2kT\,[N\ln N - (N-n)\ln(N-n) - n\ln n]$$

Simplification of this on a similar line will provide us

$$n = N \exp\left(-\frac{E_p}{2kT} \right) \tag{5}$$

The above discussion suggests us the following:

(i) Concentration of different point imperfections are found to be different at a given temperature.

(ii) Energy favours the formation of Schottky imperfections in general, although Frenkel imperfections are also formed.

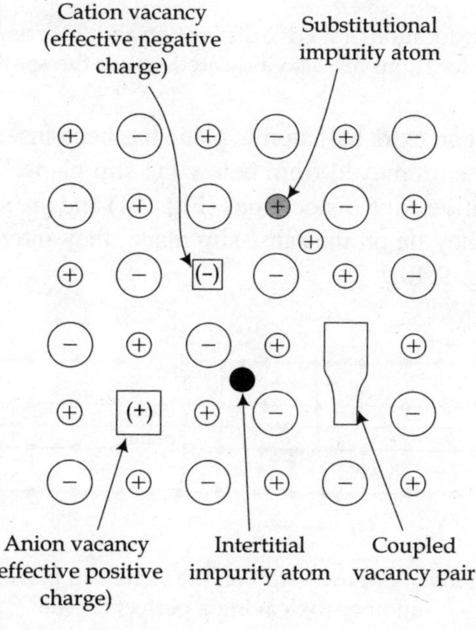

Fig. 9.2   Point imperfections in an ionic solid

(iii) Formation of Schottky imperfections increases the crystal volume (without any change in mass) and hence decreases the density of the crystal. On the other hand, formation of Frenkel imperfections leaves the volume/density of the crystal unchanged.

9.3 THE GEOMETRY OF DISLOCATIONS

One-dimensional (or line) imperfections are called dislocations. They are produced when one region of a crystal surface is slipped with respect to other region and the dislocation line represents the boundary between the two regions. Two extreme type of dislocations are distinguished as the edge and the screw dislocation. The geometry of the lattice irregularities described by the two extreme dislocations is shown in Fig. 9.3.

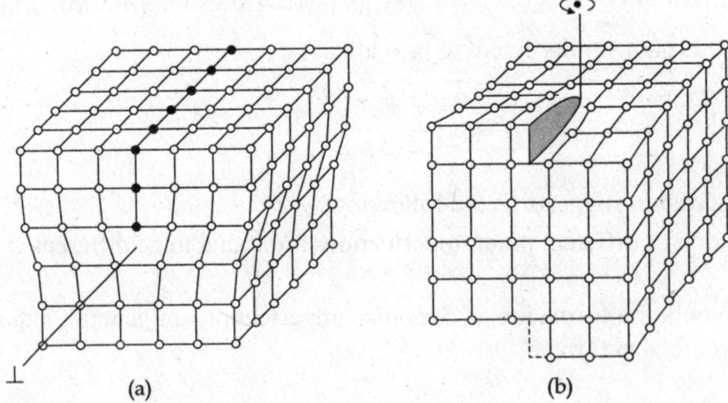

Fig. 9.3 Geometry of simple dislocations. (a) Edge dislocation and (b) screw dislocation. The lines normally used to represent the dislocations are also indicated, as are the symbols for them, ⊥ and ↻.

Figure 9.3a suggests that an extra half atomic plane has been inserted from above the slip plane or an extra half plane has been removed from below the slip plane. The two situations are distinguished as positive and negative edge dislocations (Fig. 9.4) and are symbolically represented as ⊥ and ⊤, respectively. When they lie on the same slip plane, they interact and annihilate each other leaving a perfect crystal (Fig, 9.4b).

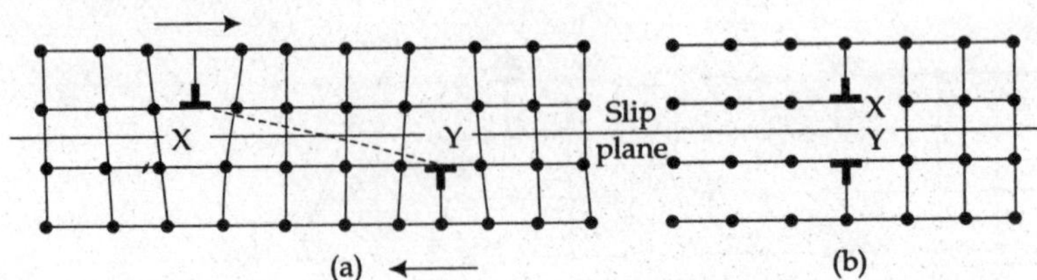

Fig. 9.4 Two edge dislocations of opposite sign on the same slip plane (a) attract and annihilate one another (b) leaving a perfect crystal

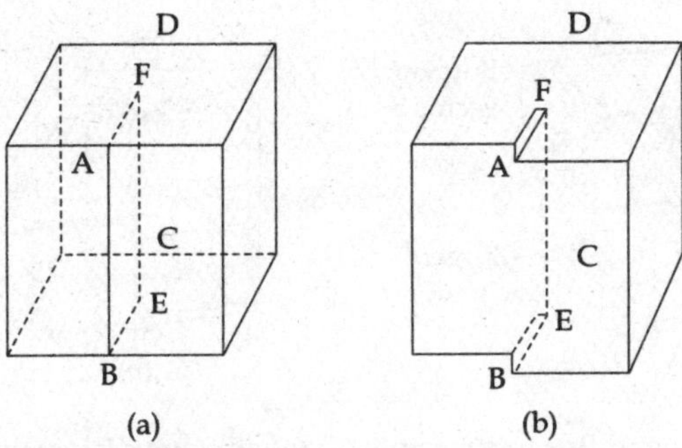

Fig. 9.5   The formation of screw dislocation.

Figure 9.5 illustrates the formation of a screw dislocation where the right half is pushed down with respect to the Burgers vector b so that the dislocation line is parallel to it. Like edge dislocation, the screw dislocation may be either right handed or left handed. They are symbolically represented as ↻ or ↺ (Fig. 9.3b) depending on whether the slip vector is parallel or anti parallel to the dislocation line.

A dislocation may either be perfect or imperfect. When the Burgers vector of a dislocation is an integral multiple of the lattice translation, the dislocation is called full or perfect dislocation. On the other hand, when the Burgers vector of the dislocation is a fraction of the lattice translation, the dislocation is a called partial or imperfect dislocation. Formation of a perfect and an imperfect dislocation is shown in Fig. 9.6 in a simple cubic lattice.

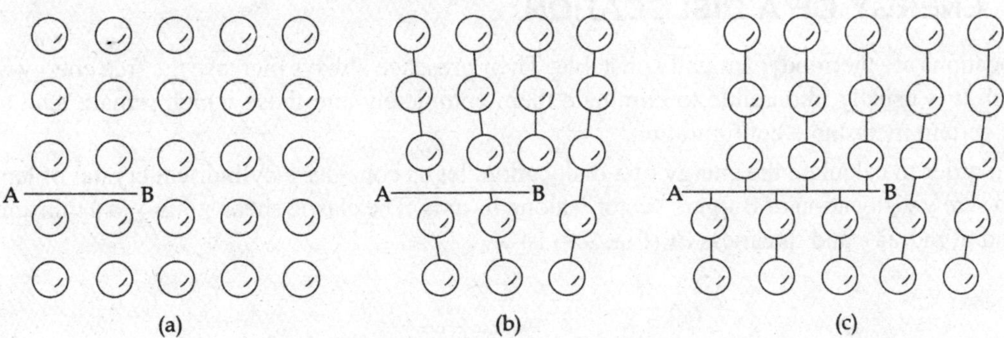

Fig. 9.6   (a) Crystal is cut and displaced along *AB* to form either. (b) Perfect dislocation, and (c) an imperfect dislocation

9.4 BURGERS VECTOR AND BURGERS CIRCUIT

One property of a dislocation is its Burgers vector b, which describes both magnitudes as well as direction of slip. The Burgers vector is usually defined from a Burgers circuit as shown in Fig. 9.7. A Burgers circuit is a sequence of lattice vectors, from one lattice point to the next such that it forms a loop. This circuit around a dislocation fails to close the loop by the Burgers vector, whereas a

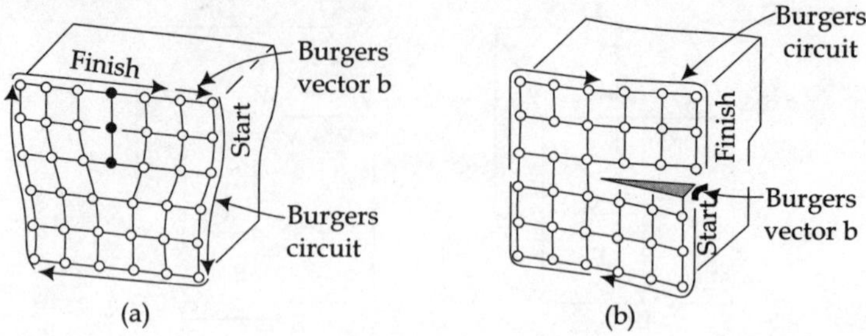

Fig. 9.7   Burgers vectors and Burgers circuits of: (a) an edge dislocation; (b) screw dislocation

similar circuit in a perfect lattice would complete the loop. The Burgers vector of a pure edge dislocation is normal to the dislocation line, that of pure screw dislocation is parallel to the dislocation line and that of mixed dislocation makes an angle (lies between 0 and 90°) with the dislocation line. The Burgers vector of a dislocation is always the same and independent of position of the dislocation. The modulus of the Burgers vector is called dislocation strength.

The Burgers vector of an edge (or mixed) dislocation and the dislocation line define the slip plane. Under the normal circumstances such dislocations are forced to more on the slip plane. On the other hand, the Burgers vector and the dislocation line of a pure screw dislocation, are parallel and do not define a unique plane. The screw dislocation is thus free to move on any of the several planes in which the Burgers vector lies. Both these processes take place without considerable activation energy.

9.5   ENERGY OF A DISLOCATION

Dislocations are thermodynamically unstable. Their presence always increase the free energy of the crystal. It is usually impossible to eliminate them completely and those which remain tend to assume certain metastable configuration.

In order to calculate the energy of a dislocation, let us consider a cylindrical crystal of length l with a screw dislocation of Burgers vector b along its axis. The elastic shear strain γ in a thin annular section of radius r and thickness dr (Fig. 9.8) is

$$\gamma = \frac{b}{2\pi r} \tag{7}$$

where $b = |b|$

The elastic energy per unit volume dE/dV, of the annular region is

$$\frac{dE}{dV} = \frac{1}{2}\,\tau\gamma = \frac{1}{2}\,G\gamma^2 = \frac{G}{2}\left(\frac{b}{2\pi r}\right)^2 \tag{8}$$

where G is the elastic shear modules. The volume of the annular region is $dV = 2\pi r l dr$. This gives us the elastic energy per unit length as

$$dE = \frac{lGb^2}{4\pi} \cdot \frac{dr}{r} \tag{9}$$

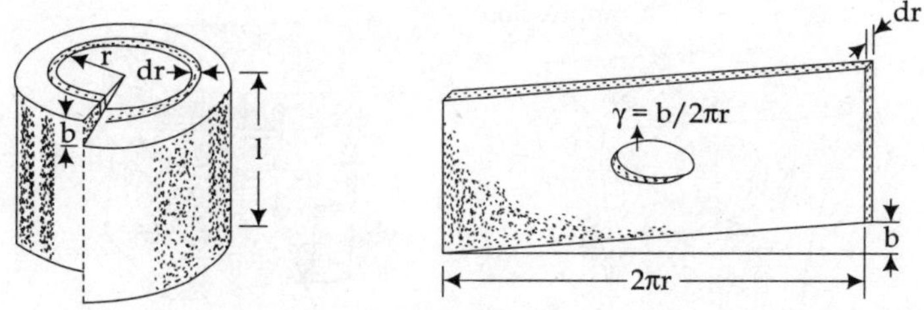

Fig. 9.8   Geometrical model for the calculation of shear strain around a screw dislocation

The strain energy due to the presence of this dislocation can be computed by integrating both sides of eq. 9 within the limit r_o to R. Therefore, we have

$$E = \int_{r_o}^{R} \frac{lGb^2}{4\pi} \cdot \frac{dr}{r} = \frac{lGb^2}{4\pi} \ln\left(\frac{R}{r_o}\right) + E_o \qquad (10)$$

where E_o is the real strain energy inside the core.

If we choose the limits $r_o = 0$ or $R = \infty$, in each case the integral becomes infinite. The difficulty in choosing $r_o = 0$ is that the Hooke's law is not valid for the high strain at the dislocation core. The value of $R = \infty$ is also unrealistic because at large values of r the strain field of one dislocation is cancelled by those of the other dislocations. It has been shown that if r_o is taken as b, the value of E_o is only a fraction of the total energy and can be neglected. Since the energy is relatively insensitive to R/r, the ratio usually adopted is $\ln\left(\dfrac{R}{r_o}\right) = 4\pi$. Under these approximations, the energy of a screw dislocation is given as

$$E \simeq lGb^2 \qquad (11)$$

Similarly, the energy of an edge dislocation is given approximately as

$$E = \frac{1}{1-v} \cdot \frac{lGb^2}{4\pi} \cdot \ln\left(\frac{R}{r_o}\right) + E_o \simeq \frac{lGb^2}{1-v} \qquad (12)$$

where v is the Poisson's ratio.

If $v = 1/3$, the energy of an edge dislocation is 3/2 times that of screw dislocation of the same length. Since the energy of a dislocation (edge or screw) is proportional to b^2, the most stable dislocations are those which have minimum Burgers vector (i.e. those in close packed directions). Equations 11 and 12 also show that the energy of a dislocation is proportional to its length, i.e. the line energy is equivalent to line tension as the surface energy is equivelent to surface tension. Thus a curved dislocation will have a line tension T, a vector acting along the line so that

$$T = \frac{\partial E}{\partial l} = Gb^2 \qquad (13)$$

Figure 9.9 illustrates the geometry of the stress field around edge and screw dislocations.

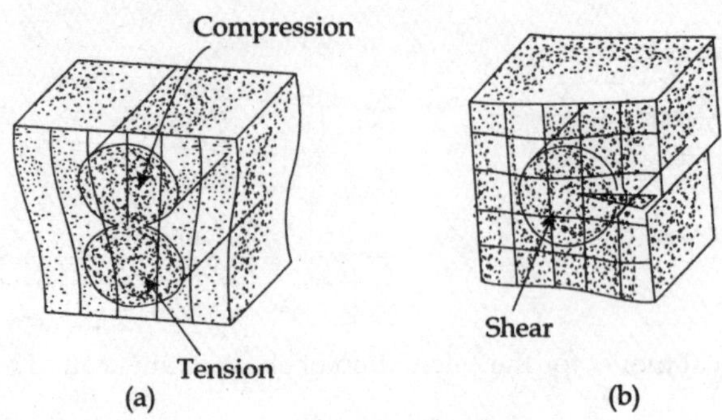

Fig. 9.9 Geometry of stress and strain fields surrounding (a) an edge dislocation (b) a screw dislocation

9.6 SLIP PLANES AND SLIP DIRECTIONS

Plastic deformation due to slip in a real crystal occurs as a result of dislocation motion. Therefore, those dislocations which require the lowest energy or stress will move most readily. Further, as the energy of a dislocation is proportional to the square of the Burgers vector, dislocations of the shortest Burgers vector, i.e. those along the close packed directions, will require the least stress. The slip planes within which slip occurs are usually the most densely packed planes, which are also most widely separated. The combination of a slip direction and the plane containing it, is defined as the slip system. For example, the combination of (111) and $[\bar{1}10]$ form a slip system but not (111) and [110], as the direction [110] does not lie on the (111) plane. The observed slip system in some *fcc*, *bcc*, *hcp* metals and *fcc* ionic crystals are listed in Table 9.1. The first listed plane in each case is either the closed packed planes or the most commonly observed planes. However, other planes may also become active at elevated temperatures. The axial ratio(c/a) also plays an important role in determining whether some other system are possible or not. For an ideal packing of spheres in *hcp*, the axial ratio is equal to 1.632. However, this ratio is not found in any *hcp* metal. For example, in Zn and Cd, the ratio has higher values than the ideal case, while the metals like Mg, T, Be, Zr etc. have smaller values. When c/a ratio in low, the (0001) plane loses the distinction of being the plane of highest density. Further, high temperatures and smaller axial ratios make other planes active for slip, particularly $\{10\bar{1}1\}$ pyramidal planes and $\{10\bar{1}0\}$ prism planes.

9.7 DISLOCATION REACTIONS

Different dislocations can react and produce new dislocations. For example two parallel dislocations of Burgers vector b_1 and b_2 situated either on the same slip plane or on two intersecting slip planes can react to form a third dislocation of Burgers vector b_3 and is given by the equation

$$b_1 + b_2 = b_3 \tag{14}$$

Reaction among larger number of dislocations are also possible, such as

$$b_1 + b_2 \rightleftharpoons b_3 + b_4 \tag{15}$$

In this case, the energies of two sides of dislocations are different.

Table 9.1 Observed slip systems in Crystals

Structure	Slip plane	Slip direction	No. of slip systems	Pictorial new of slip system
FCC Cu, Al, Ni, Pb, Au, Ag, γFe, ...	{111}	<1$\bar{1}$0>	$4 \times 3 = 12$	
BCC αFe, W, Mo, βBrass	{110}	<$\bar{1}$11>	$6 \times 2 = 12$	
αFe, Mo, W, Na	{211}	<$\bar{1}$11>	$12 \times 1 = 12$	
αFe, K	{321}	<$\bar{1}$11>	$24 \times 1 = 24$	
HCP Cd, Zn, Mg, Ti, Be, ...	(0001)	<11$\bar{2}$0>	$1 \times 3 = 3$	
Ti	{10$\bar{1}$0}	<11$\bar{2}$0>	$3 \times 1 = 3$	
Ti.Mg	{10$\bar{1}$1}	<11$\bar{2}$0>	$6 \times 1 = 6$	
NaCl, AgCl	{110}	<1$\bar{1}$0>	$6 \times 1 = 6$	

A certain amount of energy is always liberated when dislocations react. Let Q be the quantity proportional to the energy liberated, and taking in to account the fact that the energy of a dislocation is proportional to square of the Burgers vector, we can write

$$b_1{}^2 + b_2{}^2 = (b_1 + b_2)^2 + Q \tag{16}$$

where
$$Q = -2b_1 b_2 \tag{17}$$

Equation 17 can be accepted as a criterion for the stability of a dislocation. That is, two dislocations attract mutually if their Burgers vectors form an abtuse angle or repel if the two vectors form an acute angle. Equilibrium results, if the two vectors are perpendicular to each other.

Frank introduced the following criterion for the stability: That two parallel dislocations of Burgers vectors b_1 and b_2 may combine to form a third dislocation of Burgers vector b_3 only if the following reaction is satisfied

$$b_1^2 + b_2^2 > b_3^2 \tag{18}$$

Similarly, a dislocation of Burgers vector b_1 may dissociate into two parallel dislocations of Burgers vector b_2 and b_3 if

$$b_1^2 > b_2^2 + b_3^2 \tag{19}$$

It is easy to visualize that both the crystallographic direction and the magnitude of a Burgers vector are related to distance between atomic sites. In Miller index notation na[hkl] having components nah, nak, nal along the three axes is used to describe a Burgers vector whose direction is [hkl] and the magnitude is na $(h^2 + k^2 + l^2)^{1/2}$. Therefore,

$$b^2 = n^2a^2(h^2 + k^2 + l^2) \tag{20}$$

where a is the length of the cell edge and n is a number. In *fcc* crystals, slip occurs along <110> close packed directions and mostly on {111} closed packed slip planes. A perfect or a unit dislocation in this case having a Burgers vector such as $(a/2)[110]$, of magnitude $a/\sqrt{2}$ connects a cube corner to a nearby face center. In *bcc* crystals, the Burgers vector for a slip occuring from the cube corner to cube center along the close packed direction <111> is $(a/2)$ [111], with its magnitude equal to $a\sqrt{3}/2$. Hexagonal close packed crystals normally slip along <11$\bar{2}$0> directions having a Burgers vector $(a/3)$ [11$\bar{2}$0] of magnitude a, where a is the lattice translation in the (0001) plane.

Slip process can be understood better in a closes packed plane of atoms in an *fcc* or *hcp* crystals as shown in Fig. 9.10. An atom sitting at O on the next (upper) layer, can make a perfect displacement to P, Q, or R. Hence if this atoms is moved from $O \rightarrow Q \rightarrow P$ by two successive displacements with Burgers vectors OQ and OR (equivalent to QP), respectively, the effect is as if one dislocation with vector OP has passed through. It is thus possible for three such dislocations to meet at a node as shown in Fig. 9.10b. In Miller Index notation, this can be represented as

$$\frac{a}{2}[10\bar{1}] + \frac{a}{2}[0\bar{1}1] + \frac{a}{2}[\bar{1}10] = 0$$

or
$$\frac{a}{2}[10\bar{1}] + \frac{a}{2}[0\bar{1}1] = \frac{a}{2}[1\bar{1}0] \tag{21}$$

The above billiard ball model (Fig. 9.10) immediately shows that atoms prefer to slip, not along the perfect lattice vectors such as OP, but along zig-zag path such as $O \rightarrow S \rightarrow P$. Thus a perfect dislocation with Burgers vector OP splits into two partial dislocations with Burgers vectors OS and SP, respectively. For example, such a dislocation on (111) plane in an *fcc* crystal is

$$\frac{a}{2}[\bar{1}10] = \frac{a}{6}[\bar{1}2\bar{1}] + \frac{a}{6}[\bar{2}11] \tag{22}$$

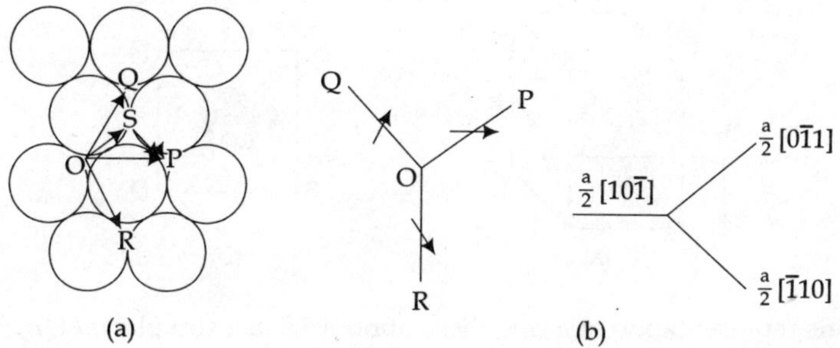

Fig. 9.10   Burgers vectors (a) On a close packed plane (b) meeting at a dislocation node

Using $\dfrac{a}{6}$ as the unit length, this reaction can be written in a more simplified form as

$$[\bar{3}\,30] = [\bar{1}\,2\,\bar{1}] + [\bar{2}\,11] \tag{23}$$

The two dislocations $[\bar{1}\,2\,\bar{1}]$ and $[\bar{2}\,11]$ are partial dislocations, known as Shockley partials.

It is necessary in any proposed dislocation reaction to ensure that the sum of the components of the Burgers Vector on both sides of the reaction is the same. Therefore, for the reaction (22), we can verify that

$$\frac{a}{2}\,[\bar{1}\,10] - \frac{a}{6}\,[\bar{1} + \bar{2},\, 2 + 1,\, \bar{1} + 1] = \frac{a}{2}\,[\bar{1}\,10].$$

9.8   DENSITY OF DISLOCATIONS

We discussed above the edge and screw dislocations each having a single dislocation line sweeping across a slip plane and giving rise to a small displacement of the order of a few Angstroms. However, the observed plastic flow (deformation) in crystals appears to involve a large number of dislocations sweeping across many slip planes. In orders to determine the density of dislocations, let us consider an element ds of the dislocation line ABC as shown in Fig. 9.11. Let dl be the displacement of the element in the slip plane along the direction normal to ds, h is the height of the crystal and A is the area of the slip plane, then the area swept by the line element is

$$dl\;ds$$

This corresponds to an average displacement of upper part of the crystal with respect to the lower part by an amount

$$\frac{(dlds)}{A}\,b \tag{24}$$

The total increase in the average displacement or the amount of slip during time dt can be obtained by integrating eq. 24 over the whole dislocation line and the corresponding increase in the shear strain is obtained if the expression is divided by h, the height of the crystal. Hence,

$$d\gamma = \frac{b}{Ah}\int (dl)\,ds = \frac{b}{V}\int (dl)\,ds = \frac{bs}{V}\,\overline{dl} \tag{25}$$

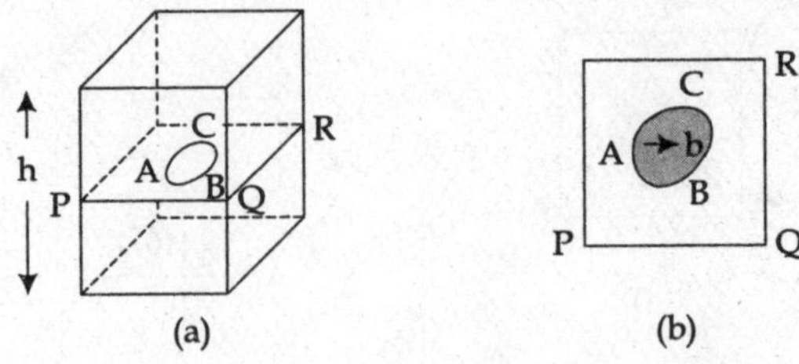

Fig. 9.11 Schematic representation of a ring dislocation *ABC* in a slip plane *PQR*. Slip has occurred only accros the hatched area

where $Ah = V$, volume of the crystal and $\overline{dl}$ is the average displacement and s is the total length of all dislocations. Further, increase in the strain per second due to the motion of the element ds is

$$\frac{d\gamma}{dt} = \frac{bs}{V}\frac{\overline{dl}}{dt} = b\rho\overline{v} \tag{26}$$

where $\dfrac{s}{V} = \rho$, the dislocation density and $\overline{v}$ is the average velocity of the element.

9.9 OBSERVATION OF DISLOCATIONS

Dislocations and other crystal imperfections like staking faults, grain or low angle boundaries can be made to observe directly or can be estimated (using indirect method) by different techniques based on distructive or non distructive methods. In the following, we provide a brief discussion of some of the important and widely used techniques.

(a) Method Based on Growth Spirals

Freshly grown crystals of some compounds such as SiC, CdI_2, etc. are found to exhibit spirals on their faces. These spirals are due to the growth of the crystal around screw dislocations. The step height of the spiral is equal to the length of the Burgers vector which is normal to the plane of observation. The sense of winding the spiral gives the sign of the dislocation. The step can be easily measured by multiple beam interferometry. This method can be used to determine the density of screw dislocations.

(b) Method Based on Etch Pits

Although, this is a distructive method but most practical and widely used method to study the dislocation density. Since the lattice surrounding the dislocation is in a highly strained state, it is preferentially attacked by a suitable etchant. This dissolves away surface atoms and leaves behind etch pits which can be observed optically.

Etching method can be used equally well to a freshly grown crystals, a cleaned surface or a polished surface. The type of the etchant depends on the nature of the crystal material. The results of an etching technique depend on the following: (a) the nature of the etchant (b) the impurities present in the crystal (c) the treatment of the crystal prior to etching and (d) the surface preparation.

(c) Optical and Electron-Optical Methods

When dislocations are piled up one above the other, they give rise bifringence (double refraction). This can be used to study the deformation. When a polarized light is made to fall on a transparent crystal, it splits into ordinary and extraordinary rays. The measurement of double refraction in any crystal provides the difference between the greatest and the least refractive indices.

Thin films are used for electron microscopic studies. Because of its very high resolution, it is possible to directly observe the dislocations.

(d) Decoration Method

There are only a limited of number of crystals which are transparent to light or infra-red radiation. The dislocations in these crystals are normally not visible. However, it is possible to decorate the dislocations by inducing precipitation along the line of the dislocation. The tendency of the foreign (impurity) atoms is to migrate towards the dislocation and form a pattern of the foreign atoms segregated along the dislocation line. The effect produced is similar in appearance to a row of beads along a fine thread. The position of the dislocation is revealed by the scattering of light at the "beads" or precipitates, and can be observed through an optical microscope. Electron microscopy can also be used. Sometimes, a combination of decoration and etching techniques is also used.

In general, the decoration method involves the heating of the crystal before examination and hence the method is restricted to study the high temperature deformation structures. However, this method is widely used to study the defects in alkali halide crystals.

(e) *X*-ray Diffraction Topography

X-ray diffraction topography is a non-destructive technique and can be used to study dislocations, stacking faults, grain and low angle boundaries in bulk crystals of suitable thickness (the optimum thickness t of the crystal in the direction of the beam corresponds to $\mu t \simeq 1$ where μ is the linear absorption coefficient). Because of the lower resolution than the electron diffraction, it is applicable to the study of crystals with relatively low dislocation densities $\sim 10^6$ cm^{-2}. However, the greater penetration of *X*-rays make this technique more useful to study the bulk crystals.

From the discussion of *X*-ray diffraction by a crystal, we know that the intensity of diffraction from crystal planes is maximum when all atoms in the unit cell occupy their normal positions. When some imperfections are present, the atoms in the unit cell change positions and hence there is a change in the intensity of diffraction. It is because of the fact that the deformed regions of a crystal diffracts differently from the undeformed regions. Therefore, if a point to point variation of diffracted intensity from a crystal is recorded, one can easily find the point to point variation in the perfection of the crystal. This is the basic principle of *X*-ray diffraction topography.

The topographic study requires a point focus and well collimated (~50cm long) monochromatic *X*-ray beam narrowed through a slit S_1. The crystal is properly aligned with the help of the turn-table (which provides the rotation around vertical axis) and the vertical circle geometer head (which provides the rotation around horizontal axis). The dectector (scintillation or proportional counter) is set to receive the diffracted beam at an angle 2θ w.r.t the direct beam. The slit S_2 is used to stop the direct beam reaching the film and allow only the diffracted beam. A typical schematic diagram exhibiting various parts of a Lang camera is shown in Fig. 9.12.

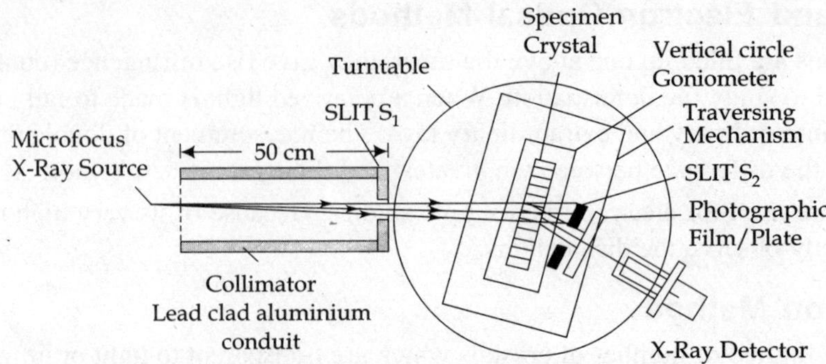

Fig. 9.12   A schematic diagram of an X-ray diffraction topography set up based upon the Lang technique

If the crystal is fixed, the topograph will reveal the defect distribution only along the path of the *X*-ray through specimen, and it is then called "section topograph." However, by moving the crystal and the photographic film/plate together, the entire crystal can be explored, and it is then called "projection topograph." The recorded topograph is then analysed to know the nature of defects present in the crystal.

From basic definition of edge and screw dislocations, we know that the Burgers vector is $\perp$ to dislocation line for edge dislocation and $||$ to dislocation line for screw dislocation. Now, let us take the case of two dislocations separately.

Case I: Screw Dislocation

In this case, the crystal planes which are $\perp$ to b have maximum deformation and provide maximum diffraction contrast, whereas the planes containing b have negligible deformation and provide minimum diffraction contrast. Now, if g is defined as the diffraction vector acting normal to crystal planes from which diffraction is observed, then the contrast criterion for observation of dislocation can be mathematically written as

$$g \cdot b = 1 \text{ for maximum contrast}$$

and
$$g \cdot b = 0 \text{ for minimum contrast}$$

Case II: Edge Dislocation

In this case, b is $\perp$ to dislocation line, therefore the contrast criteria becomes

$$g \cdot b = 1 \text{ for maximum contrast}$$

$$g \cdot b = 0 \text{ for minimum contrast}$$

and
$$g \times \hat{n} = 0 \text{ for minimum contrast}$$

where $\hat{n}$ is a unit vector along dislocation line. The two minimum contrast criteria must be satisfied at the same time to ensure minimum contrast.

9.10 SURFACE IMPERFECTIONS

As the name indicates, the surface imperfections are two dimensional in geometrical sense. They refer to regions of distortions that lie about a surface having a thickness of a few atomic diameters. The surface imperfections include grain boundary, tilt and twist boundary, twin boundary and stacking faults.

Grain Boundary

The most obvious manifestation of geometrical defects extending over whole surface is the appearance of grains and grain boundaries in polycrystalline materials. A typical polycrystalline solid consists of a number of interlocking crystals or grains, oriented randomly. The atoms along the boundary regions are being pulled by each of the two grains to join its own configuration. They can join neither grain due to the opposing forces and take up an equilibrium position. The boundary between the adjacent grains, therefore, must have a structure that some how conform to the structures and orientations of both the grains and can be compared to a non crystalline material. Actually, the atoms along the boundary represent a transition between the two adjacent misoriented crystalline regions. The crystal orientation changes sharply at the grain boundary. The angles between the crystalline orientations of nearby grains are often large (greater than 10-15°), the boundary between the grains in such cases is known as high angle boundary (Fig. 9.13).

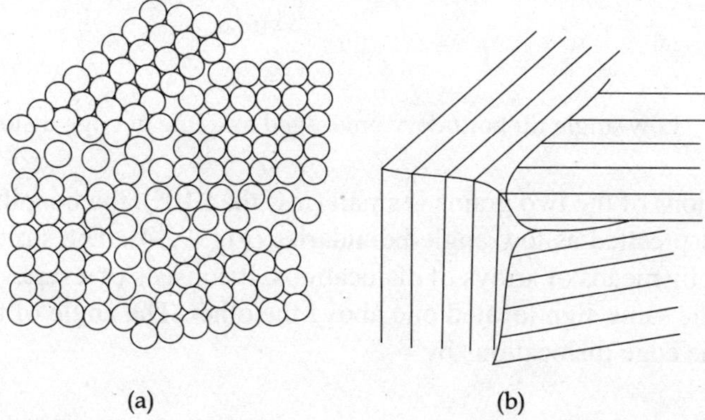

(a)   (b)

Fig. 9.13   The atomic arrangements at grain boundaries

The average number of nearest neighbours for an atom near the boundary of a close packed crystal is 11, whereas in the interior of the crystal it is 12. On an average, one bond out of twelve bonds is broken at the boundary. The grain boundary between two crystals, which have different crystalline arrangements or differ in composition, is known as interphase boundary or simply an interphase.

Tilt and Twist Boundary

Tilt boundary may be defined as a boundary between two adjacent perfect regions in the same crystal that are slightly tilted, with respect to one another. In other words, when the angle between

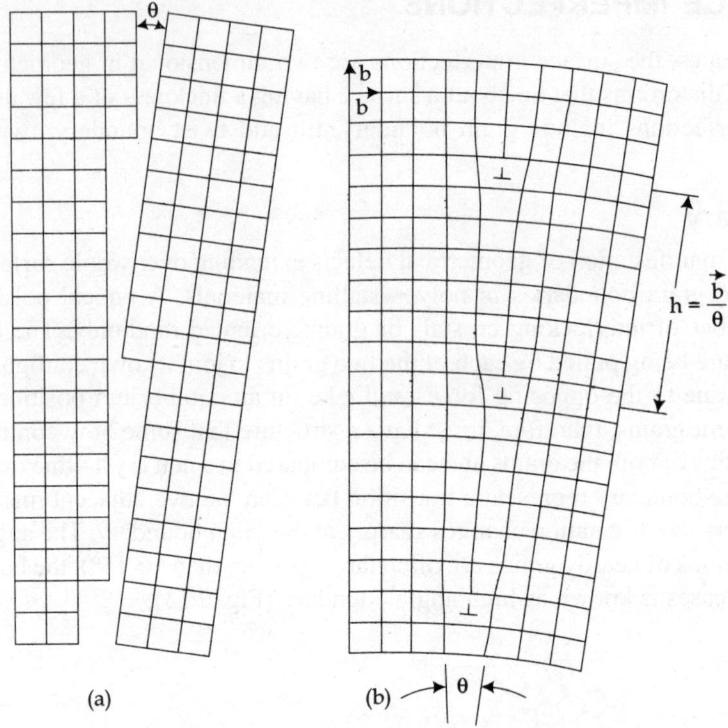

Fig. 9.14   Low-angle tilt boundary envisaged as a line of edge dislocations

the crystallite orientations of the two grains is small (less than 10°), the boundary is said to be a tilt boundary. They are also called as low angle boundaries (Fig. 9.14). The structure of such boundaries can be described by means of arrays of dislocations. It consists of a series of equally of spaced edge dislocations of the same sign located one above the other. The angle of tilt θ is related to the Burgers vector b of the edge dislocations by

$$\tan \theta = \frac{b}{h}$$

where h is the vertical average spacing between two neighboring edge dislocations. For small angle of tilt

$$\theta = \frac{b}{h}$$

If the two parts of the crystal are rotated through a small angle, about an axis which is perpendicular to the grain boundary (rather than about one that lies in the boundary as in the case of tilt boundary), a twist boundary is formed. Twist boundary consists of a set of screw dislocations, in contrast to the tilt boundary which consists of a set of edge dislocations.

If the misoriented single crystal sections are identical but are joined together in such a way that the boundary acts as a reflecting plane, the pair of crystals constitute a twin and the resulting boundary is said to be a twin boundary. In such cases the atomic arrangement on one side of the boundary is a mirror reflection of the arrangement on the other side (Fig. 9.15).

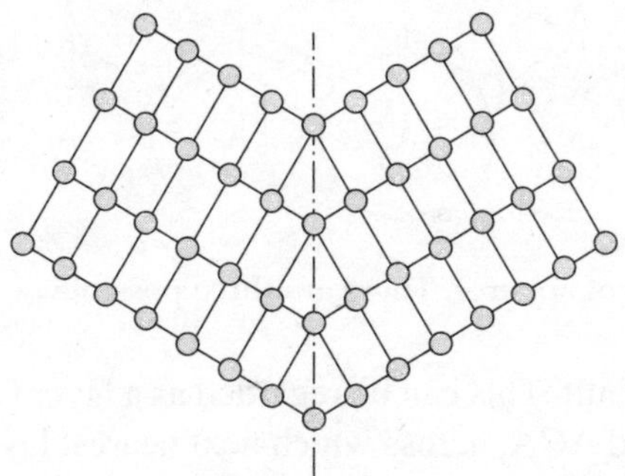

Fig. 9.15   A twin plane

There are several ways in which twinning can be produced. Growth twins are formed during the growth of a crystal. The transformation between high and low temperature modifications of a crystal may take place via twinning. Twinning produced during plastic deformation of a crystal is called deformation twinning. The simplest example of twinning in an *fcc* structure can be represented by the sequence

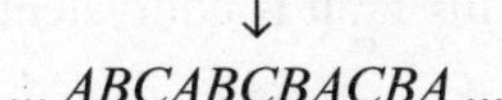

$$... ABCABCBACBA ...$$

where the arrow indicates the twin plane and the center of the fault. The occurrence of twinning is common especially in metals with *bcc* or *hcp* structure.

Stacking Faults

Stacking faults are observed in close packed structures. A close packed structure is generated by stacking close packed layers on top of one another with the restriction that no two adjacent layers are in the same orientation. Given a layer *A*, a close packing can be extended by placing the next layer such that its atoms occupy either *B* or *C* sites (Fig. 9.16). Here *A*, *B* and *C* refer to the three possible layer positions in a projection normal to the close packed layers. In the following, we shall discuss various ways of producing stacking faults in *fcc* and *hcp* crystals.

(i) Stacking Faults in *FCC* Crystals

Stacking faults may be produced in *fcc* crystals in the following ways.

(a) By removing a close packed plane

This can be achieved by diffusion of a sheet of vacancies into the plane and by collapsing together the adjoining planes to eliminate the layer of void between them. For example, removing a *B* layer, the cubic close packed sequence becomes

$$... ABCA \,|\, CABC ...$$

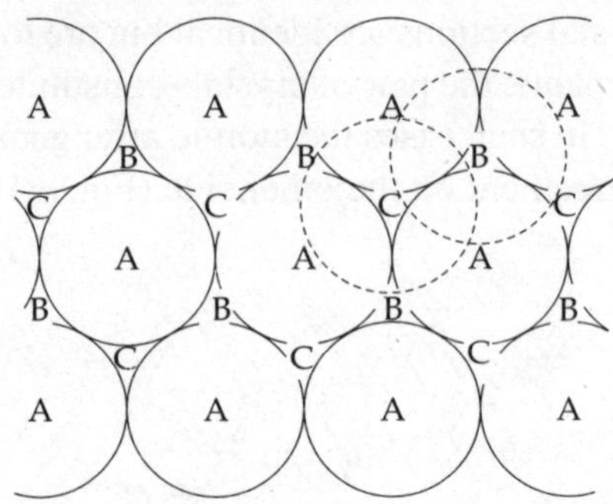

Fig. 9.16 The close packing of spheres. There are three possible positions for a layer: *A*, *B* and *C*

This is called an intrinsic fault. This can be regarded as a layer CACA of *hcp* or as two overlapping twin boundaries, CAC and ACA, across which next nearest layer are wrongly stacked. In such cases, the normal stacking sequence is maintained in the crystal on either side of the fault right up to the fault plane.

(b) By inserting an extra plane

This can be achieved by diffusion of interstitial atoms into the space between two close packed planes. If say, a B layer is introduced as in the sequence

$$\downarrow$$
$$... ABCBABC ...$$

This is called an extrinsic fault. This fault is equivalent to two neighbouring twin boundaries *BCB* and *BAB*.

(c) By slip on a close packed plane

This can be produced by displacing all atoms above the reference plane by shearing operation on {111} planes. Let us consider an *fcc* crystal and suppose that the plane *A* as shown in Fig. 9.17 is a reference plane, and the next plane above it is a *B* plane. If the *B* planes and all other planes above it are displaced by the vector (1/6) $[\bar{2}11]$, the *B* plane moves into *C* position, and the planes above it transform in cyclic order, $A \rightarrow B \rightarrow C \rightarrow A$, relative to the reference plane. The shear displacement is represented by the arrows in the reaction

$$... ABCABCABC ...$$
$$\downarrow\downarrow\downarrow\downarrow$$
$$CABCA$$

giving ... *ABCACABCA* ... . The resultant fault is equivalent to an intrinsic fault as obtained in (a).

(ii) Stacking Faults in *HCP* Crystals

In *hcp* crystals, close packed plane is the (0001) plane. This basal plane is also the most frequently observed glide plane in *hcp* crystals but unlike *fcc* crystals this does not correspond to twin plane. Referring again to the hard ball model (Fig. 9.17), one can easily verify that there are two kinds of intrinsic faults and one extrinsic fault in *hcp* structures. The mode of formation of these faults are as follows:

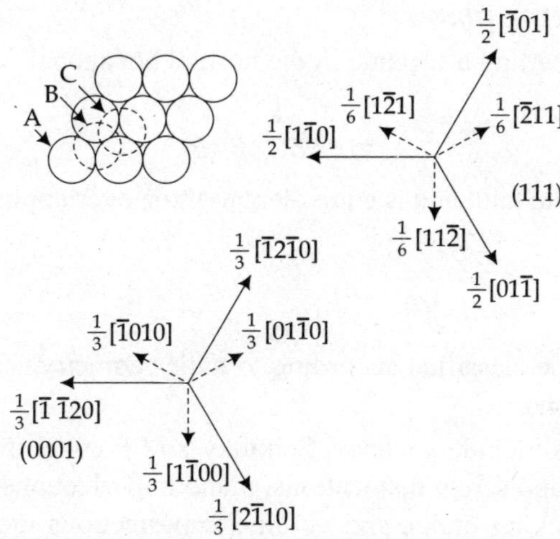

Fig. 9.17 Projection normal to the (111) plane showing three type of stacking positions *A*, *B* and *C*. *fcc* and *hcp* notations are also presented

(a) By removing a close packed plane and then shear

Removal of a close packed plane in an *hcp* structure is done in the same way as was done in the cubic case. The resulting stacking becomes

$$... ABABA|ABAB ...$$

Here, either side of boundary has the same plane (i.e., *A*) which violates the rule of close packing. Moreover, this is unstable also. In order to acquire the stable configuration, the layer above the boundary must shear.

Therefore, shearing them by the vector $\frac{1}{3}[\bar{1}100]$, the *A* plane moves into *C* position and the planes above it transform in anticyclic order, $A \rightarrow C \rightarrow B \rightarrow A$, relative to *A* plane, yielding the stacking sequence

$$... ABABA|CACA ...$$

This fault is equivalent to cubic layer *BAC*.

(b) Simply by shear

Another intrinsic fault can be produced in the *hcp* structure by directly shearing all the planes above a reference plane by the vector $\frac{1}{3}[\bar{1}100]$, i.e. the sequence

$$... ABAB|ABA ...$$

after shearing yield a sequence

$$... ABAB|CACA ...$$

Similarly, this fault is equivalent to two overlapping cubic layers *ABC* and *BCA*.

(c) By inserting an extra plane

This can be achieved by inserting a *C* plane in the normal hexagonal sequence ABAB ... such as

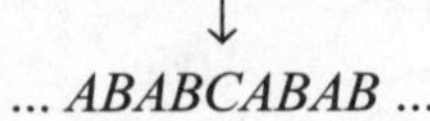

$$... ABABCABAB ...$$

This is called an extrinsic fault and is equivalent to three overlapping cubic layers i.e. *ABC* and, *BCA* and *CAB*.

9.11 SUMMARY

1. Imperfections can be classified according to their geometry, such as point, line, plane and volume imperfections.
2. Point imperfections include vacancy, Schottky and Frenkel imperfections, line imperfections include edge and screw dislocations, plane imperfections include grain, tilt and twist boundaries and stacking faults, and volume imperfections include foreign particle inclusions, voids or pores.
3. Although the energy of formation of Schottky and Frenkel imperfections are different, yet both point imperfections are found to be present in all solids, their proportions may be different.
4. In extreme cases, the Burgers vector b is $\perp$ to dislocation line in edge dislocation and $\parallel$ in screw dislocation.
5. Dislocations are thermodynamically unstable and their presence always increase the free energy of the crystal.
6. Since the energy of a dislocation is proportional to square of the Burgers vector, the dislocation with the shortest Burgers vector will require minimum energy for displacement.
7. The slip plane within which slip occurs are usually close packed planes.
8. The combination of a slip direction and the slip plane containing it is defined as a slip system.
9. Two dislocations of Burgers vector b_1 and b_2 can react to form a third dislocation of Burgers vector b_3 only if the condition $b_1^2 + b_2^2 > b_3^2$ is satisfied. Similarly, b_1 may dissociate into b_2 and b_3 only if $b_1^2 > b_2^2 + b_3^2$.
10. Dislocations and other crystal defects can be made to observe by a number of different techniques and their densities can be estimated.

9.12 DEFINITIONS

Burger circuit: A sequence of connected unit lattice translation vectors forming a circuit which would close in perfect lattice but fails to close when taken around a dislocation. The vector neccessary to complete the circuit around a dislocation is the Burgers vector of that dislocation.

Burgers vector: The vector by which the lattice on one side of an internal surface is displaced relative to the lattice on the other side as a dislocation moves along the surface; it is a property of the dislocation.

Climb: The movement of a dislocation along any internal surface other than one of its slip planes.

Critical resolved shear stress: The resolved stress on an active slip system, at which the slip is initiated.

Decoration: Decoration is a process of inducing precipitation along the line of the dislocation which are normally not visible. The resulting effect looks like a row of beads in a fine thread.

Dislocation: A line imperfection which can be visualized as the boundary between a region of an internal surface over which slip has occurred and another region over which no slip has occurred. The dislocation is called edge when the Burgers vector is perpendicular to the dislocation line. It is called screw when the Burgers vector is parallel to the dislocation line.

Etch pits: The emergence of prism or pyramid like shapes at dislocation sites on the surface of crystals when they are subjected to etching.

Frenkel imperfection: A point imperfection in which a cation vacancy is associated with an interstitial cation in an ionic crystal.

Grain boundary: A surface imperfection which separates crystal blocks of the crystal structure but different orientations in a polycrystalline aggregate.

Interstitial impurity atom: A point imperfection in which a foreign atom fits in an interstitial position in between the host atoms.

Low angle boundary: A surface imperfection separating two misoriented regions of a crystal; the angle of misorientation is small (a few degree or less).

Partial dislocation: A dislocation having a Burgers vector not equal to the lattice translation.

Schottkey imperfection: A point imperfection in which a cation vacancy is associated with an anion vacancy in an ionic crystal.

Self interstitial: A point imperfection in which an atom of the same species as those of the host material is squeezed into an interstitial position between the host atoms.

Slip: The parallel movement of two adjacent crystal planes relative to one another.

Slip plane: Any crystallographic plane containing both the Burgers vector and the dislocation line.

Slip system: The combination of a plane and a direction lying in the same plane along which slip occurs.

Stacking fault: A surface imperfection which results from the stacking of a one atomic plane on another out of sequence, so the lattices on both sides of the fault have the same orientation but are translated by less than a lattice translation with respect to one another.

Substitutional impurity atom: A point imperfection in which a foreign atom occupies a site which would be occupied by a host atom if the crystal were perfect.

Tilt boundary: A low angle boundary in which the misorientation is a rotation about an axis lying in the boundary.

Topograph: The process of mapping a section/surface of a crystal containing relatively low imperfections by moving the crystal and the film/plate together to record a point to point variation of diffracted intensity.

Twin boundary: A surface imperfection separating two regions of a crystal which are mirror images of each other with respect to the plane of the boundary.

Twist boundary: A low angled boundary in which the misorientation is a rotation about an axis normal to the boundary.

REVIEW QUESTIONS AND PROBLEMS

1. State and explain various imperfections that are possible in crystalline solid.
2. Classify the imperfection according to their dimensions. Give a brief description of each of them.
3. Obtain an expression for the concentration of vacancy in a monoatomic solid taking into account only configurational entropy.
4. Obtain an expression for the concentration of Frenkel imperfection in an ionic solid taking into account only configurational entropy.
5. Through neat diagrams explain the formation of edge and screw dislocations, calculate their energies.
6. Illustrate diagrammatically the Burgers circuit containing edge and screw dislocations.
7. What do you understand by slip plane, slip direction and slip system? Find all the members of slip system in fcc metal.
8. What are various surface imperfections? Discuss the case of low angle grain boundary and the formation of staking faults in *hcp* and *fcc* crystals.
9. The average energy required to create a vacancy in some monoatomic solid is 1.1 eV. calculate the ratio of the vacancies at 1200 K and 600K. **Ans.** 4.12×10^4
10. In a certain material, the ratio of the number of vacancies to the number of atoms is 2.58×10^{-13} at 600K. Calculate the average energy required to create a vacancy in the material.

    Ans. 1.5 eV
11. What do you understand by dislocation reaction? State Frank's stability criterion. Show both graphically and analytically that the first two dislocations add to give a third dislocation in the reaction

$$\frac{1}{6}\,[21\bar{1}] + \frac{1}{6}\,[121] \to \frac{1}{2}\,[110]$$

12. The energy required to remove a pair of ions (Na^+ and Cl^-) from NaCl is ~2 eV. Calculate the approximate number of Schottky imperfections in the NaCl crystal at temperature.

    Ans. $3.5 \times 10^5 cm^{-3}$
13. The average energy required to create a Frenkel imperfection in an ionic crystal $A^{2+}B^{2-}$ is 1.4 eV. Calculate the ratio of Frenkel imperfections at 300K and 600K in 1 gram of crystal.

    Ans. 7.5×10^5

14. The Burgers vector of a mixed dislocation is $\frac{1}{2}$ [110]. Find dislocation line lies along the [112] direction. Find the slip plane on which this dislocation lies. Also find the screw and edge components of the Burgers vector.

$$\textbf{Ans.} \ (110), \ \frac{1}{6} \ [112] \ \text{and} \ \frac{1}{3} \ [11\bar{1}]$$

15. A single crystal of copper contains a low angle tilt boundary on (010) plane and the tilt axis parallel to the [001] direction. Calculate the tilt angle, if the average spacing of the dislocations in the boundary is 1.5×10^{-6} m. **Ans.** 1.7×10^{-4} rad.

CHAPTER 10

DIFFRACTION METHODS

10.1 INTRODUCTION

We know that scattering is a fundamental property of the atoms and is treated as a microscopic phenomenon. On the other hand, the diffraction (by crystals) is a result of many microscopic scattering events and is treated as a macroscopic phenomenon. We can use X-rays (waves) or a beam of electrons/neutrons (particles) as probes for the structural study of crystalline materials because their wavelengths are of the order of interatomic spacing. Available literature suggests that most of the diffraction work has been done by using X-rays only. However, in the recent past, the use of synchrotron radiation is found to be on the rise.

Neutrons scatter appreciably only from the atomic nuclei. On the other hand, electrons and X-rays scatter from the electron cloud of the atoms. They have their own advantages and limitations but they are often complementary and supplementary to each other.

In this chapter, we intend to describe various methods based on X-ray diffraction in somewhat detail, while the description of electron/neutron diffraction will be brief.

10.2 PRODUCTION OF *X*-RAYS

X-rays are produced when a beam of highly accelerated charged particles such as electrons is allowed to strike a metal target. In the process, electrons suffer rapid deceleration and the energy lost by them is emitted in the form of electromagnetic radiations, i.e. $\Delta E = h\nu$. Such processes give "white radiations" or continuous X-rays. This phenomenon is similar to the generation of heat when a sudden break is applied on a moving wheal. This is the reason why the X-ray in German is called "Bremsstrahlung". Fig. 10.1 shows continuous X-ray spectra corresponding to a tungsten target at several different accelerating potentials, which tails off gradually at higher wavelength side, but terminate abruptly at shorter wavelength limit λ_{swl} (or λ_{min}). As the accelerating potential increases, both the maxima in the intensity distribution and the λ_{min} shift towards the shorter wavelength side. The position of λ_{min} is obtained when the entire kinetic energy of the electrons for the given electrode potential is radiated in the form of a single photon. If m is the mass of the electron, V is the accelerating potential and e is the charge of the electron, then

$$\Delta E_{max} = eV_{max} = \frac{1}{2}mv_{max}^2 = h\nu_{max} = \frac{hc}{\lambda_{min}}$$

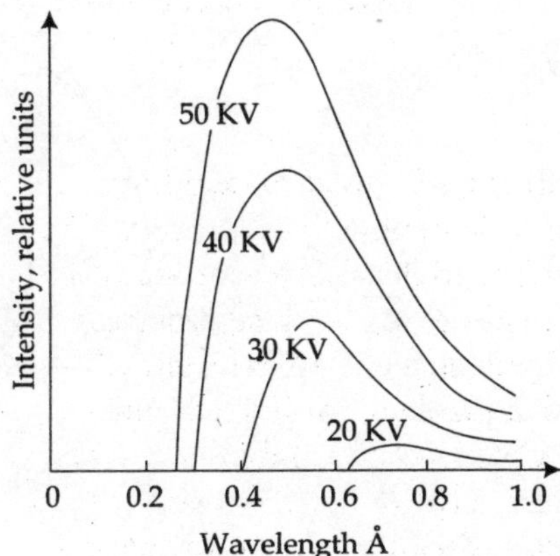

Fig. 10.1   X-ray spectra of tungsten at various accelerating potentials

or
$$\lambda_{min} = \frac{hc}{eV_{max}} \tag{1}$$

Substituting the values of constants, we obtain

$$\lambda_{min} = \frac{6.63 \times 10^{-34} \, Js \times 3 \times 10^{8} \, ms^{-1}}{1.6 \times 10^{-19} \times C \times V} = \frac{1.24 \times 10^{-6}}{V} \; \text{V–m} \tag{2}$$

The nature of the X-ray spectra for commonly used metal targets is slightly different from that is shown as Fig. 10.1. These spectra contain sharp peaks at definite wavelengths which appear to be superimposed on a continuous spectra otherwise. For example, Fig. 10.2, shows the complete X-ray spectrum for a copper target. The peaks in the spectrum correspond to the characteristic radiations and their wavelengths are characteristic of the metal targets. They are produced when the bombarding electrons eject some of the orbital electrons of the target atom and the corresponding vacancies are filled by the less energetic outer shell electrons (Fig. 10.3). For example, if the vacancies in the K-shell are replaced by the electrons from the L – shell, the resulting X-rays are called K_α radiations. On the other hand, if the vacancies in the K-shell are replaced by the electrons from the M-shell, the resulting X-rays are called K_β radiations, however this is a less probable transition. Fig. 10.4 shows the energy level diagram and possible transitions that can take place. They are:

$$E_{K\alpha_1} = W_K - W_{L\text{III}} \text{ due to } K \text{ to } L_{\text{III}} \text{ transition}$$

and
$$E_{K\alpha_2} = W_K - W_{L\text{II}} \text{ due to } K \text{ to } L_{\text{II}} \text{ transition}$$

where $\lambda_{K\alpha_1} = 1.54051$ Å and $\lambda_{K\alpha_2} = 1.54433$ Å, for copper target.

First X-ray tube was fabricated by W.C. Roentgen in 1895 when he discovered X-rays and later by W.D. Collidge in 1913. However, today these tubes have only historical importance. The modern tubes are made from tough glass with beryllium windows and specpure metal target and sealed under high vacuum. The essential parts of a modern sealed X-ray tube is shown in Fig. 10.5. A heated tungsten filament (acting as cathode) provides the electron beam, which under a large

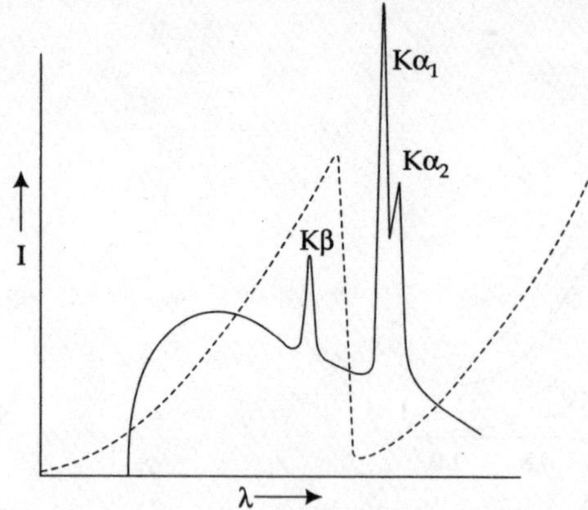

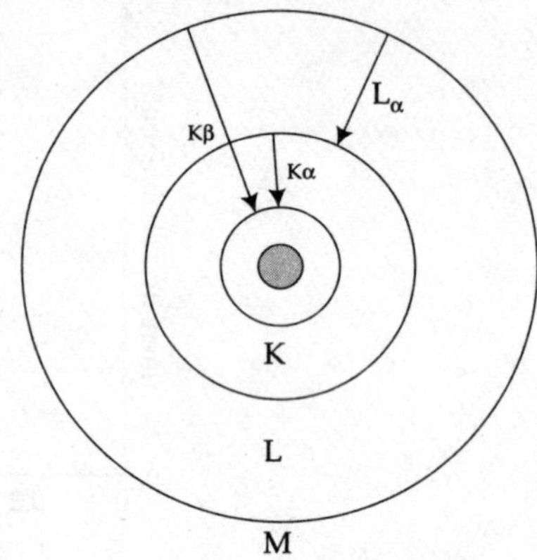

Fig. 10.2 Complete X-ray spectrum, I versus λ curve: solid line shows the characteristic lines superimposed on a background of white radiation. Dotted line shows *K* absorption edge suitable for use as a β-filter

Fig. 10.3 Electron orbits in an atom, showing transitions associated with the production of characteristic *X*-rays.

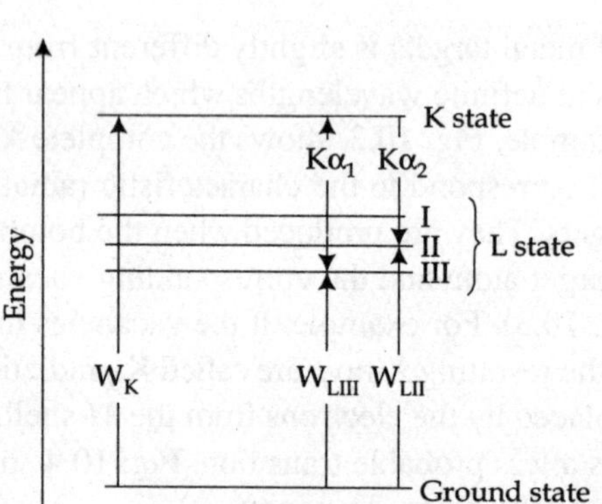

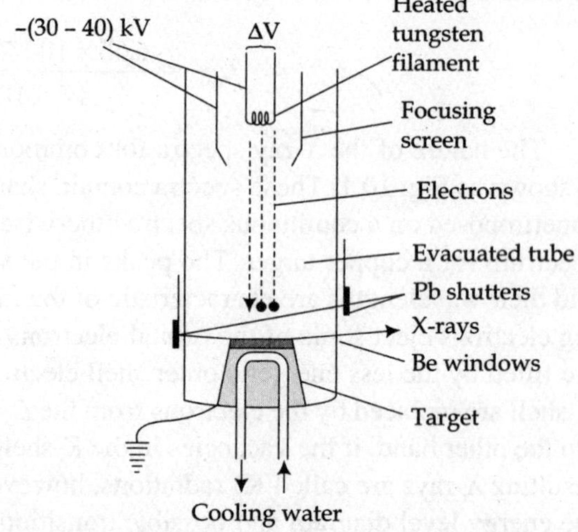

Fig. 10.4 Energy level diagram showing the allowed transitions

Fig. 10.5 Schematic diagram of essential parts of an *X*-ray tube

negative potential (~50 KV) is accelerated towards the target (acting as anode) to give reasonable amount of *K*-radiations. The tube is evacuated to prevent the electrons from losing their energy by colliding with the gas molecules before they reach the target. The negatively charged screen surrounding the filament serves to focus the electron beam into the target. The *X*-rays produced leave the tube through the beryllium windows, which when not in use are covered by shutters made of lead. Beryllium and lead are chosen because they respectively have low and high absorption coefficients for *X*-rays.

The choice of target material depends on several factors. From the engineering point of view, it should be a good conductor of heat so that it can be effectively cooled (a huge amount of heat is generated on the target for being at a large electrode potential) and of electricity (because it has to function as an anode). These requirements suggest the target material to be a metal with reasonably high melting point. Further, from crystallographer's point of view, it should emit X-rays of wavelengths of the order of 1Å. The above mentioned requirements limit the choice to the first and second – row transition metals. The most widely used target material is copper which fulfils above conditions well. Besides, there are some other metals which are used as target elements for the generation of X-rays of desired specifications. A list of commonly used target elements along with other details are provided in Table 10.1.

Table 10.1 Target elements, corresponding X-ray wavelengths and filter elements

Elements	*with at. no.*	*X-ray*	*wavelengths*	*Density*	*Optimum thickness*
Target	*Filter*	$K\alpha_1$	$K\alpha_2$		
Cr (24)	V (23)	2.2896	2.2935	0.009	0.016
Fe (26)	Mn (25)	1.9360	1.9399	0.012	0.016
Cu (29)	Ni (28)	1.5405	1.5443	0.019	0.021
Mo (42)	Nb (41)	0.7093	0.7135	0.069	0.108
Au (47)	Pd (46)	0.5594	0.5638	0.030	0.030

X-ray tubes may be permanently sealed after evacuation or may be demountable in which vacuum is maintained by continuous pumping. A sealed tube has neither rotating anode facility nor it is possible to replace any of its component if goes wrong, the whole tube has to be replaced. Also, a different tube is needed for each type of radiation. On the other hand, despite having certain advantages, demountable tubes require an extra maintenance facility and are not readily available for use. This may be one of the reasons that most crystallographers use permanently sealed tubes for their X-ray diffraction work.

10.3 X-RAY DIFFRACTION

X-rays, like other electromagnetic waves, interact with the electron cloud of the atoms. When X-rays enter into a crystal, each atom acts as a diffraction centre and the crystal as a whole acts like a three dimensional grating. Because of their shorter wavelengths, X-rays are scattered by adjacent atoms in the crystal, which can interfere constructively and give rise to diffraction effects.

However, this is possible only if the path difference between the rays diffracted by two adjacent atoms is an integral multiple of the wavelength. The diffraction pattern so obtained can tell us much about the internal arrangement of atoms in crystals. There are two different ways to understand the diffraction conditions in crystals.

Bragg's Law

In 1912, W.H.Bragg and W.L.Bragg pointed that a consequence of three dimensional periodicity of a crystal is that it can be divided into various sets of equidistant parallel planes containing identical atomic arrangements. They also discovered that the geometry of the X-ray diffraction is analogous to the reflection of light by a plane mirror. It, of course, follows that the angle between the incident and reflected rays is 2θ. Thus the words diffraction and reflection are interchangeable in Bragg's treatment. Based on these considerations Bragg derived a simple mathematical relationship known as Bragg's law.

In order to derive the Bragg's law, let us consider a set of parallel atomic planes with interplanar spacing as 'd'. A monochromatic X-ray beam of wavelength λ having common wavefront is allowed to incident on these planes at a glancing angle θ such that the rays lie in the plane of the paper. Consider two of the incoming X-rays 1 and 2 which strike the first two planes and get partially reflected at the same angle θ with the same planes as shown in Fig. 10.6. These reflected rays interfere constructively and give rise to the diffraction effect. Let AB and AD be the perpendiculars drawn from the point A on the incident and reflected ray number 2, respectively. The path difference between the two rays is

$$\Delta = (BC + CD) = 2d \sin \theta$$

For a constructive interference between the rays 1 and 2, the path difference must be an integral multiple of the wavelength, i.e.,

$$2d \sin \theta = n\lambda \tag{3}$$

where $n = 0, 1, 2, \ldots$ represents the order of reflection. This equation is known as Bragg's law. The diffraction takes place for those values of d, θ, λ and n which satisfy Bragg's law. For n = 0, $\theta = 0$ means no reflection can be observed experimentally corresponding to this. For the given values of d and λ, higher order reflections ($n = 1, 2, \ldots$) appear at higher values of θ. The highest possible order can be determined by the condition

$$(\sin \theta)_{max} = 1$$

Using this in eq. 3, we obtain another condition

$$\frac{n\lambda}{2d} \leq 1$$

This indicates that λ must not be greater than twice the interplanar spacing, otherwise no diffraction will occur.

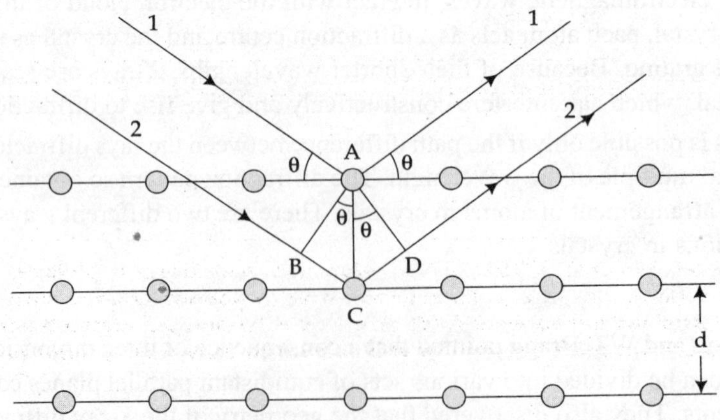

Fig. 10.6   Bragg's reflection of X-rays from the atomic planes

The Laue Equations

Although the Bragg's equation is a result of the fundamental concept of periodic arrangement of atoms in crystal planes but in true sense it does not refer to the actual arrangement of the atoms associated with them. On the other hand, Von Laue treated the phenomenon of diffraction in a more general way by considering the scattering of X-rays from individual atoms in the crystal followed by their recombinations to obtain the directions of diffraction maxima. Since crystals are periodic in all three dimensions, three equations are obtained. They are known as Laue equations. One can verify that the Bragg's equation is not only equivalent but also a consequence of the Laue equations.

To derive Laue equations, let us first consider a one dimensional row of atoms with interatomic distance 'a' (Fig. 10.7). Suppose AB is the incident plane wave front making and angle α_0 with a-row of atoms, and CD is the diffracted plane wave front leaving at an angle α with the same row of atoms. Then the path difference between the two consecutive rays is

$$\Delta = (AC - BD) = a(\cos \alpha - \cos \alpha_0)$$

and a diffracted beam is observed only if

$$a(\cos \alpha - \cos \alpha_0) = e\lambda \tag{4}$$

where $e = 0, 1, 2, ...,$ is any integer giving the order of diffraction.

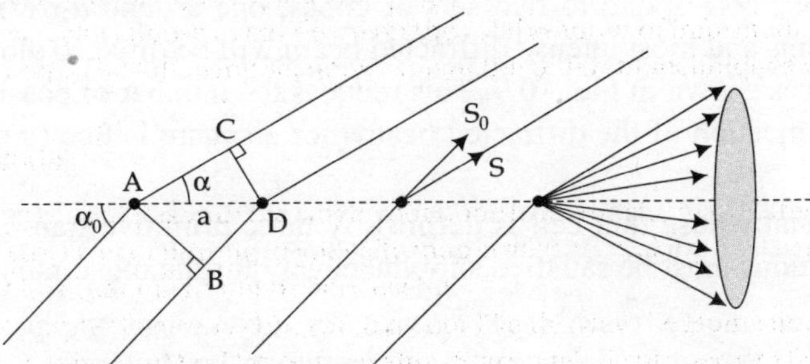

Fig. 10.7 Diffraction of X-rays by one row of atoms

This equation will be satisfied by all the diffracted beams lying on the concentric cone with respect to the a-row of atoms and has the semiapex angle α(Fig. 10.7). Thus for any given angle of incidence there will be a series of concentric cones surrounding the a-row of atoms, where each cone represents various orders of diffraction. The direction of the primary beam is indicated by the downward arrow, while the higher orders of diffraction are indicated by the numbers (Fig. 10.8).

If S_0 and S be the unit vectors in the directions of incident and diffracted beams and 'a' is the translation vector along x-axis, then equation 4 can be written in the vector notation as

$$a \cdot (S - S_0) = e\lambda \tag{4'}$$

In a plane lattice with an interatomic spacing 'a' in one direction and 'b' in another, two quations of the type 4 or 4' must be satisfied simultaneously for the intense diffracted beams to occur.

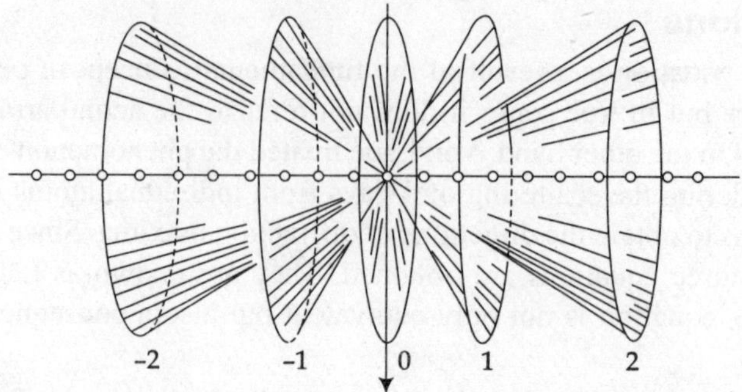

Fig. 10.8 Diffraction cones around a row of atoms. Direction of primary beam is indicated by the arrow. Numbers indicate the order of diffraction

They are:

$$a(\cos \alpha - \cos \alpha_0) = a \cdot (S - S_0) = e\lambda$$

$$b(\cos \beta - \cos \beta_0) = b \cdot (S - S_0) = f\lambda \tag{5}$$

where β_0 and β and f, respectively have the same meaning for the b-rows of atoms as α_0, α and e for a-row of atoms.

The two equations correspond to two sets of cones, one around a-rows of atoms and other around b-rows of atoms and most intense diffracted beam will be directed along the intersection of the two sets of cones as shown in Fig. 10.9. This reduces the number of possible diffracted beams. The stereographic projection of the diffracted beams for a square lattice ($a = b$, $\phi = 90°$) is illustrated in Fig. 10.10.

Finally, in a crystal whose unit cell is defined by three primitive translations a, b, and c the following three equations must be satisfied simultaneously for the most intense diffracted beam to occur. They are:

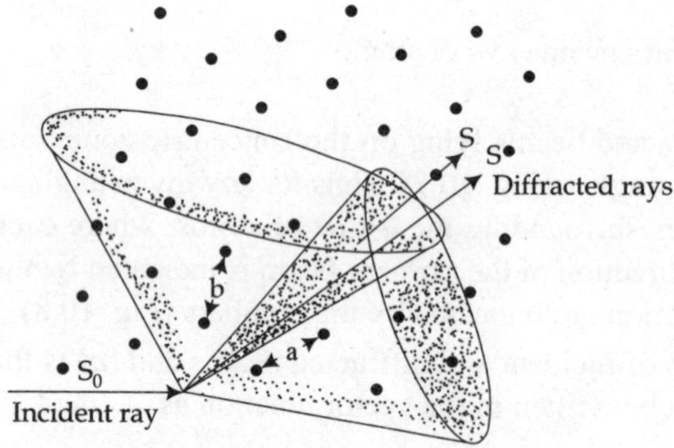

Fig. 10.9 Intersection of two diffraction cones, one each for a- and b-rows of atoms

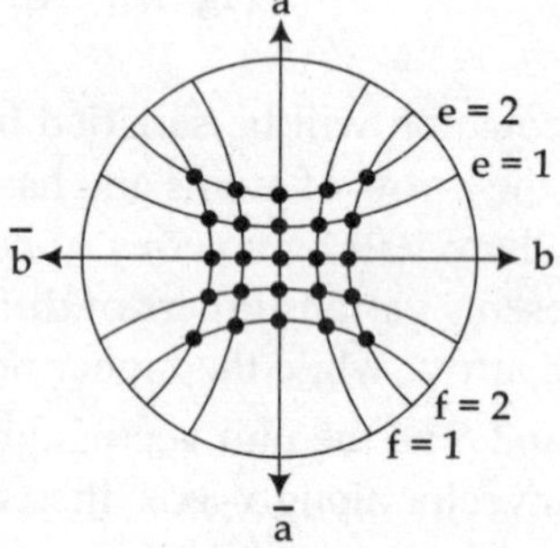

Fig. 10.10 Steriographic projection of diffracted beams from a two dimensional square lattice. Cones are concentric with a and b axes. Intersections give the direction of most intense diffracted beams.

$$a(\cos \alpha - \cos \alpha_0) = a.(S - S_0) = e\lambda$$

$$b(\cos \beta - \cos \beta_0) = b.(S - S_0) = f\lambda \qquad (6)$$

$$c(\cos \gamma - \cos \gamma_0) = c.(S - S_0) = g\lambda$$

where γ_0, γ and g, respectively have the same meaning for the c-rows of atoms as (α_0, α, e) and (β_0, β, f) for $a-$ or $b-$ rows of atoms. These three equations together are called the Laue equations. In this case, there are three sets of cones, one each around $a-$, $b-$, and $c-$ rows of atoms and the most intense diffracted beam will be directed along the intersection of all three sets of cones. This further reduces the number of possible diffracted beams.

It is to be noted that for a monochromatic X-ray beam (fixed λ) and arbitrary direction of incidence (fixed S_0), it is not possible to find a direction S which satisfies the Laue equations simultaneously and hence in general no diffraction should be observed. This actually means that the diffraction will occur only for particular angle of incidence.

In order to prove the equivalence of Bragg's equation with the Laue equations and whether the former is a special case of the later, let us consider the case of a simple cubic lattice ($a = b = c$) and assume the angle between the incident and diffracted beams to be equal to 2θ. Then squaring and adding the three Laue equations, we get

$$a^2[\cos^2 \alpha + \cos^2 \beta + \cos^2 \gamma + \cos^2 \alpha_0 + \cos^2 \beta_0 + \cos^2 \gamma_0 - 2(\cos \alpha \cos \alpha_0$$
$$+ \cos \beta \cos \beta_0 + \cos \gamma \cos \gamma_0)] = \lambda^2[e^2 + f^2 + g^2] \qquad (7)$$

From solid geometry, we know that in an orthogonal coordinate system, α, β, and γ also satisfy the following conditions:

$$\cos^2 \alpha + \cos^2 \beta + \cos^2 \gamma = \cos^2 \alpha_0 + \cos^2 \beta_0 + \cos^2 \gamma_0 = 1$$

and
$$\cos \alpha \cdot \cos \alpha_0 + \cos \beta \cdot \cos \beta_0 + \cos \gamma \cdot \cos \gamma_0 = \cos 2\theta$$

Making use of these results, eq. 7 reduces to

$$2(1 - \cos 2\theta) = \lambda^2(e^2 + f^2 + g^2)$$

or
$$\sin^2\theta = \frac{\lambda^2}{4a^2}(e^2 + f^2 + g^2) \qquad (8)$$

Similarly, squaring the Bragg's equation, we have

$$\sin^2\theta = \frac{n^2\lambda^2}{4a^2}(h^2 + k^2 + l^2) \qquad (9)$$

where $d = \dfrac{a}{\left(h^2 + k^2 + l^2\right)^{1/2}}$ for a cubic system (sec. 2.10).

Now, comparing eqs. 8 and 9, we find that

$$e = nh, f = nk \text{ and } g = nl$$

This indicates that $e\,f\,g$ defined in the Laue's treatment actually represents the nth order diffraction from a set of (hkl) planes in the Bragg's treatment.

10.4   DIFFRACTION CONDITION AND BRAGG'S LAW

We know that the Bragg's treatment of X-ray diffraction is based on simple considerations and can be conviniently applied to high symmetry lattices such as cubic, tetragonal and orthorhombic. However, its application to low symmetry lattices such as monoclinic and triclinic is fairly involved. It also fails to provide information about the intensity of scattering from a special distribution of electrons within the unit cells.

In the following, we shall have a relook at the diffraction condition in terms of reciprocal lattice concept discussed in section 7.6. Accordingly, let us represent the incident and reflected waves by wave vectors k and k' instead of simply considering the X-rays of wavelength λ interacting with the atoms in a set of planes. Fig. 10.11, illustrates the Bragg's reflection in terms of wave vectors. The vector triangle provides us

$$\Delta k = k' - k \tag{10}$$

where the change Δk in k is only in the direction perpendicular to the (hkl) planes. Also for elastic scattering

$$|k| = |k'| = \frac{2\pi}{\lambda} \tag{11}$$

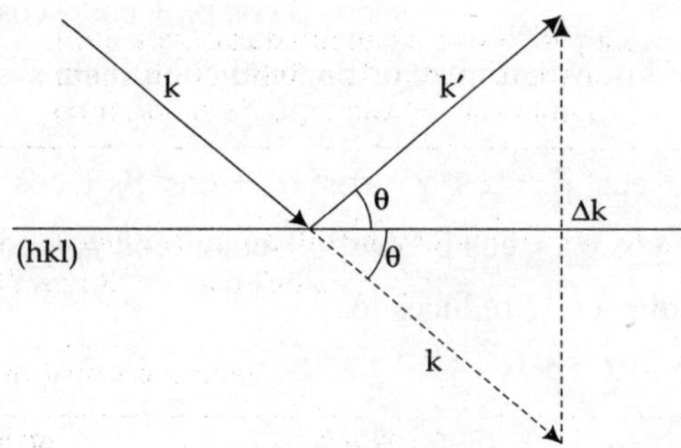

Fig. 10.11   Illustration of Bragg's reflection

The change Δk, according to the reciprocal lattice concept is just associated with G_{hkl} or with the unit vector $\hat{n}$. Therefore, from Fig. 10.11, we can write

$$\Delta k = (k' - k) \equiv 2 \sin \theta |k| \hat{n}$$

$$= \left[\frac{4\pi \sin \theta}{\lambda} \right] \hat{n}$$

$$= \left[\frac{4\pi \sin \theta}{\lambda |G_{hkl}|} \right] G_{hkl}$$

$$= \left[\frac{2d_{hkl} \sin \theta}{\lambda} \right] G_{hkl} \qquad (12)$$

where $d_{hkl} = \dfrac{2\pi}{|G_{hkl}|}$.

When the combination of λ, θ and d_{hkl} is appropriate to satisfy the Bragg condition, then

$$\Delta k = G_{hkl} \qquad (13)$$

Thus the set of points generated by the reciprocal lattice vector G_{hkl} is the array of reciprocal lattice points produced by diffraction of a crystal. The spacing between them are inversely proportional to the interplanar spacing in a real lattice. Now, construct a sphere of radius $|k'| = |k| = \dfrac{2\pi}{\lambda}$ with center at O (Fig. 10.12). If this sphere passes through any point of the reciprocal lattice, then the diffraction is possible corresponding to this wavelength or else otherwise. This sphere of diffraction is called Ewald sphere or sphere of reflection. Now, substituting Δk from eq. 13 into eq. 10, we have,

$$G = k' - k$$

or
$$k' = k + G \qquad (14)$$

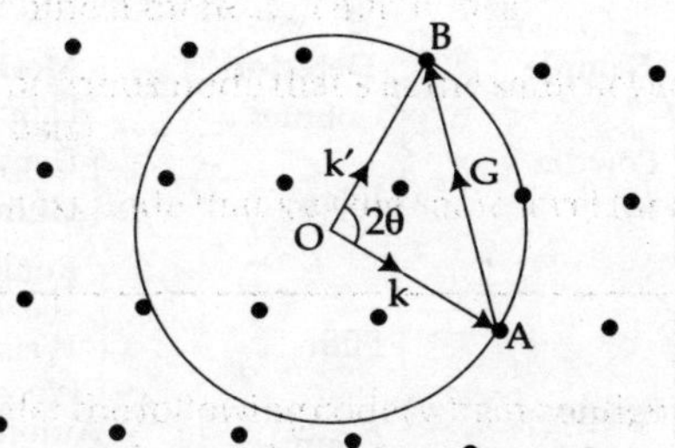

Fig. 10.12   Ewald construction

Eq. 14 Suggests that: (i) the scattering changes only the direction of k, and (ii) the scattered wave differ from the incident wave by a reciprocal lattice vector G.

Squaring eq. 14, we obtain

$$k'^2 = (k + G)^2 = k^2 + G^2 + 2k.G$$

Making use of eq. 11, we are left with

$$G^2 + 2k.G = 0$$

This is the Bragg's law in vector form and is quite useful in the study of band theory.

10.5   THE *X*-RAY DIFFRACTION EXPERIMENTS

When reduced to basic essentials, *X*-ray diffraction experiment requires an *X*-ray source, the sample under investigation and a photographic film or a detector to record the diffraction beam as shown in Fig. 10.13. Within this broad framework, there are three variables which govern the different *X*-ray diffraction techniques. They are:

- (i) radiation—monochromatic or of variable λ.
- (ii) sample—single crystal, polycrystal (powder) or a solid peice (amorphous).
- (iii) detector—photographic film or radiation counter.

These are summarized for the most important experimental techniques in Fig. 10.14. With the exception of the Laue method, monochromatic radiation is almost always used.

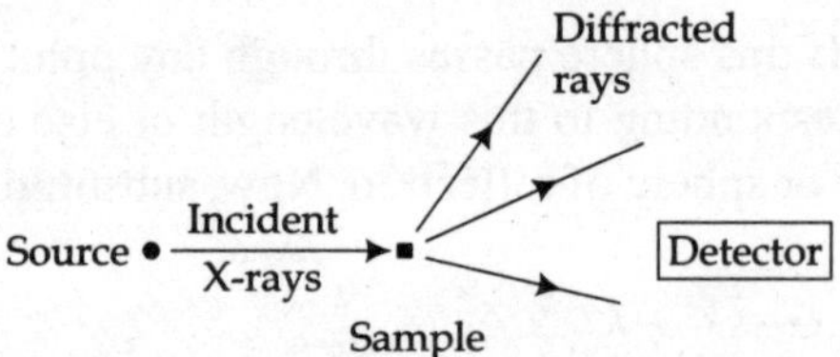

Fig. 10.13   The X-ray diffraction experiment

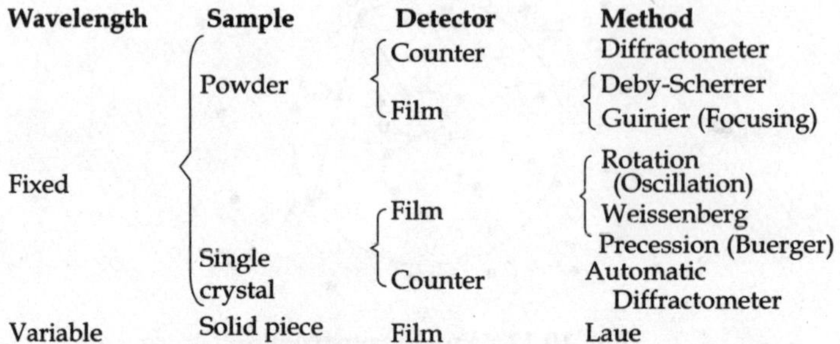

Fig. 10.14   The different X-ray diffraction techniques

10.6   THE POWDER METHOD

The powder method is the most convenient and quickest methods for obtaining diffraction data and is readily applicable to all crystalline materials. Since the diffraction data are unique for a substance, this method can be used for identification of materials (just as finger prints of a person is unique and is used to identify a human being).

A powder camera (also known as Debye-Scherrer camera) is cylindrical in shape (Fig. 10.15). It consists of a specimen holder and its adjusting system at the center. Further, it has two holes on opposite sides of the circular rim, one for collimator and the other for direct beam catcher containing a fluorescent screen. Powdered specimen prepared in the form of a thin wire with the help of

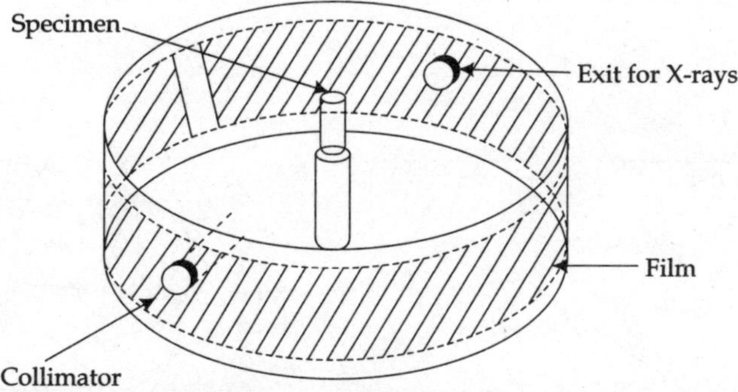

Fig. 10.15   Schematic representation of the powder method

some adhesive material or filled in a thin walled capillary tube is placed on the sample holder. The specimen is set correctly on its axis with the help of plunger so that it is free from any eccentric motion during its rotation. Then the camera is taken to the dark room where a thin strip of *X*-ray film is punched by a punch machine to make two holes (to fit the collimator and the direct beam catcher) and wrapped in a black paper (to protect it from stray light and unwanted *X*-rays) is loaded in the cylindrical cassette in such a way that the two ends of the film meet midway between the entry and exit points of the *X*-ray beam. This particular film placement is known as Straumanis and Ievins setting and has two special features: (i) it allows to record a complete diffraction pattern ranging from $\theta = 0$ to $\theta = 90°$ on the same film, and (ii) the pair of arcs situated around $\theta = 0$ and $\theta = 90°$ enable us to locate the low and high angle positions on the film accurately.

A monochromatic *X*-ray beam is allowed to fall on the specimen containing a large number of tiny crystallites arranged randomly in all possible orientations. Therefore, for each set of planes, there exist some crystallites which are oriented in a manner to satisfy the Bragg condition and give rise to diffraction effect. The locus of the diffracted beam will lie on a cone with half appex angle 2θ (Fig. 10.16). In a similar way, other sets of lattice planes give their diffraction cones on the *X*-ray film. These cones are recorded in the form of pair of arcs on the *X*-ray film. When the Bragg angle is 45°, the corresponding cone opens out into a circle whose intersection with the Ewald sphere gives rise to a straight line and is recorded at the mid point of the film situated between the incident and exist points. For $\theta > 45°$ back reflections are obtained as shown in Fig. 10.16c. After completion of exposure, the film is processed (developed and fixed) in the dark room and dried for indexing. A typical schematic powder pattern is shown as Fig. 10.17.

Indexing of Powder Lines

Before we start any measurement on the film, it is necessary to ensure $\theta = 0$ and $\theta = 90°$ positions, i.e. to distinguish the low and high angle diffraction lines. This can be done by simply observing the following two characteristic features on the film: (i) The background intensity arising mainly due to the scattering of radiation from the air molecules present in the camera is found to be maximum near $\theta = 0$, i.e. around the exit hole. (ii) Splitting of the diffraction lines as a result of resolution of CuKα radiation into its components Kα_1 ($\lambda = 1.5405$ Å) and Kα_2 ($\lambda = 1.5443$ Å) at higher angle

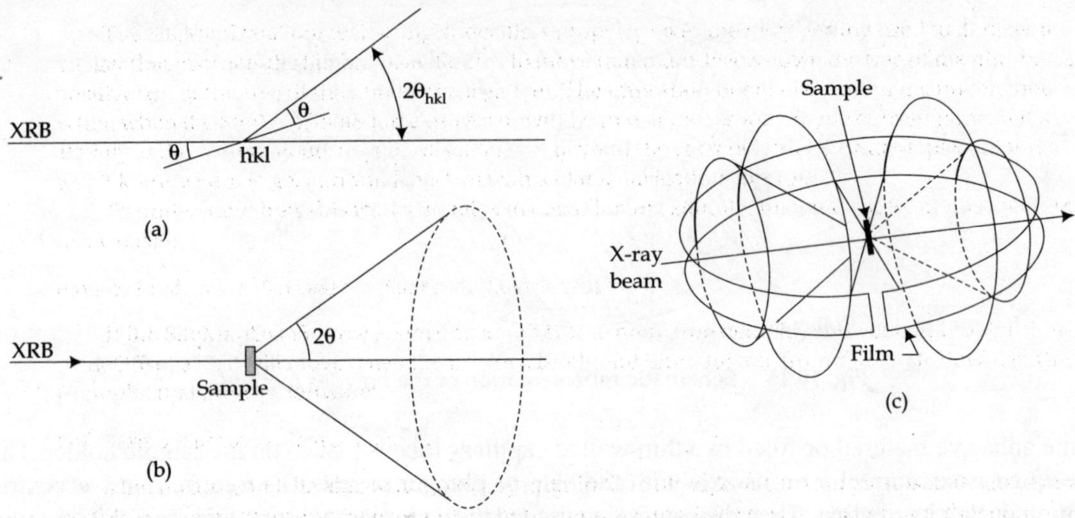

Fig. 10.16 Formation of powder pattern

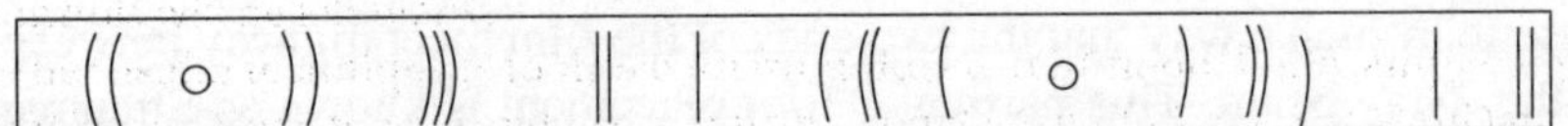

Fig. 10.17 Schematic powder pattern

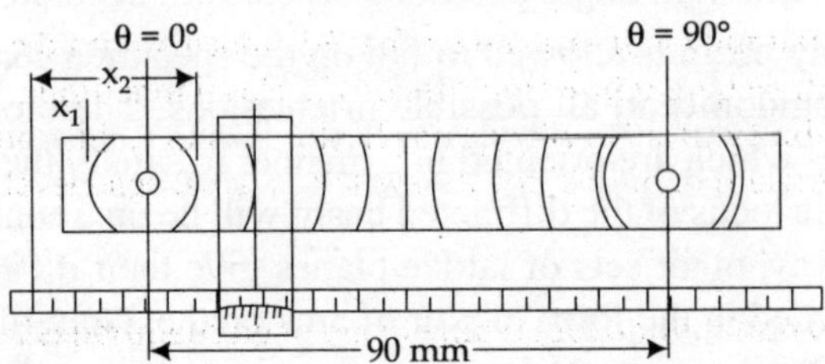

Fig. 10.18 Measurement of linear distances of arc on a schematic powder pattern

side. For the usual camera diameter (57.3 mm), these doublets appear to be a pair of closely spaced lines merging with each other to form a thick line.

After ensuring the low and high angle positions, the film is placed over the film reader in such a way that it is truly parallel to the scale of the measuring device. Cross hair of the telescope is adjusted vertically at the center of each arc to record the linear distance with respect to any given reference to determine the arc length (Fig. 10.18). Put all the readings in a tabulated data sheet as shown in Fig. 10.19. Knowing the linear distances between the pair of arcs, various diffraction angle θ's can be calculated for a known camera radius R. They are related to each other (Fig. 10.20) according to

Line No.	x_2	x_1	Check $x_2 + x_1$	Arc length $S = x_2 - x_1$	Correction $\frac{P}{100} \times S$	Corrected arc length S'	$\theta = \frac{S'}{4}$	$\sin \theta$	$d = \frac{\lambda}{2\sin\theta}$	$Q = \frac{1}{d^2}$	hkl

Fig. 10.19   Data sheet in tabulated form

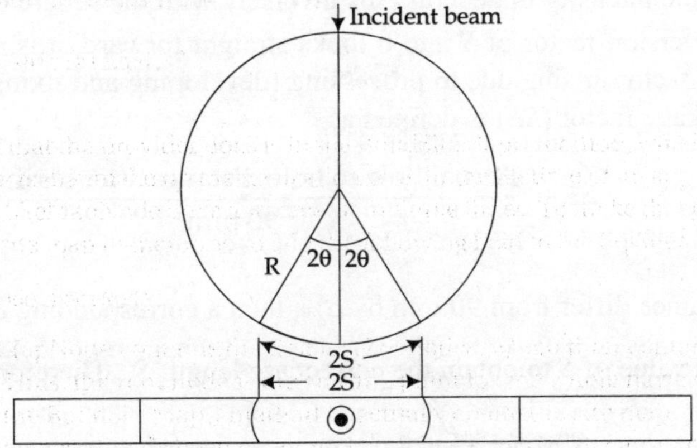

Fig. 10.20   Relation between the Bragg angle and the linear distance between pair of arcs

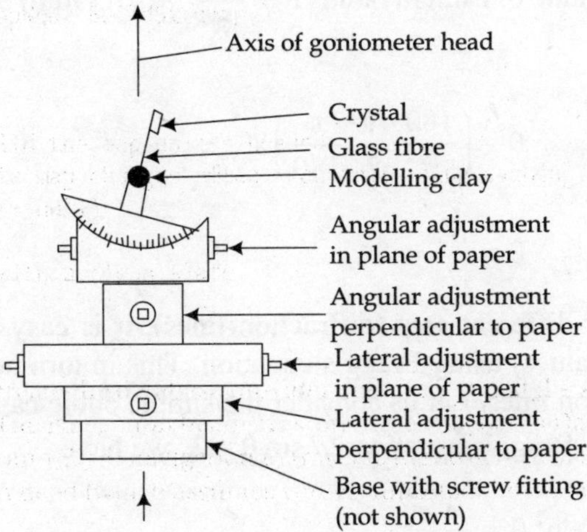

Fig. 10.21   Basic parts of gonimeter head. A key is provided to fit the square pins for moving the arcs

$$4\theta = \frac{2S}{R} \text{ or } \theta = \frac{S}{2R} \text{ radians}$$

or
$$\theta = \left(\frac{180}{\pi} \cdot \frac{1}{2R} \right) S \text{ degrees} \qquad (15)$$

It is interesting to note that $180/\pi = 57.3$. Thus for a camera diameter $2R = 57.3$ mm, $\theta = S$, so that $1°$ in θ corresponds to 1 mm in S. In view of this simple conversion factor, camera diameter is usually taken as 57.3 mm or a multiple of it. For better resolution and greater accuracy, it is sometimes desirable to use a camera diameter of $2 \times 57.3 = 114.6$ mm. However, in this case exposure time is more because the intensity in general falls inversely with the square of the distance.

Although the conversion factor of S into θ looks straight forward, it is necessary to take into account the shrinkage factor arising due to processing (developing and fixing) of the film for correct results. The shrinkage factor (SF) is defined as

$$SF = \frac{\text{Actual hole to hole distance after shrinkage}}{90 \text{ mm}} \qquad (16)$$

Let the actual distance differ from 90 mm by p%, then a corresponding correction of $\dfrac{p}{100} \times S$ must be added to each value of S to obtain the correct arc length S'. Therefore,

$$S' = S + \frac{p}{100} \times S \qquad (17)$$

Now substituting the value of camera radius $R = \dfrac{180}{\pi}$ $(= 57.3 \text{ mm})$ and the corrected arc length S' in eq. 15, we obtain

$$\theta = \left(\frac{180}{\pi} \cdot \frac{1}{2} \cdot \frac{\pi}{180} \right) S'$$

or
$$\theta = \frac{S'}{2} \qquad (18)$$

After knowing θ's for different arcs (diffraction lines), it is easy to determine the different interplanar spacings (d – values) using Bragg's equation. This in turn helps in determining the hkl values for various diffraction lines. Let us consider the simple cubic case for the purpose. Squaring and arranging the Bragg's equation $2d \sin \theta = \lambda$, we have

$$\sin^2 \theta = \frac{\lambda^2}{4d^2}$$

where $d = \dfrac{a}{\left(h^2 + k^2 + l^2 \right)^{1/2}}$ for a simple cubic crystal. Therefore,

$$\sin^2 \theta = \frac{\lambda^2}{4a^2} \, (h^2 + k^2 + l^2) = \frac{\lambda^2}{4a^2} \, N \tag{19}$$

where $N = (h^2 + k^2 + l^2)$ is an integer.

Now, for a monochromatic X-ray beam, λ is fixed and for a given sample, the lattice parameter 'a' is fixed. Thus for a fixed λ and crystal system, $\dfrac{\lambda^2}{4a^2}$ is a constant quantity. This implies that $\sin^2 \theta$ (or θ) in eq. 19 is proportional to N, a number. Some of the reflections in all cubic crystals (in general valid for other crystal system also) are found to be absent in a systematic manner.

These systematic absences give rise extinction rules, following which all possible *hkl* values for different cubic systems can be obtained. They are summarized in Table 10.2. The ratios of $(h^2 + k^2 + l^2)$ values for allowed reflections of all cubic crystals obtained by using the extinction rules are given as,

SC   :   1:2:3:4:5:6:8:9:10

BCC :   1:2:3:4:5:6:7:8:9

FCC :   3:4:8:11:12:16:19:20

DC   :   3:8:11:16:19

A comparison of these ratios with the observed ratios of $\sin^2 \theta$ values is made to identify the cubic crystal structures.

10.7 THE LAUE METHOD

The Laue method is one of the most convenient and quickest methods of determining the orientation and symmetry of a crystal or of individual grains in an aggregate and hence is widely used to orient crystals in various solid state experiments. However, since white radiation is used to get the Laue photograph (where different orders of reflection may superpose on a single spot), this method is hardly used for crystal structure determination.

A suitable crystal (as perfect as possible) is selected by examining the same under polarizing microscope. It is mounted on a glass fibre parallel to the principal axis using some adhesive (like shellac, glue or even quick fix) material. The fibre alongwith the crystal is then mounted on a goniometer head with the help of modelling clay is shown in Fig. 10.21. The crystal is then aligned properly on the goniometer head with the help of four arcs (two mutually perpendicular arcs are for linear adjustment). The goniometer head with the aligned crystal is then placed on a single crystal

Table 10.2   Extinction rules for cubic crystals

Crystals	*Allowed reflections*
SC	for all values of $(h^2 + k^2 + l^2)$
BCC	for even values of $h + k + l$
FCC	when h, k, and l are all odd or all even,
DC	when h, k, and l are all odd or all even, $(h + k + l)$ should be divisible by four

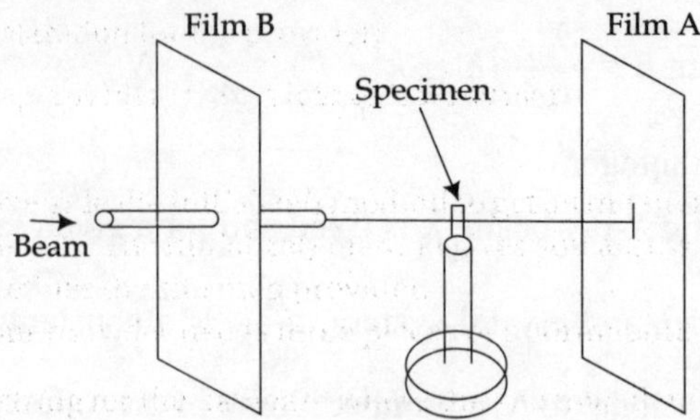

Fig. 10.22   Schematic representation of the Laue technique

X-ray diffraction camera. The goniometer head is held fixed such that the incident *X*-ray beam (continuous radiation) is normal to one face of the crystal. A flat plate film cassette is placed either in between the source and the crystal for recording the back reflection or behind the crystal to record the transmission Laue spots. A schematic representation of the Laue technique is shown in Fig. 10.22. The crystal to film distance in both the cases is normally kept small and is fixed between 3-5 cm. This increases the number of reflections intercepted by the film and reduces the exposure time. The two arrangements differ from one another in one respect, i.e. to intercept the direct incident beam a small lead cup is fixed at the center of the cassette for recording the transmission Laue photograph, whereas in the back reflection a hole is made at the same point of the cassette to fit the collimator for the *X*-ray beam to pass.

Indexing of Laue Photographs

A close relationship between gnomonic projection and reciprocal lattice (section 7.5) indicates the usefulness of the former for indexing the Laue photographs. Fig. 10.23 shows a relationship among the reflecting plane of a crystal, its normal, and the direction of the reflected beam. Here, it is assumed that the photographic film and the plane of projection are coincident and represent the vertical line *PR*. If the crystal to film distance is taken to be unity, then the distance *r*, the point of intersection of the reflected beam with the vertical line from the center of the photographic plate (through which the direct beam passes), is

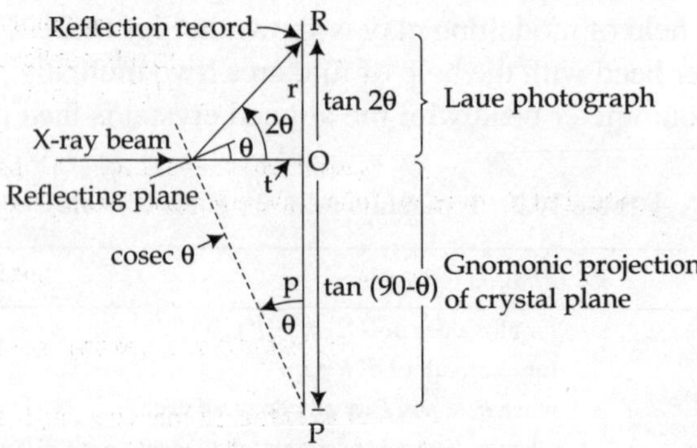

Fig. 10.23   Relationship among the reflecting plane, its normal and the direction of the diffracted beam

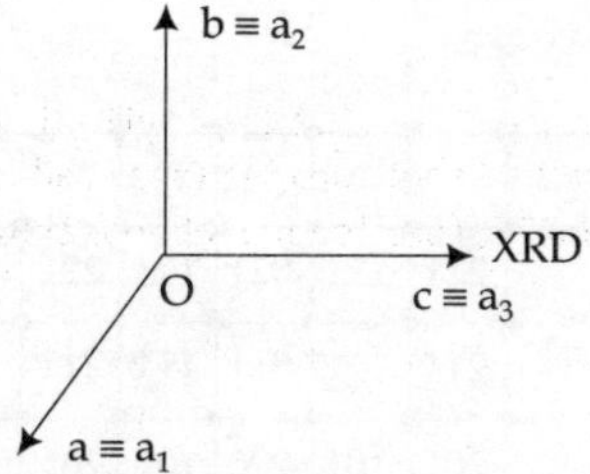

Fig. 10.24   Crystal axes and X-ray beam

$$r = \tan 2\theta$$

or
$$2\theta = \tan^{-1} r \qquad (20)$$

Similarly, the distance p, the point of intersection of normal to the reflecting plane (hkl) with the vertical line from the center of the photographic plate, is

$$p = \tan(90 - \theta) = \cot\theta \qquad (21)$$

The distances r and p are related inversely. Combining eqs. 20 and 21, we obtain

$$p = \cot\frac{1}{2}\,\tan^{-1} r \qquad (22)$$

This equation is derived by assuming the crystal–to–film distance to be unity. However, it is often more convenient to place the film at some other distance, say D and yet measure p in a plane at unit distance from the crystal. In this case, the eq. 22 is modified as

$$p = \cot\frac{1}{2}\,\tan^{-1}\left(\frac{r}{D}\right) \qquad (23)$$

With the help of this equation, it is easy to transform the location of each spot of the Laue photograph to the gnomonic projection of the plane whose reflection produces it.

In order to determine the indices of poles, it is necessary to know the orientation of the crystal w.r.t the X-ray beam before hand. Now, for simplicity consider a cubic crystal (where $a = b = c$) aligned in such a manner that the X-ray beam is parallel to c-axis, while b-axis is lying in a vertical plane and a-axis in a horizontal plane as shown in Fig. 10.24. Since b is in the vertical plane, the $ho1$ poles are regularly spaced along the central horizontal line as shown in Fig. 10.25. Similarly, $ok1$ poles lie along the vertical central line. The indices of other hkl poles can be read easily from the figure. To find out the indices of other poles proceed as follows: Consider a pole (Fig. 10.25) labelled as R. Here, $h = 2$ and its k index apparently is $-\dfrac{1}{2}$, then taking $l = 1$ and clearing the fraction

we obtain $\left(2, -\dfrac{1}{2}, 1\right) \rightarrow (4\bar{1}2)$. This is the indices of R. Similarly, other points can be indexed.

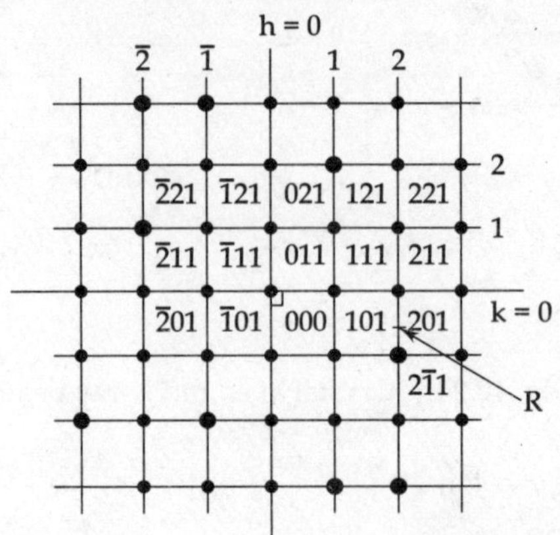

Fig. 10.25 Indexing or Laue spots of a simple cubic crystal

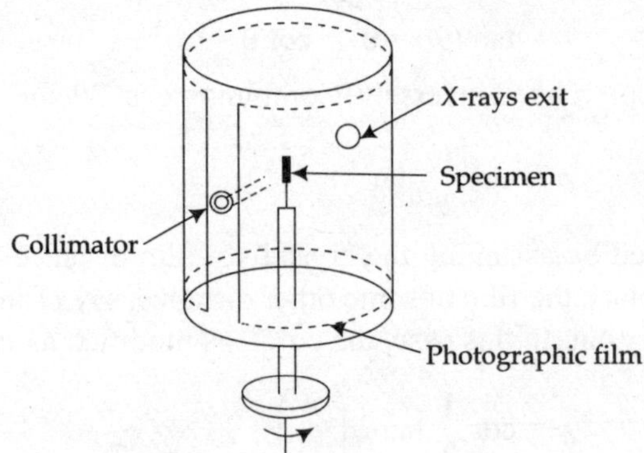

Fig. 10.26 Schematic representation of rotating crystal technique

10.8   THE ROTATION/OSCILLATION METHOD

The crystal setting is done exactly as in the case of Laue method. However, in this case instead of keeping the crystal stationary it is either rotated or oscillated (usual oscillation range is 15°) by coupling the gonimeter head with a motor having suitable gear arrangement (Fig. 10.26). A monochromatic beam of X-ray is allowed to fall on the crystal through a collimator of about 1 mm internal diameter. The X-ray reflections are recorded on an X-ray film loaded in a cylinderical camera (usually of 3 cm radius) whose axis is coincident with the axis of rotation (or oscillation). A direct beam stopper is kept at the exit point of the X-ray beam. A schematic view of the rotating (oscillating) technique is shown in Fig. 10.26. A typical rotation/oscillation photograph after unrolling and processing (developing and fixing) the film looks like as shown in Fig. 10.27. The interpretation of rotation/oscillation photograph is presented below.

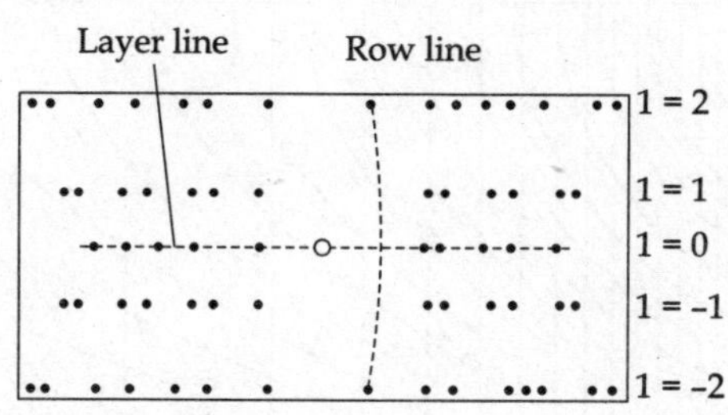

Fig. 10.27   A typical rotation photograph

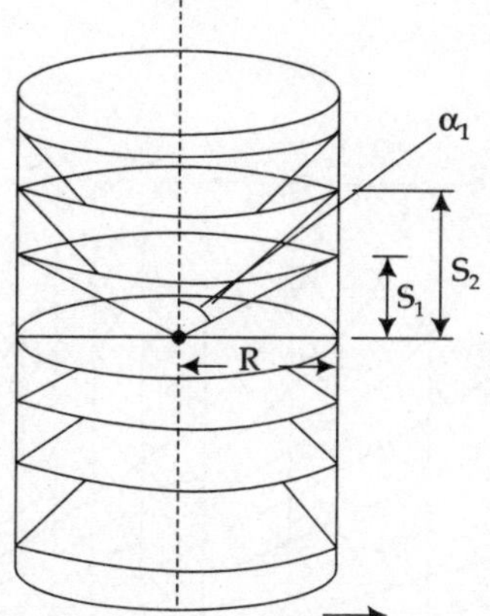

Fig. 10.28   Formation of layer lines from cones of the diffracted rays

Interpretation of Rotation/Oscillation Photographs (Formation of Layer Lines)

In deriving the Laue equations (sec. 10.3), we observed that all the diffracted beams lie on a series of concentric cones (representing different orders of diffraction) surrounding the row of atoms. These cones are actually responsible for the formation of layer lines in a rotation/oscillation photograph (Fig. 10.28). All the reflections on the zero layer line come from the planes which are parallel to the axis of rotation/oscillation and hence the corresponding Miller index is zero. The position of various reflections on this layer line depends on the d-spacings of the planes in question (i.e. on the other Miller indices). The planes that have the largest d-spacing diffract X-rays at low Bragg angles and the corresponding reflections are near to the centre of the film (exit point of the direct X-ray beam corresponds to the zero Bragg angle). Other layer lines are separated by $S_1, S_2, S_3, \ldots$ etc. away from the zero layer line whose Miller indices l are 1, 2, .. etc.

If the crystal is assumed to be rotated about c-axis and the incident X-ray beam is perpendicular to the axis of rotation, then from eq. 6, we obtain

$$c \cos \gamma = n\lambda \tag{24}$$

where g is replaced by n and is equal to zero for (zero), ±1 for (first), ±2 for (second) layer lines, etc. Similar equations hold for rotation of the crystal about other axes. If S_n is the separation of the nth layer line from zero layer line, γ is replaced by α and R is the radius of the camera used then from Fig. 10.28, we have,

$$\cot \alpha_n = \frac{S_n}{R} \tag{25}$$

With the help of eqs. 24 and 25, we obtain

$$S_n = \frac{(n\lambda / c)}{\sqrt{1 - \left(\dfrac{n\lambda}{c}\right)^2}} \, R \tag{26}$$

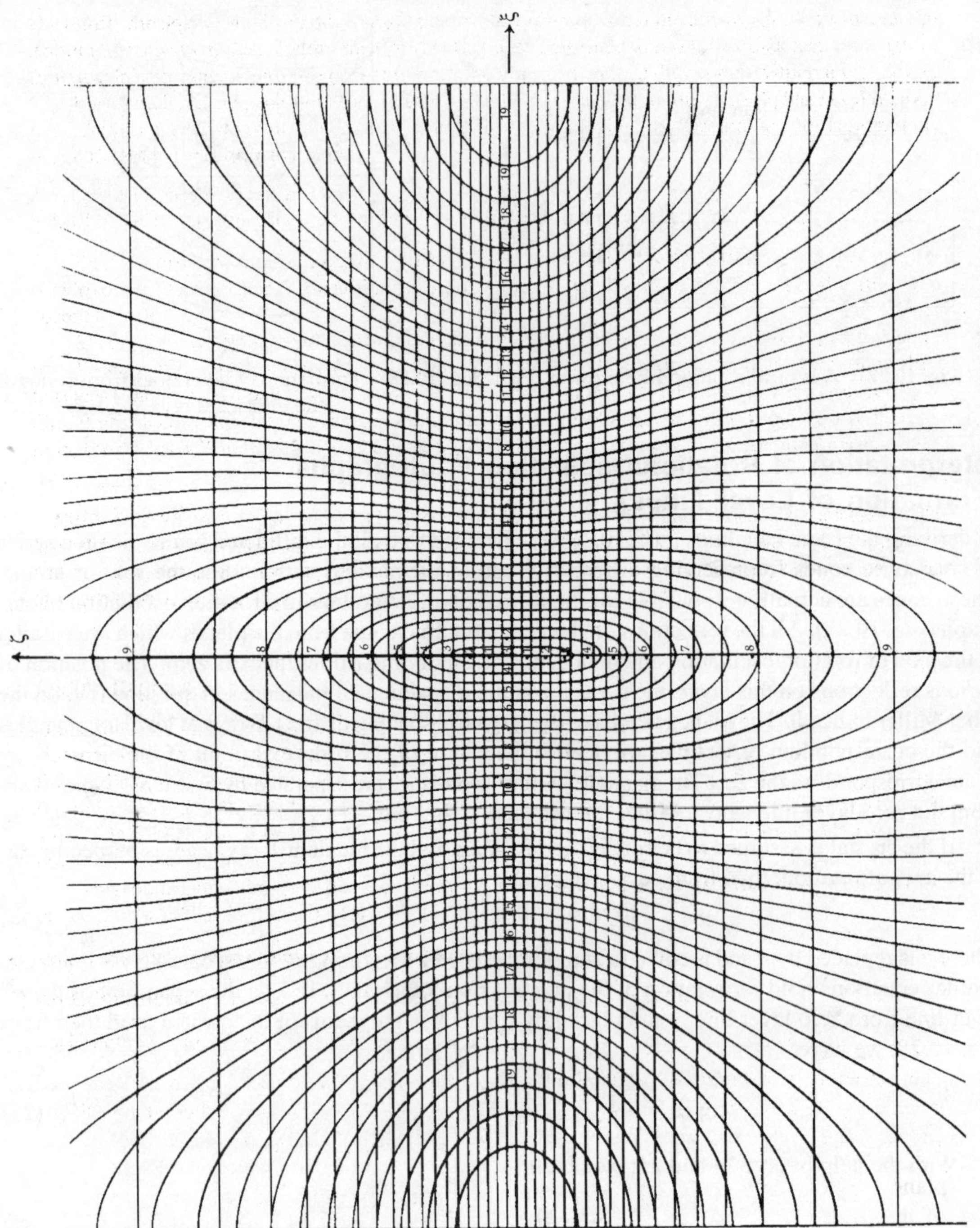

Fig. 10.29   A Bernal chart (Reproduced by courtesy of the Institute of Physics)

So that the cell dimension, c, is given by

$$c = \frac{n\lambda}{S_n}\,(R^2 + S_n{}^2) \tag{27}$$

Similarly, by taking a-axis and b-axis rotation photographs their unit cell dimensions can be obtained. However, this is a tedius and time consuming process.

As far as the rotation photograph is concerned, its left and right sides are identical because each reciprocal lattice point passes through the sphere of reflection in an identical manner on both sides of the direct beam. Similarly, the top and bottom halves of the film are also identical because of its centrosymmetic nature. Therefore, it can be said that all rotation photographs have the inherent mm symmetry and the useful information can be extracted (e.g. to determine the lattice parameters when the crystal's symmetry is known) by indexing only one quadrant of the film.

Indexing of rotation/oscillation photographs

Indexing of rotation/oscillation photograph is a process to assign the appropriate Miller indices h, k, and l to each reflection in one quadrant of the film. The indexing procedure for both rotation and oscillation photographs is quite similar. The simplest, easiest and quickest method of indexing such photographs is graphical method.

For simplicity, let us first discuss the indexing procedure of a zero layer c-axis rotation/oscillation photograph and then generalize it for other layer lines. For the purpose, let us suitably place the X-ray photograph on a Bernal chart (it is a graphical chart named after its inventor, J.D. Bernal) as shown in Fig. 10.29 (copies of the Bernal chart on transparent paper or plastic are commercially available) which enables us to know the coordinates ζ and ξ for any reflection directly, where ζ is the vertical distance of any layer line above the zero layer and ξ is the horizontal distance of any point (spot) from the axis of rotation/oscillation of the reciprocal lattice (Fig. 10.30). The value ζ directly provides the index $l = 0, 1, 2, \ldots$ for zero, first, second, etc. layer lines. On the other hand,

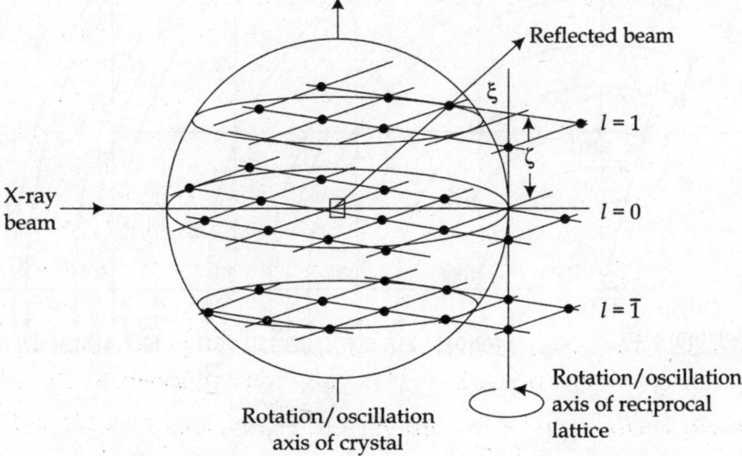

Fig. 10.30　The geometry of rotation/oscillation photographs

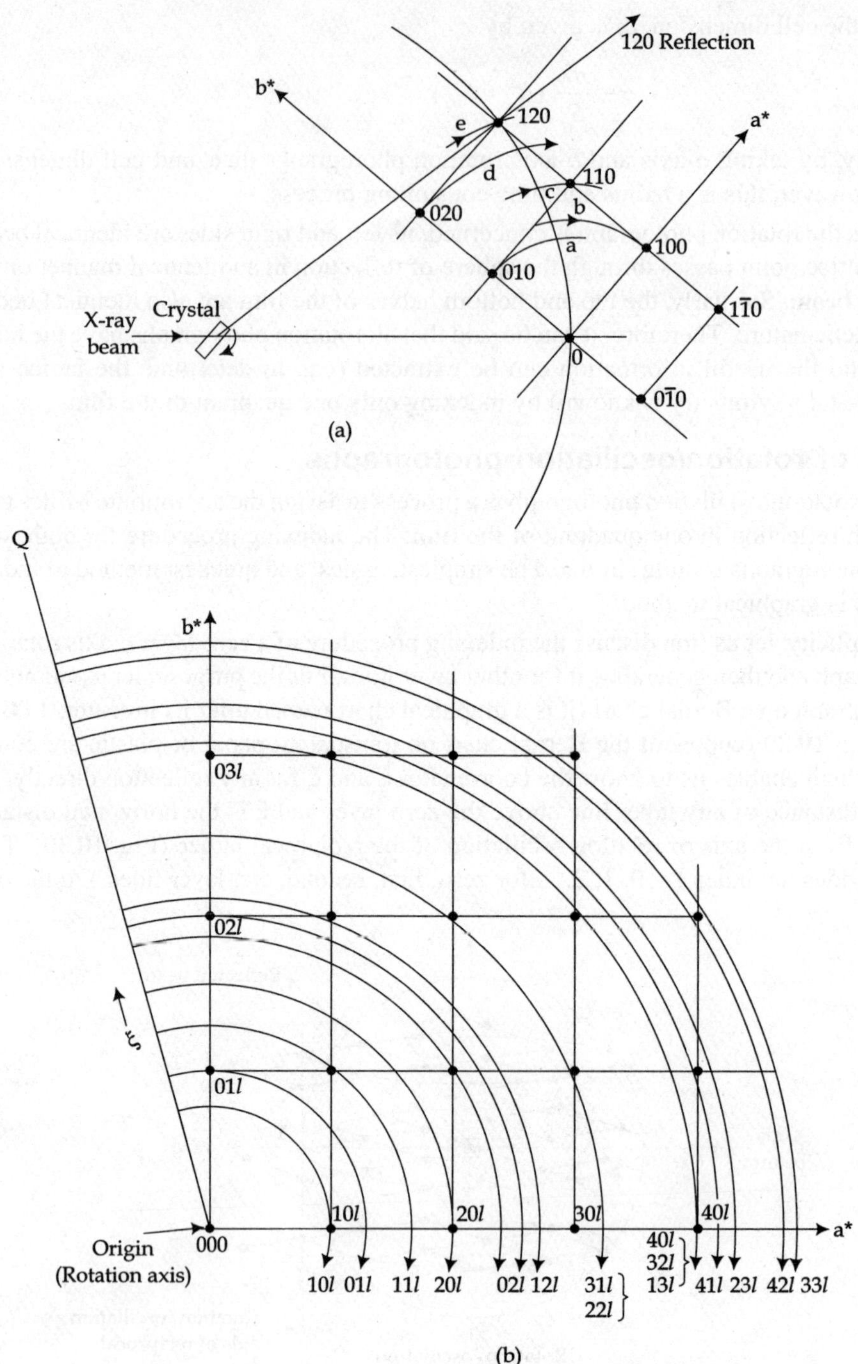

Fig. 10.31   Indexing of a rotation photograph

determining the indices *h* and *k* within a layer is generally not a straight forward process particularly for crystals with low symmetry and/or large unit cells because the information about a two dimensional reciprocal lattice layer is compressed into a one – dimensional layer line. To resolve this problem, let us draw a $a* - b*$ reciprocal lattice net (Fig. 10.31) with the help of the earlier determined lattice parameters where one reciprocal lattice ($r \cdot l$) unit is taken as 10 cm throughout in the indexing process.

To index the rotation photograph, draw a redial line from the origin on which the ξ value of each spot of the zero layer is plotted. Each redial distance is then swung one by one about the origin till its end is found to coincide with the corresponding reciprocal lattice point $[[hko]]*$. In this way, all the reflections in the zero layer line can be indexed. In a similar manner, the reflections of other layer lines can also be indexed. However, it is to be noted that the radius of the circle intersecting with the Ewald sphere decreases with the increase in ζ value (Fig. 10.32) for higher layer lines. Accordingly, the radius of the circle of *n*th layer line is given by

$$R_n = \sqrt{1 - \zeta_n^2} \tag{28}$$

For low symmetry crystals (where $c*$ is not perpendicular to $a* - b*$ net), the origin is shifted by an amount of offset w.r.t the origin of the zero layer as shown in Fig. 10.33.

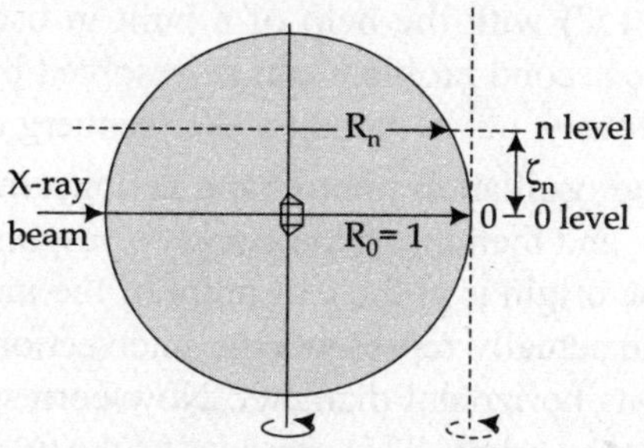

Fig. 10.32 Radius of *n*th circle

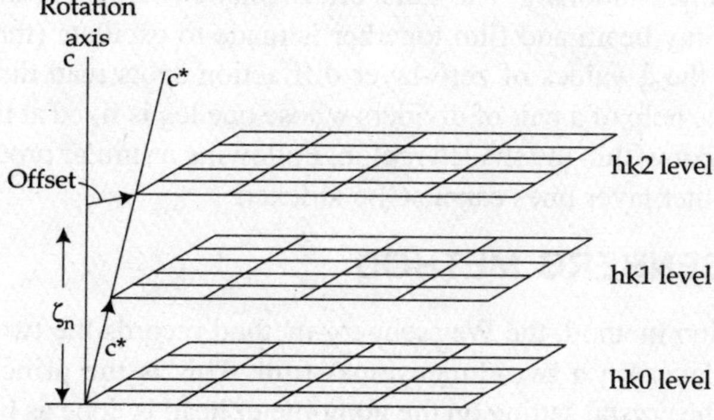

Fig. 10.33 Offset w.r.t the origin of the zero-layer for low symmetric crystals

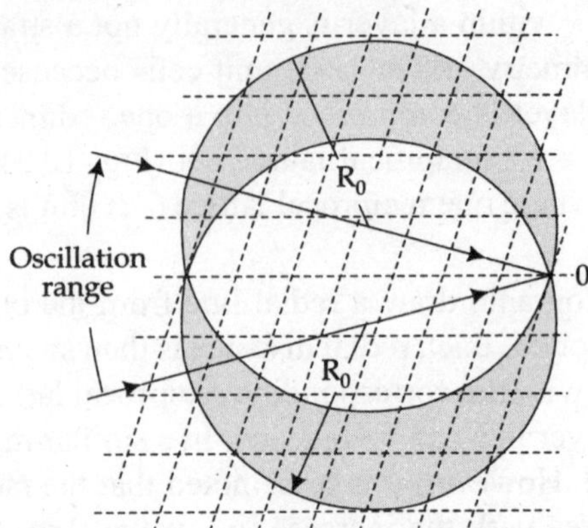

Fig. 10.34 Indexing of zero-layer 15°-oscillation photograph

The rotation photographs inherently have a couple of drawbacks. They are: (i) the actual meaning of the third cylinderical coordinate φ gets lost during a complete rotation and (ii) there is an ambiguity in the assignment of the indices h and k due to overlapping of some reflections having identical or similar ξ values. An improvement upto a certain extent can be made in the first case by oscillating the crystal (usually 15°) with the help of a built in oscillation cams or angular limit switches. On the other hand, the second problem can be resolved by moving the film linearly and oscillating the crystal simultaneously, i.e. by taking a Weissenberg or a precession photograph.

As far as the indexing of an oscillation photograph is concerned first of all prepare $a^* - b^*$ reciprocal lattice net as before – and then insert the circle corresponding to zero layer of the reciprocal lattice of the crystal whose origin is at the exit point of the incident X-ray beam as shown in Fig. 10.30. The zero layer circle actually represents the intersection of the reciprocal lattice plane with the Ewald sphere around its horizontal diameter. Now corresponding to the 15° oscillation, rotate the circle through 15° w.r.t the origin. This provides us the two extreme positions of the circle and the possible reflections will lie within the shaded crescent shaped region (Fig. 10.34) produced by the intersection of the two circles. This is equivalent to oscillating the reflection circle and keeping the reciprocal lattice net stationary. The same effect can be obtained if the crystal is assumed to be stationary and the X-ray beam and film together is made to oscillate (through the same angle) about the origin. Next, the ξ values of zero-layer diffraction spots read directly from the Bernal chart are marked with the help of a pair of dividers whose one leg is fixed at the origin and the other swept around to find a spot within the shaded region. Following a similar procedure and making use of eq. 28, the spots of other layer lines can also be indexed.

10.9 THE WEISSENBERG METHOD

Unlike rotation/oscillation method, the Weissenberg method records the two dimensional reciprocal lattice points of a layer on a two-dimensional film. This is the principal advantage of the Weissenberg method. The crystal setting on the goniometer head is done as before. However, in the Weissenberg method, the goniometer head is fixed at the end of a horizontal spindle which can be rotated through a motor having suitable gear arrangement. Horizontal linear motion can also be

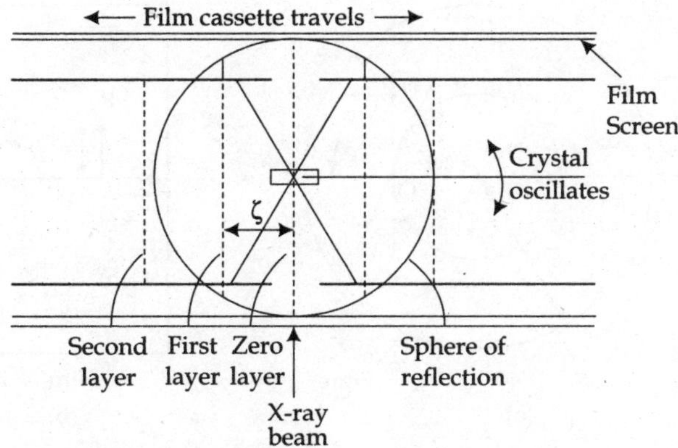

Fig. 10.35   Zero-layer equi-inclination Weissenberg setting

imparted to the crystal (spindle). With the help of a screw arrangement in order to set the crystal in the path of the incoming X-ray beam.

The Weissenberg camera is essentially a steel cylinder with a horizontal slot (also called as collimator slot) cut parallel to its axis. The standard camera has an inner diameter of 57.3 mm. The choice of this value will be clear later.

The main body of the camera consists of a double base construction. The lower base is fixed and is fitted with three lavelling screw legs for height adjustment. The lower base also provides the track for the upper movable base by means of crank and screw. The microscope, collimator holder, motor and reversible micro-switch, the spindle (the crystal holder) etc. are fitted with the movable base, which can rotate over a circular scale. The Weissenberg camera base which rests on tacks can have to and fro motion parallel to the axis of rotation/oscillation. When the camera is used for Weissenburg photographs, the motion of the camera base and the crystal oscillation is synchronously coupled by using a split nut while the two reversing switches control the range of crystal oscillation (usually 200°). When the camera is used for simple rotation photographs, the nuts and reversing switches are disengaged.

A comparison of rotation/oscillation method and Weissenberg method provides us the following:

1. In rotation/oscillation method many Laue cones are recorded on a single film at a time while in Weissenberg method only one Laue cone is allowed at a time to pass through the gap between the two layer line screens. Further, to record a desired cone, the position of the layer line screens and the gap between them need to be adjusted accordingly.

2. In rotation/oscillation method the axis of rotation/oscillation is vertical, while the axis of rotation/oscillation in Weissenberg method is horizontal (Fig. 10.35). This facilitates the equi-inclination setting by varying the angle between the axis of rotation/oscillation and X-ray beam in the range 0° to 30°.

3. Rotation/oscillation photographs of the same crystal depend only on camera diameter while Weissenberg photographs of the same crystal depend on camera diameter and coupling (of film translation to crystal rotation/oscillation) mechanism.

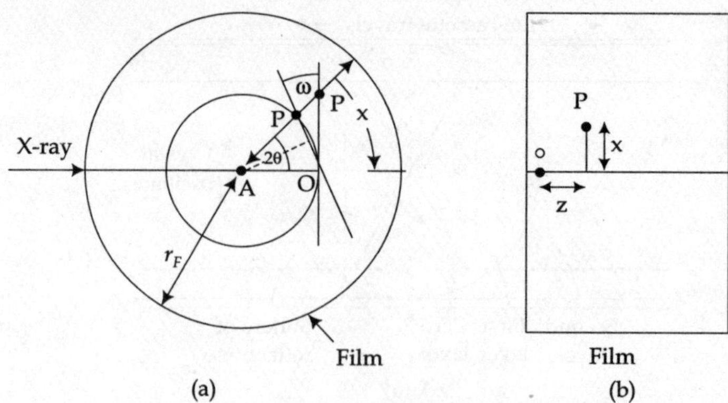

Fig. 10.36   (a) Zero-layer Weissenberg geometry (b) the coordinates x and z (of the *r. l.* point P) on the film

As a result of rotatory motion of the crystal and translatory motion of the film in a Weissenberg set up, there are two instrumental constants. To determine them, let us see the condition of forming reflection in a zero layer and the corresponding Weissenberg photograph (Fig. 10.36), where Fig. 10.36a shows the central reciprocal lattice row (line) as tangent to the sphere of reflection at the exit of X-ray beam. Further, from Fig. 10.36a it is clear that the distance x (the distance of diffraction spot from the origin of the reciprocal lattice net) is proportional to the angle of diffraction, i.e.

$$2\theta = C_1 x \tag{29}$$

where C_1 is the first instrumental constant and can be evaluated for maximum (optimum) diffraction angle 360°, then the proportion is written as

$$\frac{2\theta}{x} = \frac{360°}{\text{Camera circumference}} = \frac{360°}{2\pi r_F}$$

or

$$2\theta = \left(\frac{360°}{2\pi r_F}\right) x \tag{30}$$

where r_F is the camera (or film) radius. From eqs. 29 and 30, we obtain

$$C_1 = \frac{360°}{2\pi r_F} \tag{31}$$

To keep the conversion from x to 2θ simple, let us suppose the value of $C_1 = 2°/\text{mm}$. From eq. 31, this provides us the diameter of the camera, i.e. $2r_F = 57.3$ mm. This is the standard Weissenberg camera diameter, others may be an integral multiple of this.

Further, the coupling of the crystal rotation/oscillation and the film translation is such that a rotation of an angle ω in the crystal produces a translation z in the film (Fig. 10.36b) and their ratio is written as

$$\frac{\omega}{z} = C_2$$

or

$$\omega = C_2 z \tag{32}$$

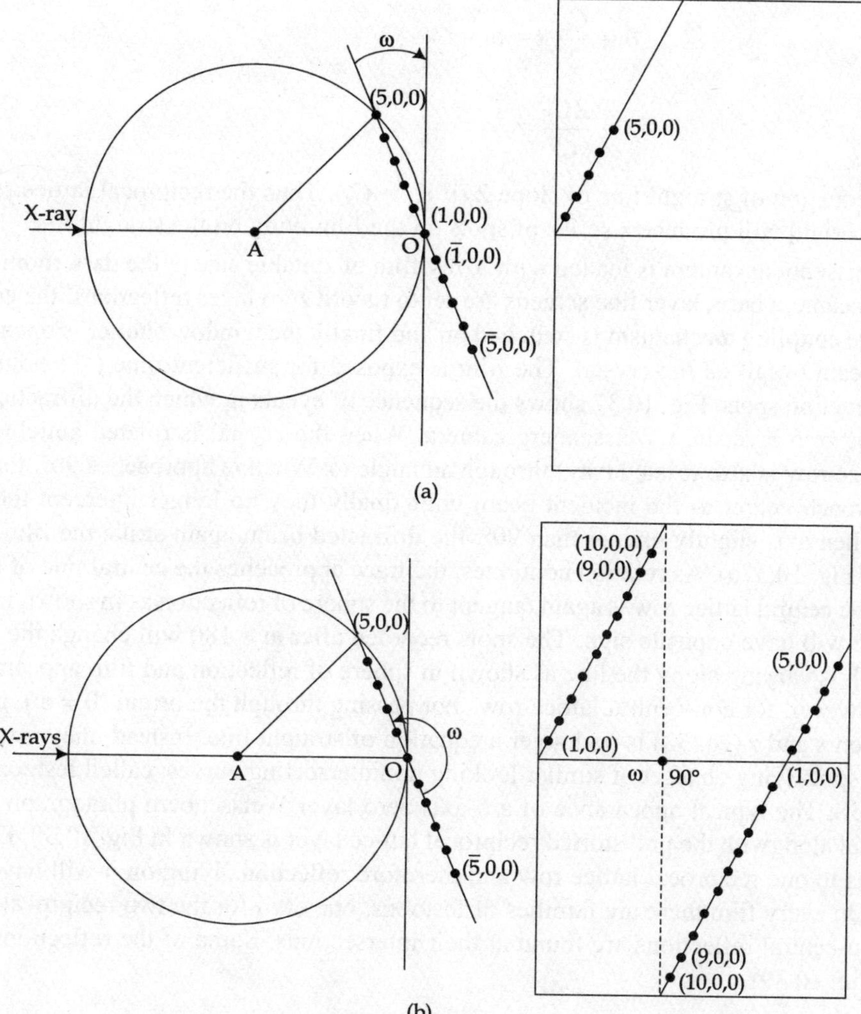

Fig. 10.37   (a) Sequence of diffraction spots for zero-layer Weissenberg setting before ω approaches 90°
(b) Showing the events during complete range of oscillation

where C_2 is the second instrumental constant. It is desirable to measure the angles ω and 2θ of a Weissenberg photograph by means of the same scale to avoid any confusion in its indexing. With this choice, all the Weissenberg photographs have similar appearance except for the magnification. Weissenberg photographs with $C_1 = C_2$ are said to have undistorted scale.

Figure 10.36a shows that a normal to the zero level reciprocal lattice row passing through A bisects the angle 2θ. Thus, for this particular level, θ is equal to ω because the sides of the two angles are perpendicular. Therefore, from eqs. 29 and 32, we can write

$$\theta = \frac{C_1}{2} \, x = \omega = C_2 z$$

or
$$x = \frac{2C_2}{C_1} \, z \qquad\qquad (33)$$

which is a equation of straight line of slope 2 (if $C_1 = C_2$). Thus the reciprocal lattice row passing through the origin will produce a series of spots on the film lying on the straight line.

The Weissenberg camera is loaded with X-ray film of suitable size in the dark room and then it is placed on camera base, layer line screens are set to record zero layer reflections, the collimator is replaced, the coupling mechanism is switched on and finally the window shutter is opened to allow the X-ray beam to fall on the crystal. The film is exposed for sufficient time (~ 1 hour) to record intense diffraction spots. Fig. 10.37 shows the sequence of events in which the diffraction spots are recorded for zero layer in a Weissenberg camera. When the crystal is rotated anticlockwise, the central lattice row is also rotated (say) through an angle ω. When ω approaches 90°, the diffracted beams approach nearer to the incident beam until finally they no longer intercept the film (Fig. 10.37a). When ω is slightly greater than 90°, the diffracted beam again strike the film, but on the lower half (Fig. 10.37b). As rotation continues, the trace approaches the central line of the film. At $\omega = 180°$, the central lattice row is again tangent to the sphere of reflection as in shown in Fig. 10.36 but its trace will have opposite sign. The spots recorded after $\omega > 180$ will change the sign of that index which is varying along the line as shown in sphere of reflection and film appearance in Fig. 10.37b. However, for non-central lattice rows not passing through the origin ($\theta \neq \omega$), the relationship between x and z (eq. 33) is no longer a equation of straight line. Instead, they give rise to the diffraction spots lying on a set of similar looking nonintersecting curves, called festoons as shown in Fig. 10.38. The typical appearance of a b-axis zero layer Weissenberg photograph of a monoclinic crystal alongwith the undistorted reciprocal lattice layer is shown in Fig. 10.39. Each festoon corresponds to one reciprocal lattice row and therefore reflections lying on it will have one index common. On every film there are families of festoons, one each for the two reciprocal lattice axes and the non-central reflections are found at their intersections. Some of the reflections have been indexed (Fig. 10.39).

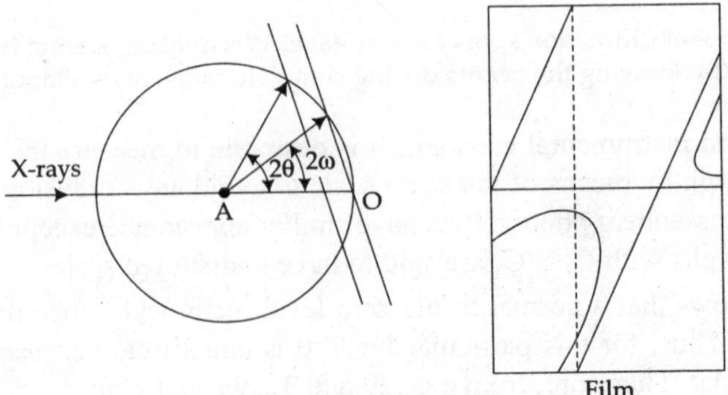

Fig. 10.38   Sequence of events for non-central lattice rows ($\theta \neq \omega$) in the form of festoons

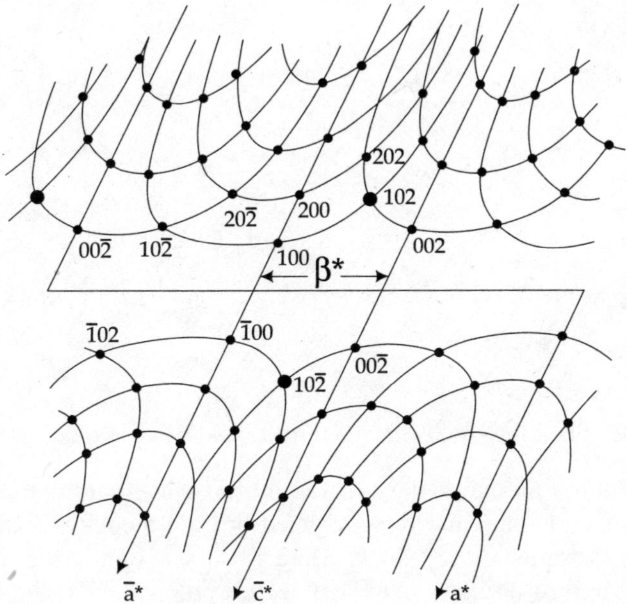

Fig. 10.39 Zero-layer Weissenberg photograph of a monoclinic crystal

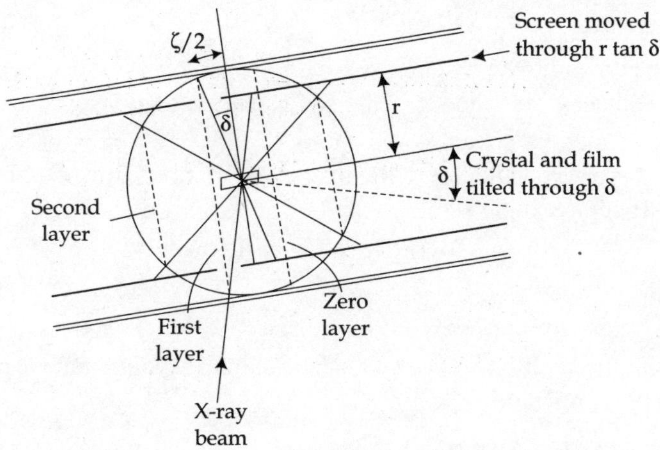

Fig. 10.40 Equi-inclination Weissenberg setting

Equi-inclination Setting

In order to record the Weissenberg diffraction data of a particular Laue cone (usually higher level), it is necessary to adjust the layer line screens such that the diffracted beams from the desired layer only can pass through the gap between the screens. There are several ways to achieve this and record upper level Weissenberg photographs. However, the equi-inclination setting (Fig. 10.40) is the most advantageous. For example, the obvious advantage over the normal beam method is that

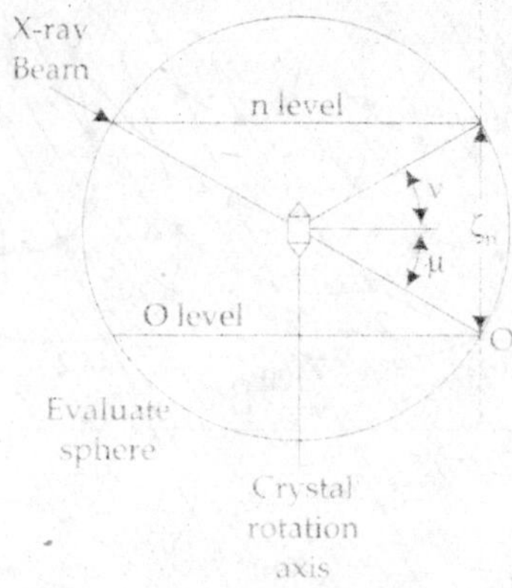

Fig. 10.41   Showing equally inclined direct and diffracted beams (i.e. $\mu = v$)

the origin of the n-level lies on the Ewald sphere, so that the central blind region is eliminated. In this case, the direct (X-ray) beam and the diffracted beam are equally inclined to the reciprocal lattice layer (i.e. $\mu = v$) as shown in Fig. 10.41. Here, μ is the complement of the angle between the incident beam and the axis of oscillation, while v is the complement of the semi apex angle of the Laue cone. For eqi-inclination setting, the angle μ is given by

$$\mu = \sin^{-1}\left(\frac{\zeta_n}{2}\right) \tag{34}$$

where $\zeta_n = n\,\zeta_1$ is the perpendicular distance between zero and n^{th} layers in reciprocal lattice units. If $\zeta_n = 1$ (the radius of the sphere), then $\mu = 30°$, the usual inclination angle.

In the zero layer setting, the crystal is set symmetrically between the gap of layer line screens and the diffracted rays are perpendicular to the axis of oscillation. However, for higher layers, this is not the case and hence layer line screen must be shifted (with respect to zero layer setting) to select the desired cone. The shift, S_n towards the crystal tip to collect the nth layer diffracted beam is given by

$$S_n = r_s \tan \mu \tag{35}$$

where r_s is the radius of the layer line screen. Once the angle of inclination is known, the shift in the layer line screen can be estimated.

Indexing of Weissenberg Photographs

Zero layer and equi-inclination Weissenberg photographs can be easily indexed by properly placing them on a Weissenberg chart (Fig. 10.42) and directly reading the Cartesian coordinates of diffraction spots after properly identifying the straight lines as the reciprocal lattice axes. Following this, any reflection can be indexed by determining the constant indices of the two festoons on which it lies (Fig. 10.39). For example, the reflections $0kl$ will always lie between b* ($0k0$) and c* ($00l$) axes,

the reflections $0\bar{k}l$ between $b*$ and $\bar{c}*$ axes, the reflections $0k\bar{l}$ between $\bar{b}*$ and $c*$ axes and so on. For other details one can refer any standrad text.

Camera Diameter 5.73 cm   2° Rotation per mm Travel

Fig. 10.42   A Weissenberg chart

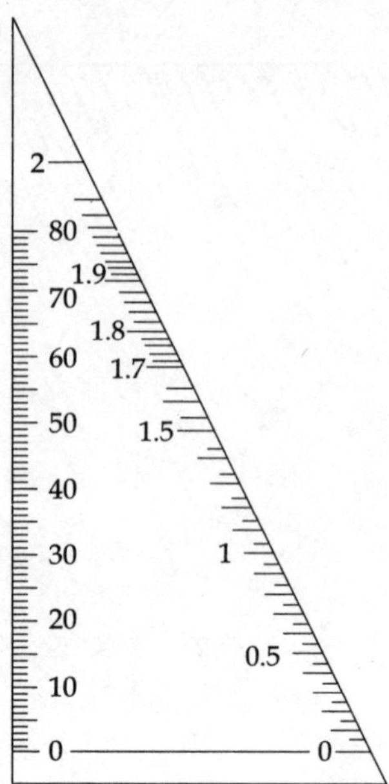

Fig. 10.43   Triangle calibrated in reciprocal lattice

The Weissenberg photographs are useful in many different ways such as accurate alignment of crystals, measurement of unit cell parameters, determination of space groups etc. However, at present, as a part of the indexing process we shall confine our discussion to the measurement of unit cell parameters only.

The cell parameter (called the repeat distance) along the direction of axis of rotation/oscillation is essentially determined by rotation/oscillation photographs only and hence its value is the same as given by eq. 27. However, the measurement of other two cell parameters is made on the zero layer photograph. The most straight forward method for obtaining them is to measure their reciprocal lattice constants and then convert them into their corresponding direct lattice constants. The distance $\xi(x_i)$ of any reflection from the reciprocal lattice origin along the central lattice line can be made by using a triangular scale (Fig. 10.43) calibrated in r.l.u for 57.3 mm diameter camera. However, for accurate value of lattice constants, the film is placed on a measuring device such that the hairline is parallel to the central line of the film, and the perpendicular distance $(2y_n)$ between the corresponding reflections on the axial line is measured (Fig. 10.44). The relationship among ξ, y, R and θ are depicted in Fig. 10.45, where R is the camera radius. From the figure, we obtain

$$\theta = \frac{y/2}{R} \text{ in radian}$$

and

$$\frac{\xi}{2} = \sin\theta = \sin\left(\frac{y}{2R}\right)$$

or

$$\xi = 2\sin\theta = 2\sin\left(\frac{y}{2R}\right)$$

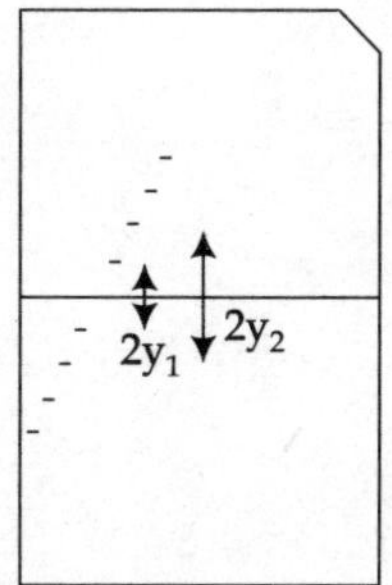

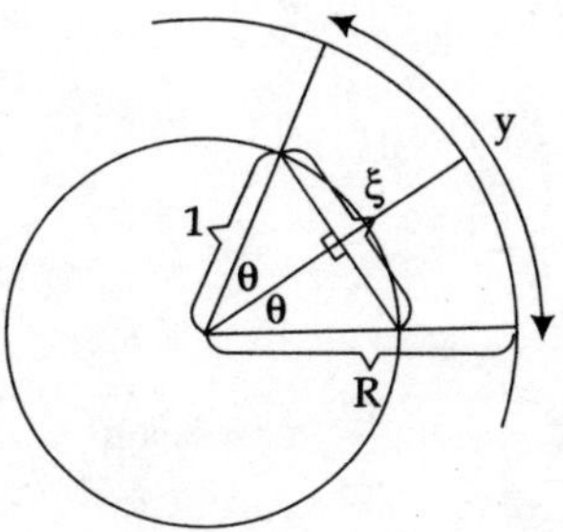

Fig. 10.44   Distance measured on Weissenberg film and calculate interplanar spacings

Fig. 10.45   Geometry showing relation y, R, and θ

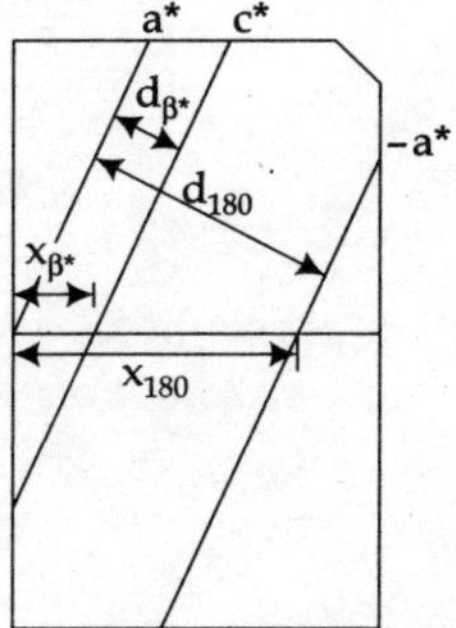

Fig. 10.46   Diagram showing distances measured in determining reciprocal interaxial angles

For n^{th} reflection (in degrees), we can write

$$\xi_n = 2\sin\frac{2y_n}{4R} \times 57.3 \text{ (in degrees)} \tag{36}$$

where $\xi_1 = \dfrac{\xi_n}{n}$.

Now, we know that in general, the Weissenberg camera diameter $2R = 57.3$ mm. Accordingly, eq. 36 reduces to

$$\xi_n = 2\sin y_n \tag{37}$$

where y in mm = angle in degrees.

Therefore, the cell parameter along the given axis a or c will be

$$a \text{ or } c = \frac{n\lambda}{\xi_n \sin\beta} = \frac{n\lambda}{2\sin(y_n)\sin\beta} \tag{38}$$

where $\sin\beta = \sin\beta^*$.

The angle between the $a^* - c^*$ axes is determined by the ratio of the distance between the two axes and that between the two traces of single axis as shown in Fig. 10.46. That is

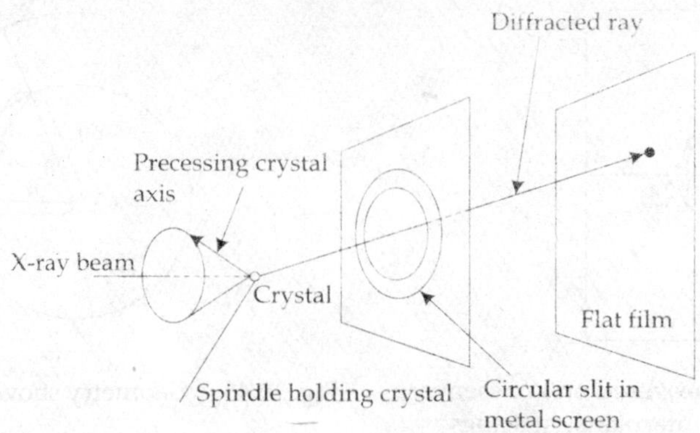

Fig. 10.47 Principle of the precession camera

$$\beta^* = \frac{d_{\beta^*}}{d_{180}} \times 180^\circ = \frac{X_{\beta^*}}{X_{180}} \times 180^\circ \tag{39}$$

These distances can be measured either perpendicularly to the axial lines or along the central line of the film. The latter has the advantage because most of the cameras have $X_{180} = 90$ mm and therefore eq. 39 reduces to

$$\beta^* = 2X_{\beta^*} \tag{40}$$

where X_{β^*} is in mm. Consequently, the angle between the a-c (real) axes is given by

$$\beta = 180 - 2X_{\beta^*} \tag{41}$$

10.10 THE PRECESSION METHOD

The Weissenberg method gives us a distorted picture of a reciprocal lattice plane. On the other hand, the precession method produces an undistorted picture of a reciprocal lattice plane from which the angles and distances can be read off more or less directly. In the precession method, a crystal is first oriented with one of its axes parallel to the incident X-ray beam (i.e. a family of reciprocal lattice planes is made perpendicular to the X-ray beam), the crystal is tilted to an angle μ (generally about 20-30°) with the beam and then made to precess about it as illustrated in the form of a line diagram in Fig. 10.47. The crystal, the layer screen (a metal plate with suitable circular slit) and the film are coupled together to move identically along the direction of the incident beam so that during their motion: (i) the film is always parallel to the reciprocal lattice plane, (ii) a constant distance between the crystal and film is maintained and (iii) only a desired cone of diffracted beams (from the plane of interest) will pass through the circular opening on their way to the film. Fig. 10.48 shows the diffraction cones generated by the zero and first layer reciprocal lattice planes and the layer screen used to stop the diffracted beam from the first layer. The crystal to screen distance depends on the radius of the circular slits. If S is the crystal to screen distance and r, the radius of the slit, then from the Fig. 10.48, we have

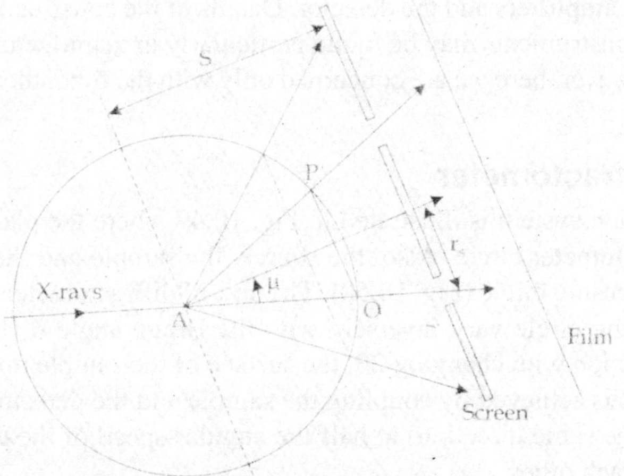

Fig. 10.48 Diffracted cones generated by the zero and first reciprocal lattice planes. The screen is set at a distance from the crystal S so that only the diffracted beams originated from the zero level are let through and therefore produce a signal on the film

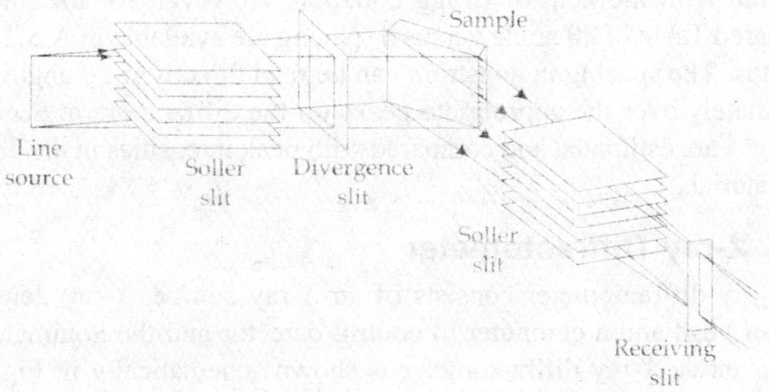

Fig. 10.49 A typical X-ray diffraction system

$$\cot \mu = \frac{S}{r_s}$$

or $$S = r_s \cot \mu \tag{42}$$

10.11 X-RAY DIFFRACTOMETERS

The photographic methods of recording diffraction patterns are now supplemented in almost every X-ray laboratory by the X-ray diffractometer. The advantages of this method are considerable, e.g. the intensities can be measured directly. Also, the work and time required are very less, especially if only a certain part of the intensity chart is required.

Apart from the *X*-ray generator and the *X*-ray tube, the diffractometer includes an *X*-ray diffraction goniometer, several amplifiers and the detector. Details of the construction and operation of the commercially available instruments may be found particularly in manufacturers' literature and vary for different makes. However, here we are concerned only with the fundamental principles common to all.

Powder *X*-ray Diffractometer

A typical *X*-ray diffraction system is illustrated in Fig. 10.49 where the plane of the sample lies at the center of the diffractometer circle. Also, the source, the sample and the detector all lie on the circumference of the focusing circle (Fig. 10.50). The size of diffractometer circle remains constant while that of the focusing circle vary inversely with the Bragg angle θ (Fig. 10.50). In order to preserve the focusing action with changing 2θ, the surface of the sample must remain tangential to the focusing circle. This is achieved by coupling the sample and the detector such that the plane of the sample rotates (in the same direction) at half the angular speed of the detector, i.e., retaiting a $\theta/2\theta$ relationship with each other.

The ideal powder sample should be homogeneous, with a grain size of about 1 to 25 microns. There should be no preferred orientation, or any micro/macro crystalline strain.

In order to interpret the diffractogram (it is a plot of intensity of diffracted radiation versus diffraction angle), the first step to determine the interplanar spacing d. Earlier in the photographic method, it was found with the help of Bragg equation. However, for the interpretation of the diffractogram, prepared Table of 2θ angle versus d spacing are available in A.S.T.M data file for all common wavelengths. The spacing in angstrom can be read directly if 2θ angle is known. Record these values immediately over the appropriate peaks on the diffractogram. Secondly, the relative intensities of the peaks are estimated and compared with peak intensities in the Table for identification of the given material.

Single Crystal X-ray Diffractometer

A single crystal *X*-ray diffractometer consists of an *X*-ray source, *X*-ray detector (scintillation counter) a gonimeter head and a computer to control detector and the gonimeter head. A typical fully automatic four circle *X*-ray diffractometer is shown schematically in Fig. 10.51. The term

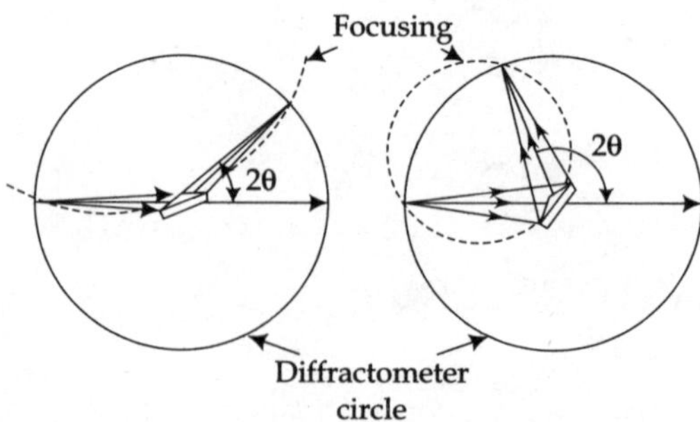

Fig. 10.50 Alignment of the sample and detector with respect to the source. (a) All lie on the circumference of the circle (b) relationship between diffractometer circle and focusing circle

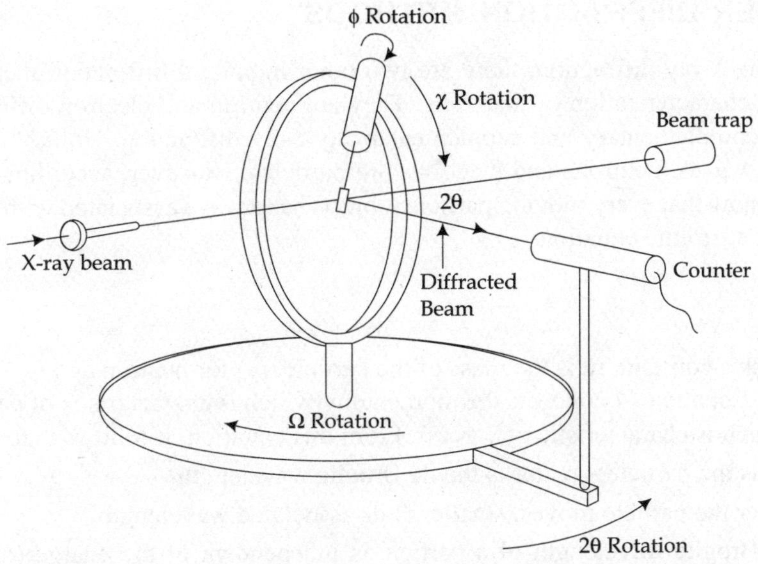

Fig. 10.51   Schematic diagram of a four-circle diffractometer

'four circle' refers to the number of rotational motions available with the system. Out of these, three rotations are associated with the crystal and one with the counter.

The crystal setting is done with the help of three angular rotations, ϕ, χ and Ω for a number of reflections (at least two) then these values are calculated for other reflections. The detector (counter) rotates about a vertical axis (also called the 2θ axis) so that the plane containing the incident and diffracted beams is always horizontal. For a particular reflection, the counter is set at a position corresponding to the Bragg angle 2θ. After initial settings, the computer (either built in or linked to the diffractometer) can supervise the process of collecting a complete set of diffraction data automatically on its own. Checks and safegaurds can be built into the system so that the errors if any can be detected and corrected.

The accuracy of diffractometer intensity is affected due to a number of factors such as non-availability of highly stabilized power supply, statistical fluctuation, etc. If in a given time, t, N counts are made then the standard error (deviation) in the number of counts is given by

$$\sigma = \sqrt{N}$$

and the fractional standard deviation is given by

$$\sigma_f = \frac{\sqrt{N}}{N} = N^{-1/2} \tag{43}$$

Therefore, to have a standard error of 1%, it is necessary to make 10^3 counts and for a standard error of 0.1%, 10^6 counts and so on. However, the number of counts which can be made is limited by the time available to collect the data. A typical count rate for a strong reflection would be about 30,000 counts per second, while 3-10 counts per second for a weak but observable reflection.

10.12   OTHER DIFFRACTION METHODS

In addition to the *X*-ray diffraction, there are two other important diffraction methods which are widely used for characterization of materials. They are neutron and electron diffraction methods. They are often complementary and supplementary to *X*-ray diffraction. Unlike *X*-rays which are electromagnetic waves, neutrons and electrons are particles. However, according to de Broglie's hypothesis we know that every moving particle (object) has a wave associated with it, whose length is given by the de Broglie equation

$$\lambda = \frac{h}{mv} \tag{44}$$

where h is Planck's constant, m is the mass of the particle (m_n for neutron and m_e for electron) and v is the velocity. Equation 44 connects the momentum (which is characteristic of particles) with the wavelength (which is characteristic of waves). From this equation, it follows that:

1. Lighter is the particle, greater is the de Broglie wavelength.
2. The faster the particle moves, smaller is its associated wavelength.
3. The de Broglie wavelength of a particle is independent of the charge (or nature) of the particle.

The wavelength of the associated wave therefore depends on the velocity of the particle, which in turn depends on its kinetic energy E, where $E = mv^2/2$. Substituting $v = \sqrt{(2E/m)}$, the eq. 44 becomes

$$\lambda = \frac{h}{m\sqrt{(2E/m)}} = \frac{h}{\sqrt{2mE}} \tag{45}$$

Neutron Diffraction

Substituting the appropriate mass value for neutron $m_n = 1.675 \times 10^{-27}$ kg, the above equation reduces to

$$\lambda = \frac{0.28}{E^{1/2}} \tag{46}$$

where λ is in angtroms and E is in electron volts. As we know that the wavelength suitable for diffraction experiment must be of the order of interatomic spacings (~ 1 Å) of crystalline solids. Therefore, the wavelength $\lambda = 1$ Å in eq. 46 yields an energy of about 0.08 *eV*. This energy is of the same order as the energy of the thermal neutrons, 0.025 *eV* (which is of the order of *kT* at room temperature, $T \sim 300$ K). Fig. 10.52a shows a typical wavelength distribution for the neutron beam from the pile of a reactor. Unlike the spectral distribution of *X*-rays from an *X*-ray tube (Fig. 10.52b), there are no sharp characteristic peaks in the reactor spectrum. The most common method for obtaining a more or less monochromatic beam of neutron is to use a crystal monochromator as shown in Fig. 10.53. The white "neutron" beam from the reactor face is allowed to fall, through a collimator on to a large single crystal of lead, beryllium or germanium, etc. oriented suitably so that the Bragg condition (eq. 3) is satisfied for chosen wavelength. The reflection angle θ is adjusted so

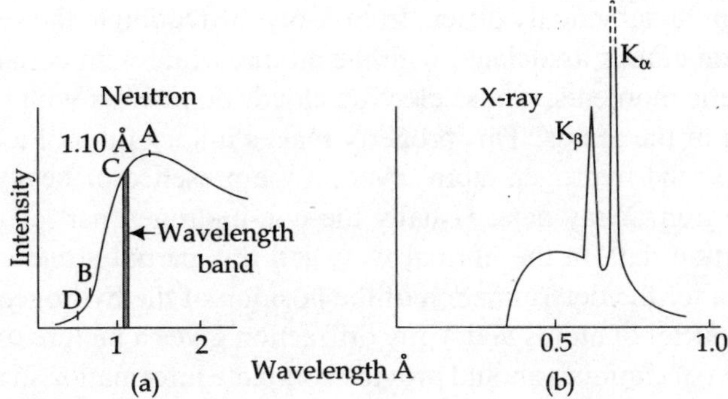

Fig. 10.52   Intensity versus wavelength distribution (a) For the neutron beam from the pile of the reactor (b) from an X-ray tube of Mo target

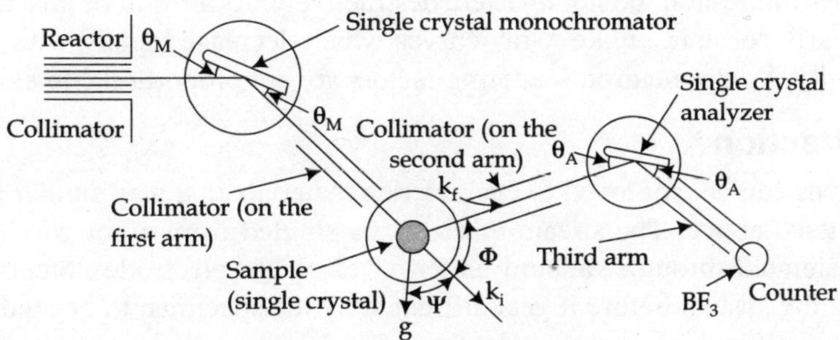

Fig. 10.53   Geometry of a triple axis neutron spectrometer

that the fraction of neutron having energies close to kT is selected. The extent of angular range over which the neutrons are to be collected is a compromise between the two cases: (i) if the range is small, the beam will be highly monochromatic but very weak and (ii) if the range is large, the beam will be stronger but less monochromatic. In any case, the beam is not very strong. A typical monochromatised neutron beam has an intensity of the order of 10^6 neutron/cm^2/sec., which is about four orders of magnitude smaller than that obtained with X-rays. Thus for neutron diffraction, the single crystal to be studied must be much larger (at least 1 mm^2 or more) than those required for X-ray diffraction. However, for crystals of very large molecules, there is an additional difficulty, i.e. the low intensity problem can not be solved simply by using bigger crystals or by increasing the counting times. Actually, the unit cells of these crystals are very large and the corresponding diffraction maxima are therefore very closely spaced, their resolution is a great problem.

The diffracted beams of neutron cannot be recorded on film like X-rays. This must be measured with the help of a counter. Therefore, the whole apparatus must be heavily shielded, making it very massive as compared to X-ray diffraction equipment. Besides, the facility of neutron beam is relatively rare. These are some of the reasons because of which the neutron diffraction is much less widely used than X-ray diffraction.

Neutron diffraction fundamentally differs from X-ray diffraction in the sense that the X-rays are scattered by the electron clouds associated with the atoms, while with certain exceptions (e.g. the atoms that have magnetic moments, whose electron clouds do interact with neutrons), neutrons are scattered by the nuclei of the atoms. This property makes it a useful tool to locate the position of light atoms (particularly the hydrogen atom) even in the presence of heavy ones, which may be difficult or impossible from X-ray data. Usually the non-hydrogen part of the structure is solved from the X-ray diffraction data in the normal way, and this partial structure is coupled with the neutron diffraction data for the determination of the position of the hydrogen atoms. Since neutron diffraction locates the nuclei of atoms and X-ray diffraction gives a picture of their electron clouds, a combination of the two techniques should provide complete information in regards to the distribution of electron clouds about the nucleus.

Because the nucleus of an atom, unlike the associated electron cloud, is negligibly small as compared to the wavelength of the scattered radiation. Consequently, the neutron waves diffracted from opposite sides of the atom do not interfere destructively. As a result of this, neutron scattering factors remain nearly constant unlike X-ray curves which decrease rapidly with increasing Bragg angle θ. Also, unlike X-rays, neutron scattering factors are not proportional to atomic numbers.

Electron Diffraction

A beam of electrons can be employed to characterize materials in a way similar to the X-rays and neutrons as discussed above. The stream of electrons emitted from a hot wire (as in thermionic emission) is accelerated through vacuum under a very large electrode potential and is passed through a collimating system before it gets diffracted by the specimen to be studied. A schematic view of the electron diffraction system is shown in Fig. 10.54.

The energy E acquired by these electrons depends on the accelerating voltage V, approximately as

$$E = eV \tag{47}$$

where e is the charge on the electron. The eq. 47 is only an approximate because at the higher voltages (usual in electron diffraction) the electron velocity begins to approach the velocity of light, so that they gain mass due to effect of relativity. However, if we ignore the relativity correction, the wavelength associated with the moving electrons can be written as

$$\lambda = \frac{h}{\sqrt{2mE}} = \frac{h}{\sqrt{2meV}} \tag{48}$$

where $m = 9.11 \times 10^{-31}$ kg, the mass of an electron. Substituting the appropriate mass value for an electron, the eq. 48 reduces to

$$\lambda(\text{Å}) = \left(\frac{150}{V}\right)^{1/2} = \frac{12.24}{(V)^{1/2}} \tag{49}$$

From the above equations, it is clear that the wavelength is roughly inversely proportional to the square root of the accelerating voltage and hence can be easily varied to get the desired wavelength electron beam. Monochromatic beam is obtained simply by applying stabilized voltages to

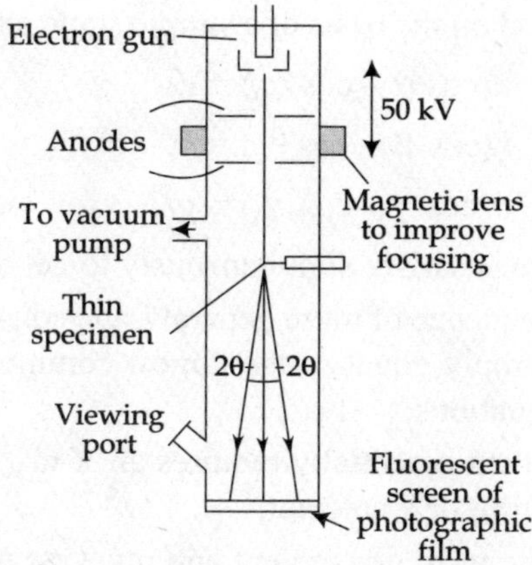

Fig. 10.54 Schematic view of the electron diffraction technique

accelerate the electrons. Calculations show that the scattering of electrons from an atom is about four orders of magnitude greater than the scattering of X-rays from the same atom. This suggest that the electron beam has a short stopping distance of about 500 Å or so and hence the specimen to be studied must be thin.

10.13 SUMMARY

1. X-rays like visible light, are a form of electromagnetic radiation having wavelengths of the order of 1 Å. X-rays are produced when a beam of fast moving electron strikes a metal target. For a given electrode potential, the wavelength of the resulting X-ray is

$$\lambda(\text{Å}) = \frac{1.24 \times 10^{-6}}{V} \text{ V-m}$$

2. X-rays may be either monochromatic (containing single wavelength also called characteristic X-ray) or continuous (containing all wavelengths also called white radiation).

3. The target material should be:

   (i) Good conductor of heat so that it can be efficiently cooled.

   (ii) Good conductor of electricity because it has to function as an anode.

   (iii) Metal which emit X-rays of wavelengths of the order of 1 Å suitable for the study of crystalline materials.

4. Each atom in a crystal acts as a diffraction center and the crystal as a whole acts like a three dimensional diffraction grating. The condition for diffraction is given by well known Bragg's law

$$2d \sin \theta = n\lambda$$

5. Laue equations are derived on the basis of a simple static atomic model. They are given by:

$$a(\cos \alpha - \cos \alpha_0) = e\lambda$$

$$b(\cos \beta - \cos \beta_0) = f\lambda$$

$$c(\cos \gamma - \cos \gamma_0) = g\lambda$$

All the three equations must satisfy simultaneously to get the most intense beam.

6. Bragg equation is a consequence of more general Laue equations. The order of diffraction n in Bragg's equation is simply equal to the largest common factor of the numbers e, f, g appearing in the Laue equations.

7. X-ray diffraction experiment essentially requires an X-ray source sample under investigation and a photographic film or a detector.

8. The powder method is the most convenient and quickest method for obtaining diffraction data. The Bragg angle θ and the distance between a pair of arcs are related through

$$\theta = \frac{S}{4}$$

9. Laue method is one of the most convenient and accurate methods for determining the orientation and symmetry of a single crystal. Gnomonic projection is extremely useful in the interpretation of the Laue photographs.

10. Rotation/Oscillation method is normally used to determine the lattice parameters of single crystals. For an a-axis rotation photograph, the separation of nth layer line (with respect to zero layer) and the lattice parameter a are given by

$$S_n = \frac{(n\lambda/a)}{\sqrt{1-(n\lambda/a)^2}} \cdot R$$

and

$$a = \frac{n\lambda}{S_n}(R_n^2 + S_n^2)^{1/2}$$

11. The Weissenberg method records the two dimensional reciprocal lattice points of a layer on a two dimensional film in the form of non-intersecting curves called festoons. This method is used to determine the lattice parameters accurately.

12. The precession method gives us an undistorted picture of a reciprocal lattice plane from which the lattice parameters (angles and distances) can be read off more or less directly.

13. X-ray diffractometers (Powder and single crystal) are based on radiation counters (proportional, scientillation or G.M) and the detector. Photographic diffraction work is supplemented in almost every X-ray laboratory by the X-ray diffractometer data.

14. Depending upon the energy of neutron from the pile of a reactor, the wavelength of the neutron beam is given by

$$\lambda(\text{Å}) = \frac{0.28}{E^{1/2}}$$

15. Ignoring the relativistic correction, the wavelength associated with the moving electron beam is given by

$$\lambda(\text{Å}) = \frac{12}{[E(eV)]^{1/2}}$$

10.14 DEFINITIONS

Absorption edge: An abrupt discontinuity in the intensity of an *X*-ray absorption spectrum at a particular wavelength.

Compton scattering: It is the elastic scattering of protons by electrons.

De Broglie Hypothesis: With every moving particle (such as electron or proton) a wave is associated whose wavelength is universely proportional to the momentum of the particle.

Ewald sphere: It is the sphere of radius $1/\lambda$ within which the diffracted beam and reciprocal lattice point intersect each other. It is also called sphere is reflection.

Linear Attenuation Coefficient: It is measure of the fractional fall in the intensity of *X*-rays per unit thickness of absorbing medium.

Mass Absorption Coefficient: It is the fractional decrease in the intensity of a beam of unit cross section in travelling unit mass of the absorbing medium.

Monochromatic radiation: It is the radiation restricted to a very narrow band of wavelength; ideally one wavelength.

Pair Production: Simultaneous production of an electron and a positron from high energy (> 1.022 *MeV*) photon.

Reciprocal lattice: A theoretical lattice associated with a crystal lattice. The reciprocal lattice points are obtained by drawing normals to each crystal plane from a common origin and lying at a distance to the reciprocal of interplanar spacing.

X-rays: Electromagnetic radiation of very short wavelength ($\sim 1\text{Å}$) lying between ultraviolet and γ-rays in the spectrum.

REVIEW QUESTIONS AND PROBLEMS

1. Explain the concept of reciprocal lattice. Derive the Bragg's condition in terms of the reciprocal lattice vectors.

2. Derive Bragg's law in *X*-ray diffraction. Give an account of powder method of crystal structure analysis.

3. What are Laue equations for diffraction of *X*-rays by a crystalline solid? Show that the Bragg's equation is a special case of the Laue equations.

4. Describe briefly the methods for crystal structure determination by *X*-ray diffraction. Explain the importance of geometrical structure factor taking the example of cubic crystals.

5. Find the geometrical structure factor for *fcc* structure in which all atoms are identical. Hence show that for *fcc* lattice no reflections can occur for which the indices are partly even and partly odd.

6. (a) What is the minimum wavelength in white radiation of X-ray if the applied voltage on the tube is 30 kV?

   (b) What is the wavelength associated with an electron of kinetic energy of 10 keV?

   (c) What is the wavelength associated with a neutron at 300 K ($K.E. = 1/2\ kT$)?

   Ans. 0.40 Å, 3.85 Å, 2.52 Å

7. Calculate the geometrical structure factor for a bcc lattice. Name some important planes which will be missing from the X-ray diffraction spectrum.

8. The K_α line from molybdenum has a wavelength 0.71 Å. Calculate the wavelength of K_α line of copper. The atomic number of molybdenum is 42 and that of copper is 29.

   Ans. 1.52 Å

9. What element has a K_α X-ray line of wavelength 0.71 Å?

   Ans. $Z = 42$ (molybdenum)

10. Electrons bombarding the anode of a Coolidge tube produce X-rays of wavelength 1Å. Find the energy of each electron at the moment of impact.

    Ans. 1.24×10^4 eV

11. The wavelength of the L_α (X-ray) line of platinum (atomic number 78) is 1.32 Å. An unknown substance emits L_α (X-ray) of wavelength 4.17 Å. Calculate the atomic number of the unknown substance. Given $b = 7.4$ for L_α.

    Ans. $Z = 47$ (silver)

12. The diamond structure is formed by the combination of two interpenetrating fcc sub-lattices; the basis being (000) and (¼ ¼ ¼). Find the structure factor of the basis and prove that if all indices are even the structure factor of the basis vanishes unless $h + k + l = 4n$, where n is an integer.

$$\textbf{Ans.}\ 1 + \exp\left[-\frac{i\pi}{2}(h + k + l)\right]$$

13. What is the attenuation of an X-ray beam after passing through a thickness of a material equal to four-half value thickness?

    Ans. 93.7%

14. The mass absorption coefficient of aluminium for X-rays of a certain energy is 0.027 m^2/kg. Calculate the half-value thickness of aluminium for these X-rays. What thickness of aluminium would alternate the X-rays by 80%? Density of aluminium is 2700 kg/m^3.

    Ans. 0.022 m

15. A beam of neutron with energies ranging from zero to several electron volts is directed at a crystal with interplanar spacing 3.03 Å. Determine the angle between the incident beam and the crystal so that the reflected neutrons will have a kinetic energy of 0.1 eV.

    Ans. 8°36′

16. Calculate the glancing angle on the plane (110) of a cube rock salt ($a = 2.81$ Å) corresponding to second order diffraction maximum for the X-rays of wavelength 0.71 Å.

    Ans. 20.9°

17. A beam of X-rays incident on a sodium chloride crystal ($a = 2.81$ Å). The first order reflection is observed at a glancing angle 8°35′. What is the wavelength of X-rays? At what angle should the second order Bragg's reflection occur?

   Ans. 0.842 Å, 17.4°

18. Sylvine (KCl) crystallizes in the form of simple cubic structure. Determine the interatomic spacing d and the glancing angle at which an X-ray of wavelength of 1.1787 Å is reflected in the third order. The density of KCl is 1990 kg/m^3 and the molecular weight is 74.6.

   Ans. 3.96 Å, 42.6°

19. A *bcc* crystal is used to measure the wavelength of some X-rays. The Bragg angle for reflection from (110) plane is 20.2°. What is the wavelength? The lattice parameter of the crystal is 3.15 Å.

   Ans. 1.54 Å

20. X-rays with a wavelength of 1.54Å are used to calculate the spacing of (200) plane in platinum. The Bragg angle for this reflection is 22.4°. What is the side of the unit cell of the aluminium crystal?

   Ans. 4.05Å

21. Using a diffractometer and a radiation of wavelength 1.54Å, only one reflection from an *fcc* crystal is observed when 2θ is 121°. What are the indices of reflection? What is the interplanar spacing?

   Ans. {111}, 0.885Å

22. The first three lines from the powder pattern of a cubic crystal have the following S values: 24.95, 40.9 and 48.05mm. The camera radius is 57.3 mm. Molybdenum K_α radiation of wavelength 0.71Å is used. Determine the structure and the lattice parameter of the material.

   Ans. DC, 5.66Å

23. The Bragg angle corresponding to a reflection for which $(h^2 + k^2 + l^2) = 8$ is found to be 14.35°. Determine the lattice parameter of the crystal. X-rays of wavelength 0.71Å are used. If there are two other reflections with smaller Bragg angles, what is the crystal structure?

   Ans. 4.05Å, *fcc*

24. Aluminium (*fcc*) has a lattice parameter of 4.05Å. When a monochromatic radiation of 1.79Å is used in a powder camera, what would be the first four S values? The camera diameter is 114.6 mm.

   Ans. 90.1, 105.0, 154.7 and 188.7 mm

25. The first reflection using copper K_α radiation from a sample of copper powder (*fcc*) has S value of 86.7 mm. Compute the camera radius.

   Ans. 57.3 mm

CHAPTER 11

FACTORS AFFECTING X-RAY INTENSITIES

11.1 INTRODUCTION

In the preceding chapter we discussed in detail the diffraction methods for measuring the X-ray intensities using photographic (film) and counter tube techniques. An X-ray diffraction pattern recorded on a film shows a considerable variation in the amount of blackening corresponding to different reflections. On the other hand, the X-ray intensities recorded from diffractometers are inherently more accurate than the film based methods. However, all these diffraction data obtained from various experimental methods are actually the raw data. Their intensities are affected by various geometrical and physical factors. They are subjected to a number of necessary corrections before they are used as an input for final crystal structure determination. The process of applying correction factors to get the corrected intensity is referred to as data reduction.

11.2 THE STRUCTURE FACTOR

We know that the directions of the diffracted beams given by Bragg's law or Laue's equations are governed entirely by the geometry and the size of the unit cell, and therefore these directions are not affected by the arrangement of atoms associated with each lattice point. However, the complexity of the atomic structure within the unit cell does affect the intensity of the diffracted beams. In other words, the intensity of the diffracted beams depends on atomic scattering factor and the position of each atom in the unit cell. Since X–rays are scattered by electrons of the atom only, therefore X-ray scattered from one part of an atom interfere with those scattered from other part at all angles of scattering (at $2\theta = 0$, all atoms scatter in phase). Further, as the electrons are distributed throughout the atomic volume, the diffracted beams are obtained as a result of combination of the scattered waves from the electrons of all atoms in the unit cell. This process involves two distinct contributions:

1. Scattering from electrons of the same atom (the atomic scattering factor or atomic form factor f).
2. The summation of this scattering from all atoms in the unit cell (the geometrical structure factor F).

The atomic scattering factor is a measure of the efficiency of an atom in scattering *X*-rays. It is defined as the ratio of the amplitude scattered by the actual electron distribution in an atom to that scattered by one electron localized at a point. Therefore, if the atoms are assumed to be points only, then the atomic scattering factor is first equal to the number of electrons present (i.e., atomic number of neutral atoms, Z). The atomic scattering factor of an atom falls off with increasing scattering angle or, more precisely with increasing value of (sin θ/λ) as shown in Fig. 11.1. On the other hand, neutrons are scattered by atomic nuclei rather than by electrons around a nucleus. Since, the nucleus is very small as compared to the size of the atom (it can be treated as a point atom), the scattering for a non-vibrating nucleus is almost independent of scattering angle.

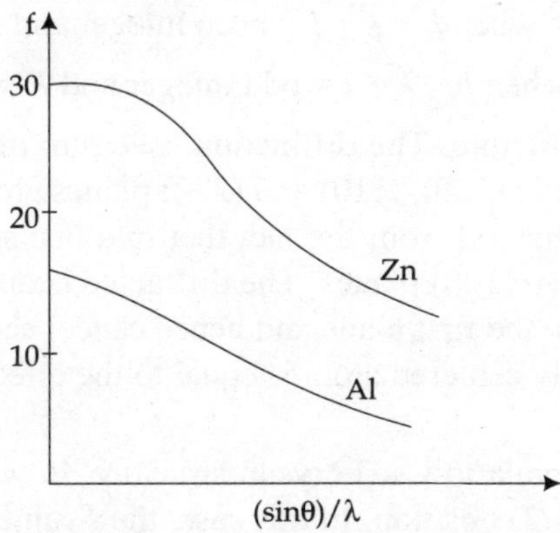

Fig. 11.1   Variation of atomic scattering factor with (sin θ/λ)

In order to know the intensity of the *X*-ray beam by one unit cell in a particular direction where there is a diffraction maximum, it is necessary to sum the waves that arise from all atoms in the unit cell. Mathematically, this involves adding the waves of the same wavelength but with different amplitudes and phases. The intensity of the diffracted beam is then obtained by squaring the resultant amplitude.

If the atomic positions in the unit cell are denoted by *x*, y and z the resultant amplitude for a given (*hkl*) reflection can be expressed as

$$F(hkl) = \sum_{j=1}^{N} f_j \exp\{2\pi i(hx_j + ky_j + lz_j)\} \tag{1}$$

or

$$F(hkl) = \sum_{j=1}^{N} f_j \cos 2\pi(hx_j + ky_j + lz_j) + i \sum_{j=1}^{N} f_j \sin 2\pi(hx_j + ky_j + lz_j) \tag{2}$$

where the coordinates *x*, *y*, *z* are expressed as fraction of the unit cell parameters. As mentioned above, the intensity of the diffracted beam is given by the square of the amplitude, i.e.

$$I \propto |F(hkl)|^2 \tag{3}$$

where

$$|F(hkl)|^2 = \left(\sum_{j=1}^{N} f_j \cos 2\pi(hx_j + ky_j + lz_j)\right)^2 + \left(\sum_{j=1}^{N} f_j \sin 2\pi(hx_j + ky_j + lz_j)\right)^2 \tag{4}$$

Let us take a practical example for better understanding. For simplicity, let us consider a bcc structure consisting of identical atoms in the unit cell. Position of the atoms in the unit cell are: (000) and (1/2 1/2 1/2). Here, for atom 1, $x_1 = y_1 = z_1 = 0$, for atoms 2 $x_2 = y_2 = z_2 = 1/2$ and $f_1 = f_2 = f$, then eq. 4 becomes

$$|F|^2 = f^2\left(\cos 2\pi(0) + \cos 2\pi\left(\frac{h}{2} + \frac{k}{2} + \frac{l}{2}\right)\right)^2 + f^2\left(\sin 2\pi(0) + \sin 2\pi\left(\frac{h}{2} + \frac{k}{2} + \frac{l}{2}\right)\right)^2$$

or
$$|F|^2 = f^2\left[\{1 + \cos \pi(h + k + l)\}^2 + \sin^2\pi(h + k + l)\right] \tag{5}$$

Also from eq. 3 and 5, one can verify that

$$I \propto |F|^2 = 4f^2 \text{ when } h + k + l = \text{even integer and } F = 2f$$
$$= 0 \text{ when } h + k + l = \text{odd integer and } F = 0$$

Metallic sodium has bcc structure. The diffraction spectrum of sodium does not show (100), (300), (111) or (221) planes, while (200), (110) and (222) planes are present. The absence of (100) and other planes could be understood from the fact that in a bcc structure, there exists a parallel plane at midway between the two (100) planes. The diffracted beam from these middle planes are out of phase by π with respect to the first plane and hence cancel each other. This is due to the fact that the effective number of body centered atom is equal to the effective number of corner atom in a *bcc* structure.

Let us apply the above formula to CsCl crystal structure, in which there is a Cs ion at (000) position and Cl ion at (1/2 1/2 1/2) position. In this case, the f values for the two ions are different and the lattice is not a body centered. The structure factor in this case yields

$$|F|^2 = \{f_{Cs} + f_{Cl} \cos \pi(h + k + l)\}^2 + \{f_{Cl} \sin \pi(h + k + l)\}^2$$

or
$$|F|^2 = (f_{Cs} + f_{Cl})^2 \text{ when } h + k + l = \text{even integer}$$
$$= (f_{Cs} - f_{Cl})^2 \text{ when } h + k + l = \text{odd integer}$$

Hence, the structure factor of CsCl structure is given as

$$F(hkl) = f_{Cs} + f_{Cl} \text{ when } h + k + l = \text{even integer}$$
$$= f_{Cs} - f_{Cl} \text{ when } h + k + l = \text{odd integer}$$

In a similar manner, with the help of above formulae the structure factors can be calculated for different *hkl* values of any given crystal.

11.3   THE LORENTZ FACTOR

We know that wherever a reciprocal lattice point intersects with sphere of reflection (Ewald sphere) a diffraction spot is produced. The lorentz factor is a measure of the amount of time that a reciprocal lattice point remains on sphere of reflection during the measuring process. Depending on the method (diffraction geometry) used to record the diffracted intensity and on the position of the different reciprocal lattice point, the time required for different reciprocal lattice point to pass through the sphere of reflection are different. In fact, the time that a reciprocal lattice point is in diffracting position depends on two factors: (i) The position of the reciprocal lattice point, and (ii) the velocity with which the reciprocal lattice point passes through the sphere of reflection.

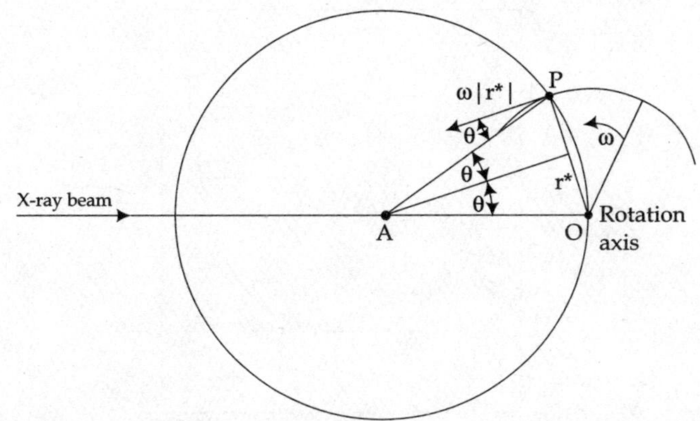

Fig. 11.2 Diffraction experiment for Lorentz correction

Let us first derive the form of Lorentz factor for some simple cases such as a zero-level rotation/ weissenberg photograph (or the equatorial reflections measured with a diffractometer) and then for a more complicated situation. Fig. 11.2 shows the Ewald sphere for a diffraction experiment in which the crystal is rotated about an axis normal to the plane defined by the incident and diffracted beams. Consequently, the reciprocal lattice is assumed to be rotated with a constant angular velocity ω; so the linear velocity of a reciprocal lattice point at a distance $d* = |1/r*|$ from the origin is

$$v = d*\omega$$

and its normal component (tangential at the point P) along the radius of the Ewald sphere is

$$v_n = d*\omega \cos\theta \tag{6}$$

The Lorentz factor can now be defined as

$$L = \frac{\omega}{v_n\lambda} \tag{7}$$

which is proportional to the time during which diffraction takes place for a given reciprocal lattice point. Writing $d*$ in terms of Bragg's law and then substituting the value of v_n in eq. 7 we have

$$L = \frac{\omega}{2\sin\theta\cos\theta}$$

or

$$L \propto \frac{1}{\sin 2\theta} \tag{8}$$

This is the simplest form of Lorentz factor. The variation of L with θ is shown in Fig. 11.3. At low Bragg angles, L is large because the reciprocal lattice point is close to the origin (of the reciprocal lattice) and at high θ, L is large because hkl point passes tangentially, or nearly so, through the surface of Ewald sphere. The Lorentz factor reduces the intensities by maximum amount at $\theta = 45°$.

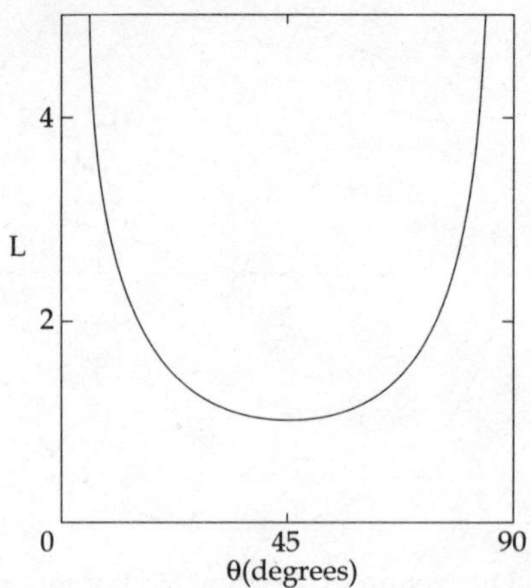

Fig. 11.3 Lorentz factor for normal-beam equatorial geometry, as a function of θ

The Lorentz factor for an equi-inclination setting Weissenberg case is given by

$$L = \frac{\sin \theta}{\sin 2\theta \sqrt{\sin^2 \theta - \sin^2 \mu}} \tag{9}$$

where μ is the inclination angle. For zero layer Weissenberg arrangement, $\mu = 0$ and eq. 9 reduces to eq. 8. The Lorentz factor for a more general case in which the axis of rotation makes an angle $90°$-μ with the incident X-ray beam and $90°$-v with the diffracted beam is given by

$$L = \frac{1}{\cos \mu \cos v \sin \gamma} \tag{10}$$

where γ is the projection on the zero layer of the Weissenberg photograph, the angle 2θ between the incident and diffracted beam or the angles μ, v, γ are the setting angles of the diffractometer. Equation 10 can be readily evaluated for each reflection as part of the computer program for reducing the intensity data to observed structure factor.

Equation 8 can be considered as a special case of the general formula in eq. 10 with $\mu = v = 90$ and $\gamma = 2\theta$. Similarly, the general formula in eq. 10 can be reduced to other simpler forms of setting in inclination geometry.

11.4 THE POLARIZATION FACTOR

We know that the characteristic radiation obtained from an X-ray tube is unpolorized and becomes partially polarized after diffraction from a crystal. The amount of polarization depends on the scattering angle of the diffracted beam. The orientation (the parallel and perpendicular components) to the electric vector E of the incident X-ray beam and the direction of the diffraction (scattered) beam is schematically shown in Fig. 11.4. The amplitude of each diffracted component is proportional to $\sin \phi_0$, where ϕ_0 is $90°$-2θ for the parrallel component and $90°$ for perpendicular component. Thus the amplitudes are in the ratio $\cos^2 2\theta : 1$. The mean intensity for the two states of polarization is defined as the polarization factor, which is expressed as

$$p = \frac{1 + \cos^2 2\theta}{2} \qquad (10)$$

For $\theta = 0$ or $90°$, $p = 1$ and the intensities of the parallel and perpendicular components are equal, meaning thereby that the diffracted beam remains unpolarized. However, for $\theta = 45°$, $p = \frac{1}{2}$ and the intensity of parallel component is zero, so that the diffracted beam is completely polarized. For all other values of θ, the diffracted beam is partially polarized as shown in Fig. 11.5.

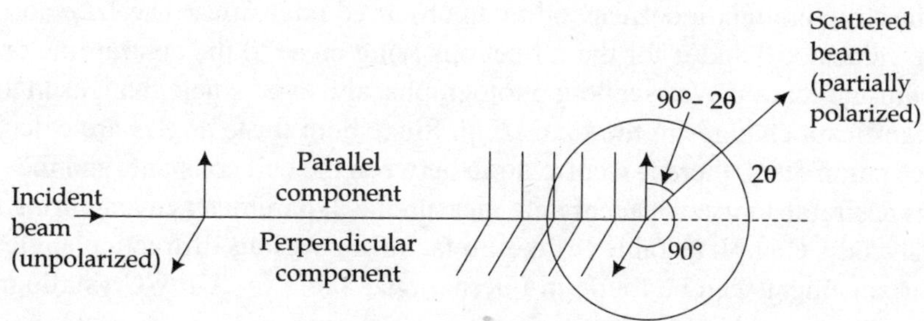

Fig. 11.4   Derivation of polarization factor for unpolarized incident beam

The polarization factor is common to all *X*-ray diffraction method except when a crystal monochromatized primary beam is used. However, in such cases the primary beam itself becomes partially, polarized, which should be taken into account as it will affect the intensity of reflection from the crystals under study.

We discussed above the Lorentz factor and the polarization factor separately. However, since both Lorentz and polarization factors depend on θ, it is a common practice to combine them into a single expression, called the Lorentz and polarization factor or simply the *Lp* factor to obtain the final structure factor, $|F|$ which is proportional to $\sqrt{I}$. Applying the Lorentz and polarization factor together, a more general form of the structure factor can be written as

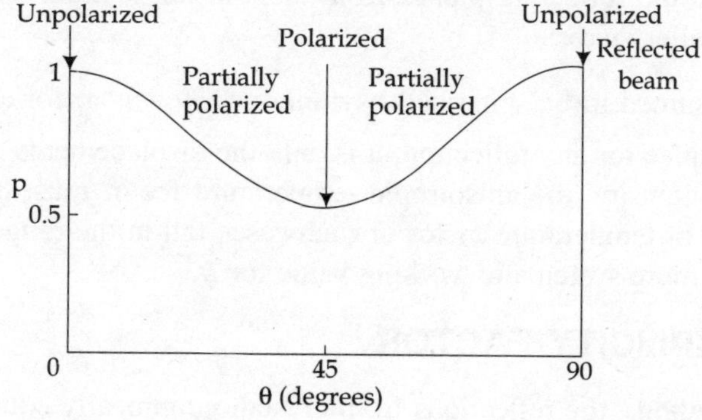

Fig. 11.5   Polarization factor for unpolarized in incident beam, as a function of θ

$$|Fhkl| = \sqrt{\frac{KI_{hkl}}{Lp}} \tag{11}$$

where L is the Lorentz factor, p is the polarization factor and K is a term depends on crystal size, beam intensity, and a number of fundamental constants for a given set of measurements. However, for data reduction calculations, instead of $|F_{hkl}|$, $|F_{\text{relative}}|$'s are estimated which is defined as

$$|F_{\text{rel}}| = K'|F_0| = \sqrt{\frac{I_{hkl}}{Lp}} \tag{12}$$

where $|F_0|$ is the structure factor corresponding to observed intensities. The $1/Lp$ correction is quite sensitive to the values of θ and μ for the reflections lying close to the central line or outer edge of zero-level or equi-inclination Weissenberg photographs, and even small changes in these quantities can produce significant changes in the final $|F_{\text{rel}}|$. Since both these angles are calculated from the measured lattice parameters, there is a correlation between the cell constants and the intensities. For this reason, it is desirable to use the accurately measured cell parameters even during the early stage of the structural study. Complete Tables of the Lp factor for various diffraction angles and different experimental arrangements can be found in International Table for X-ray Crystallography.

11.5 THE TEMPERATURE FACTOR

Normally the atomic scattering factors are derived on the basis of electron distributions of atoms at rest. However, at any temperature $T > 0K$, the atoms in crystals are always vibrating about their mean positions and the magnitude of the vibration depends upon the temperature, mass of the atom and the nature of interatomic forces binding them. In general, the higher the temperature, the greater the vibration. This causes the atoms to occupy a larger volume and make the scattering power of the real atom to fall rapidly than they would at rest. It has been shown theoretically (Debye, 1914) that the relation between the scattering factor f of a vibrating atom to that f_0 of an atom at rest is

$$f = f_0 \exp\left(-B \sin^2\theta/\lambda^2\right) \tag{13}$$

where $B = 8\pi^2 \overline{u_\perp^2}$ is known as temperature factor or temperature coefficient, $\overline{u_\perp^2}$ is the mean square displacement normal to the reflecting planes from their mean position and $\exp\left(-B \sin^2\theta/\lambda^2\right)$ is known as the Debye-waller factor.

$\overline{u^2}$ is generally assumed to be spherically symmetric. However, for an anisotropic crystal it varies with direction. Since for any reflection, it is only the displacements normal to the reflecting planes that affect the intensity, the anisotropic temperature factor must be a function of Miller indices. Typical values of temperature factor in such cases fall in the range 2-3 Å^2. However, the Wilson plot provides a more systematic working value for B.

11.6 THE MULTIPLICITY FACTOR

In X-ray diffraction methods, the reflections from crystallographically equivalent planes (in some cases such as powder method or rotation method even non-equivalent planes also) are found to superimpose. The number of planes having different orientations in the crystal but with the same *d*-spacing and intensity values, is known as the "multiplicity factor". The value of the multiplicity factor depends upon the indices of the reflecting planes *hkl* and the crystal symmetry. The multiplicity factor is evaluated by finding the maximum possible number of (*hkl*) combinations which are equivalent, taking both positive and negative values of *h*, *k*, and *l*. For example, in a cubic system, the multiplicity factor is 6, corresponding to the family of equivalent planes {100}, 12 corresponding to {110} and 8 corresponding to {111}, respectively. The highest multiplicity factor in this crystal system is 48 corresponding to a set of equivalent planes with $h \neq k \neq l \neq 0$.

For tetragonal crystals, only *h* and *k* can be interchanged as $a = b \neq c$ but negative and positive permutations are still possible, the highest multiplicity is 16. Similarly, for orthorhombic crystals where *h*, *k* and *l* cannot be interchanged as $a \neq b \neq c$ but negative and positive permutations are still possible, the highest multiplicity is only 8. In a similar manner, the multiplicity factor can be evaluated for each and every crystal class (point group) and is provided in Table 5.1 under the heading order of the group.

11.7   THE ABSORPTION FACTOR

When *X*-rays of intensity *I* are passed through a material of thickness *dx* their intensity is attenuated by an amount *dI* (Fig. 11.6). The decrease *dI*, is proportional to the original intensity *I* and thickness *dx*.

Consequently,

$$-dI \propto I dx \text{ or } -dI = \mu I dx$$

or
$$\frac{dI}{I} = -\mu\, dx \tag{14}$$

where μ is the linear absorption coefficient. Now, integrating eq. 14 and applying the boundary condition that $I = I_0$ at $x = 0$, we obtain

$$I = I_0 e^{-\mu x} \tag{15}$$

Since the amount of reduction in the intensities of incident and diffracted beams will depend on their path lengths through the crystal, each reflection is affected differently by absorption. Thus in reality, the calculation for each reflection must be carried out separately. However, alternatively let us consider a small volume element *dV* of a crystal scattering the incident *X*-ray beam into the *hkl* reflection (Fig. 11.7) where t_1 is the path length of the incident beam inside the crystal before being scattered and t_2 is the path length of the diffracted beam. The amount by which the intensity of the hkl reflection is reduced due to such absorption is given by

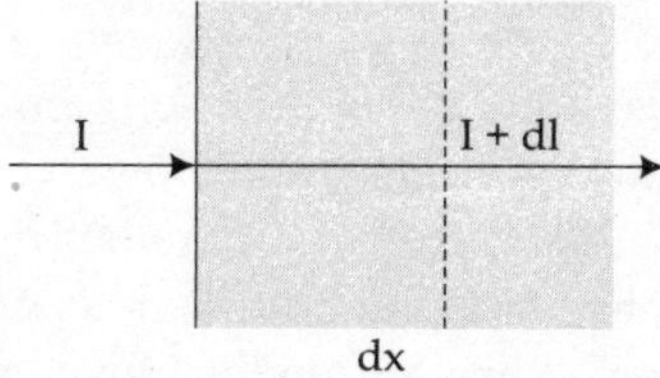

Fig. 11.6   The attenuation of an *X*-ray bream after passing through a small distance in a material

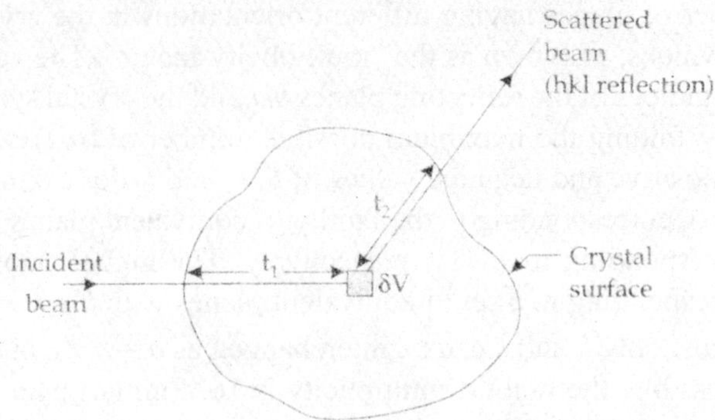

Fig. 11.7 Scattering from volume element δV of crystal

$$A_{hkl} = \frac{\int_V e^{-\mu(t_1 + t_2)}}{\int_V dV} = \frac{1}{V} \int_V e^{-\mu(t_1 + t_2)} dV \tag{16}$$

where V is the volume of the crystal. The reciprocal of A_{hkl} is the absorption factor, $A^{*}_{hkl} = \dfrac{1}{A_{hkl}}$, where A^{*} represents the factor by which the observed intensity is multiplied to obtain the corrected intensity. Thus the absorption factor A^{*}_{hkl} is given by

$$A^{*}_{hkl} = \frac{V}{\int_V e^{-\mu(t_1 + t_2)} dv} \tag{17}$$

The absorption factor A^{*} calculated on the basis of eq. 17 is extremely tedious and often the most serious source of error in X-ray work. This is because of the fact that A^{*} cannot be evaluated correctly except for some simple geometrical shapes or the materials having low linear coefficient of absorption μ. For these reasons, the practical approach is to minimise the effect of absorption by shaping the crystals to cylindrical or spherical form or to use more penetrating radiation, e.g., $MoK\alpha$, instead of carrying out the calculations. On the other hand, in neutron diffraction, the μ values are normally much smaller and hence the application of absorption correction is relatively easier.

11.8 EXTINCTION

In the preceding section, we considered the effect of absorption is to attenuate the intensity of the incident X-ray beam when it passes through a material. However, soon after the discovery of X-ray diffraction, a phenomenon usually called " extinction" rather than attenuation was realized. This pertains to the attenuation in the intensity of strong reflections due to an interference effect caused by secondary reflections of the diffracted beam from the lower crystal planes. Darwin, who first of all investigated this effect theoretically and divided it into two categories, primary extinction and secondary extinction.

The primary extinction can be easily understood with the help of Fig. 11.8. At the Bragg angle position every incident beam is observed to suffer multiple reflections from different crystal planes. It is observed that after an even number of reflections, the beam will be directed towards the primary incident beam. On the other hand, after an odd number of reflections beam will be directed towards the diffracted beam. Since each reflection introduces a phase change of $\pi/2$, the doubly reflected beam will differ in phase by π from the primary incident beam. For this reason, the doubly reflected beam, instead of strengthening the primary beam, actually weaken it by interfering destructively. All the primary beam penetrates deeper and deeper, it continuously gets attenuated in this manner. The same consideration holds for odd reflected beams which are directed towards the diffracted beam. The net result is that both incident and diffracted beams are attenuated considerably. This causes the diffracted intensity to be proportional to $|F|$ rather than $|F^2|$. A crystal for which this is true is termed as an "ideally perfect" crystal. However, we know that the crystals grown in the laboratory do contain certain degree of imperfections.

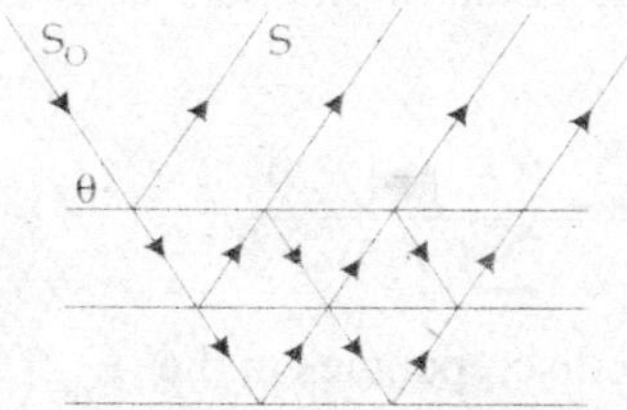

Fig. 11.8   Multiple reflections from a family of lattice planes

In order to explain extinction theoretically in crystals containing imperfections, Darwin considered an over simplified situation of an imperfect crystals as made up of disoriented mosaic blocks (Fig. 11.9). Primary extinction is said to be present when the mosaic blocks are too large. However, it is negligibly small when the mosaic blocks are very small or when the structure factor $|F|$ is very small.

A different situation is found to exist (even if the mosaic blocks are so small that there is no primary extinction) when there is no enough disorientation in the mosaic blocks. In such a case, the small perfect regions (the mosaic blocks) reflect as units and a significant fraction of the intensity is diffracted by upper mosaic blocks, so that the lower blocks receive less intensity where it decreases like an ordinary absorption. As a result, the lower blocks diffract less amount of intensity than they

Fig. 11.9   Schematic illustration of an ideally imperfect crystal made up of small disoriented mosaic blocks

are actually capable of. Therefore, the reduction in the intensity of the primary beam due to reflection from previous blocks was called by Darwin as secondary extinction. It is negligibly small when the disorientation of the mosaic blocks becomes sufficiently large or if the reflection is very weak. On the other hand, it is more pronounced for strong reflections at low $(\sin \theta)/\lambda$ values in sufficiently perfect crystals.

Various methods have been proposed for extinction correction. However, in general they are based on over simplified concept of clear cut separation into primary and secondary extinctions, a satisfactory and practical method for extinction correction has been given by W.H. Zachariasen (1967).

11.9 THE *R*-FACTOR

In order to check the correctness of a proposed or trial structure, the values of observed structure factor $|F_o|$ and calculated structure factor $|F_c|$ are compared in various ways. However, the most widely used assessment made by the crystallographers is to know a quantity called the residual factor or *R*-factor, defined as

$$R = \frac{\sum |(|F_o| - |F_c|)|}{\sum (|F_o|)} \tag{18}$$

where the numerator is the sum of all discrepancies in the $|F_{hkl}|$'s and the denominator is the sum of all observed $|F_{hkl}|$'s. When the value of R is comparatively a small fraction, the proposed structure is treated to be correct, and when it is a large fraction the structure is supposed to be incorrect.

11.10 SUMMARY

1. *X*-ray diffraction intensities are affected by various geometrical and physical factors.
2. Scattering of *X*-rays from electrons of the same atom gives rise to the atomic scattering factor (or form factor). On the other hand, the summation of this scattering from all atoms in the unit cell gives rise to the geometrical structure factor.
3. Lorentz factor is a measure of time during which the reciprocal lattice point remains on the sphere of reflection. It depends on the diffraction geometry used to record the diffracted beam.

   The simplest form of the Lorentz factor is given by

   $$L \propto \frac{1}{\sin 2\theta}$$

   For an equi-inclination Weissenberg setting,

   $$L = \frac{\sin \theta}{\sin 2\theta \sqrt{\sin^2 \theta - \sin^2 \mu}}$$

4. The characteristic radiation obtained from an *X*-ray tube is unpolarized and becomes partially polarized after diffraction from a crystal. The general form of the polarization factor is expressed as

$$p = \frac{1 + \cos^2 2\theta}{2}$$

$$= 1 \text{ for } \theta = 0 \text{ or } 90° \text{ (unpolarized)}$$

$$= 1/2 \text{ for } \theta = 45° \text{ (polarized)}$$

5. Rise of temperature causes the scattering power of a real atom to fall more rapidly than they would at rest. The relation between scattering factor *f* of a vibrating atom to that f_o of an atom at rest is

$$f = f_o \exp\left(-B \sin^2\theta/\lambda\right)$$

where *B* is known as temperature factor and the exponential term is known as Deby-Waller factor.

6. In *X*-ray diffraction methods, the reflections from crystallographically equivalent planes (in some cases such as powder method or rotation method even non-equivalent planes also) are found to superimpose and give rise to multiplicity factor. Its value depends on the indices of reflecting planes and the crystal symmetry, maximum value is 48 for cubic system.

7. When *X*-rays pass through any crystal material, the reduction in the intensity due to absorption is

$$I = I_o \, e^{-\mu x}$$

On the other hand, the absorption factor may be defined as a reciprocal of the amount by which the intensity of the hkl reflection is reduced due to absorption of incident and diffracted beams by a small elemental volume in the crystal.

$$A^*(hkl) = \frac{1}{A(hkl)} = \frac{V}{\int\limits_V e^{-\mu(t_1 + t_2)} dV}$$

8. Extinction is nothing but the attenuation in the intensity of strong reflections due to an interference effect caused by secondary reflections of the diffracted beam from the lower crystal planes. Darwin divided it into primary and secondary extinctions.

11.11   DEFINITIONS

Absorption factor: It is defined as the reciprocal of the amount by which the intensity of hkl reflection is reduced due to absorption of incident and diffracted beams by a small elemental volume in the crystal.

Data reduction: It is the process of applying various correction factors on the raw intensity data (experimental data are affected by a number of geometrical and physical factors) and make them usable as an input for crystal structure determination.

Extinction: It is the attenuation in the intensity of strong reflections due to an interference effect caused by secondary reflections of the diffracted beam from the lower crystal planes.

Lorentz factor: It is a measure of the amount of time during which the reciprocal lattice point remains on the sphere of reflection (Ewald sphere). The value depends on the scattering and diffraction geometry used to record the reflections.

Multiplicity factor: It is the result of superimposition of X-ray reflections of crystallographically equivalent planes (sometimes non-equivalent planes also) in X-ray diffraction experiments. Its values depends on the indices of reflecting planes and the crystal symmetry.

Polarization factor: It is a factor that takes into account the reduction in intensity on X-ray scattering due to the state of polarization of the incident beam.

Primary extinction: It relates to an interference process which reduces the intensity of beam as it passes through a crystal.

R-factor: An index that gives a crude measure of the correctness of a structure and the quality of the data. Its values between 0.06–0.02 are considered good.

Scattering factor: It is the scattering power of an atom for a given reflection and is expressed as the scattering power of an equivalent number of electrons located at the position of the atomic nucleus. The atomic scattering factor therefore may be defined as the ratio of the amplitude of the coherent scattered radiation from an atom to that from a single electron situated at the atomic center.

Secondary extinction: It arises when an appreciable amount of incident radiation is reflected by the first planes encountered by the beam. The deeper planes thus receive less incident intensity and therefore reflect less power than they are actually capable of. Therefore, the reduction in the intensity of the primary beam due to reflection from previous planes is known as secondary extinction.

REVIEW QUESTIONS AND PROBLEMS

1. What are the possible factors affecting X-ray intensities during data collection? Briefly discuses them.

2. What do you understand by the term "atomic scattering factor" and geometrical structure factor"? Obtain intensity expressions for *bcc* and *fcc* structures. Explain the absence of some Miller planes in these structures.

3. What do you understand by the term "Lorentz factor"? Obtain Lorentz factor for different diffraction geometry. Discuss its dependence on diffraction angle.

4. What do you understand and the term "Polarization factor"? Discuss the polarization factor in detail.

5. Explain why the Lorentz and polarization correction factors are taken together to obtain final structure factor. Discuss the case of Lp correction in detail.

6. Discuss the importance of temperature factor in structure factor calculation in detail.

7. What do you understand by the term " Multiplicity factor"? Explain the term by taking some specific cases.

8. Discuss in detail the role of absorption factor in the X-ray intensities.
9. What do you understand by the term "extinction"? Discuss its types, effects and remedies.
10. Discuss the role and importance of R-factor in crystal structure determination.

CHAPTER 12

STRUCTURE FACTORS AND FOURIER SYNTHESIS

12.1 INTRODUCTION

We shall start this chapter with the fundamental aspects of waves and their representations to understand the composition and representation of the structure factor in a better way. The structure amplitude and the phase associated with the diffracted waves (particularly in X-ray diffraction) will be attempted to understand with the help of their optical analogues.

Without going in to the mathematical details of Fourier series, we shall make use of its application in X-ray diffraction by crystals. We shall try to understand the process of Fourier synthesis of electron density in terms of structure factors and vice-versa. We shall also discuss the utility of modifying the structure factors on the basis of symmetry and lattice centering.

12.2 WAVES MOTION

Let us consider a simple harmonic wave which can be described in terms of a point P moving on the circumference of a circle at constant angular velocity. Its projection on the x-axis, the point M executes simple harmonic motion w.r.t. its mean position as shown in Fig. 12.1. This figure also shows the linear displacement of M as a function of angular displacement ϕ of the radius vector F. The maximum displacement equal to the magnitude of F is defined as the amplitude of the wave, $|F|$.

The point N (the projection of the point P on the y-axis) also executes simple harmonic motion but is 90° out of phase with M, because of the orthogonality of the axes. It is to be noted that at any time t (where the time is counted from the instant that the point P is at X),

$$OM = A = |F| \cos \phi = |F| \cos \omega t$$

and
$$ON = B = |F| \sin \phi = |F| \sin \omega t$$

Therefore, for orthogonal axes (Fig. 12.2)

$$|F| = (A^2 + B^2)^{1/2} \tag{1}$$

and
$$\tan \phi = \frac{B}{A} \tag{2}$$

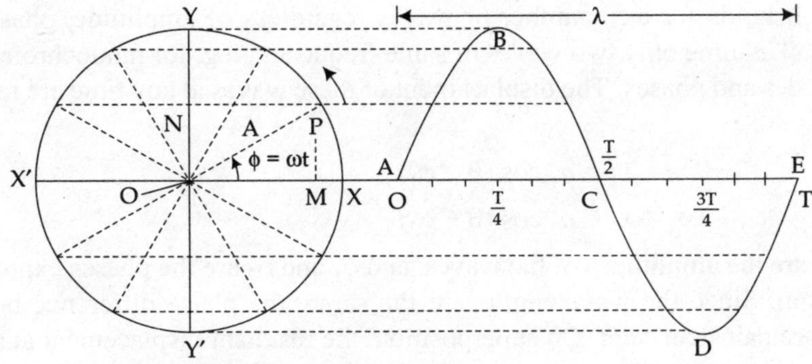

Fig. 12.1 Representation of simple harmonic wave

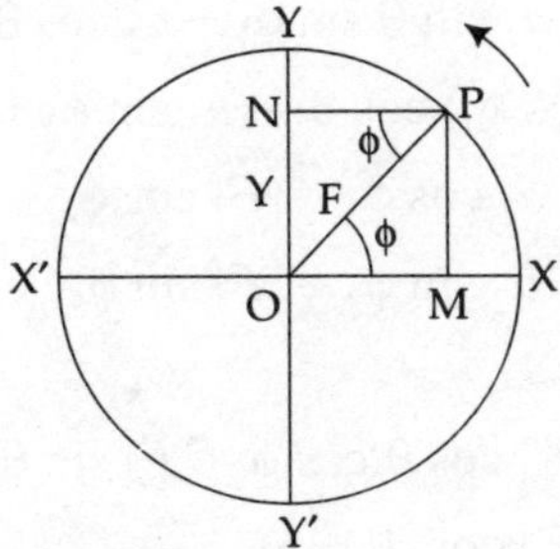

Fig. 12.2 Amplitude and phase for an orthogonal system

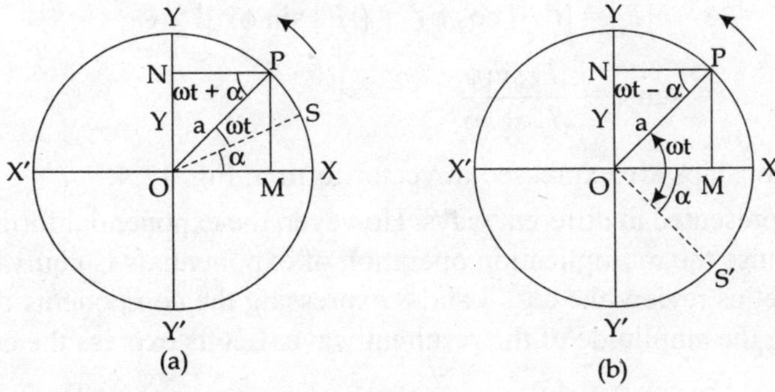

Fig. 12.3 Representation of phase angles

Here, the angle $\phi = \omega t$ is termed as the phase of the wave. However, if the time is counted from the instant that the point P is at S or S' as shown in Fig. 12.3, then the corresponding phase angles are $(\omega t + \alpha)$ or $(\omega t - \alpha)$ where α is known as epoch.

12.3 SUPERPOSITION OF WAVES

The Principle of superposition of waves states that the amplitude of the resultant wave in any direction is the algebric sun of individual waves scattered in that direction, taking into account their relative phases. It holds for any number of waves regardless of amplitude, phase or frequency. However, we shall assume only two waves of same frequency (e.g. for monochromatic X-rays) but different amplitudes and phases. The displacement of these waves at any time are represented algebraically as

$$x_1 = a_1 \cos(\theta + \phi_1) \tag{3}$$

and
$$x_2 = a_2 \cos(\theta + \phi_2) \tag{4}$$

where a_1 and a_2 are the amplitudes of the waves, and ϕ_1 and ϕ_2 are the phases expressed relative to an arbitrary origin. Since the wavelengths are the same, the phase difference between the two waves, $(\phi_1 - \phi_2)$ remains constant. On superposition, the resultant displacement at any time is

$$x_r = x_1 + x_2 = a_1 \cos(\theta + \phi_1) + a_2 \cos(\theta + \phi_2) \tag{5}$$

After expansion, eq. 5 can be written as

$$x_r = a_1 \cos\theta \cos\phi_1 - a_1 \sin\theta \sin\phi_1 + a_2 \cos\theta \cos\phi_2 - a_2 \sin\theta \sin\phi_2$$

or
$$x_r = (a_1 \cos\phi_1 + a_2 \cos\phi_2)\cos\theta - (a_1 \sin\phi_1 + a_2 \sin\phi_2)\sin\theta \tag{6}$$

If we put
$$a_1 \cos\phi_1 + a_2 \cos\phi_2 = |F| \cos\phi \tag{7}$$

and
$$a_1 \sin\phi_1 + a_2 \sin\phi_2 = |F| \sin\phi \tag{8}$$

in eq. 6, then

$$x_r = |F| \cos\theta \cos\phi - |F| \sin\theta \sin\phi$$
$$= |F| \cos(\theta + \phi) \tag{9}$$

Thus the resultant of two waves is a wave of the same frequency. The amplitude of the resultant wave is given by

$$|F| = [(|F| \cos\phi)^2 + (|F| \sin\phi)^2]^{1/2} \tag{10}$$

and
$$\tan\phi = \frac{|F| \sin\phi}{|F| \cos\phi} \tag{11}$$

The above relationships are expressed in vector form in Fig. 12.4.

Waves can be represented in different ways. However, the exponential form of representation is easier to handle because the multiplication operation of exponentials is equivalent to simple addition of exponents. Let us review the eqs. 7 and 8 expressing the components of the resultant wave and eq.10 expressing the amplitude of the resultant wave. Let us express the components as

$$|F| \cos\phi = A$$

and
$$|F| \sin\phi = B$$

Hence, the amplitude of the resulting wave can be written as

$$|F| = [FF^*]^{1/2} = (A^2 + B^2)^{1/2}$$
$$= [(A + iB)(A - iB)]^{1/2} \tag{12}$$

where $A + iB$ and $A - iB$ are the complex conjugate of each other, both A and B are real numbers, but iB is an imaginary number because $i = \sqrt{-1}$ is an imaginary number.

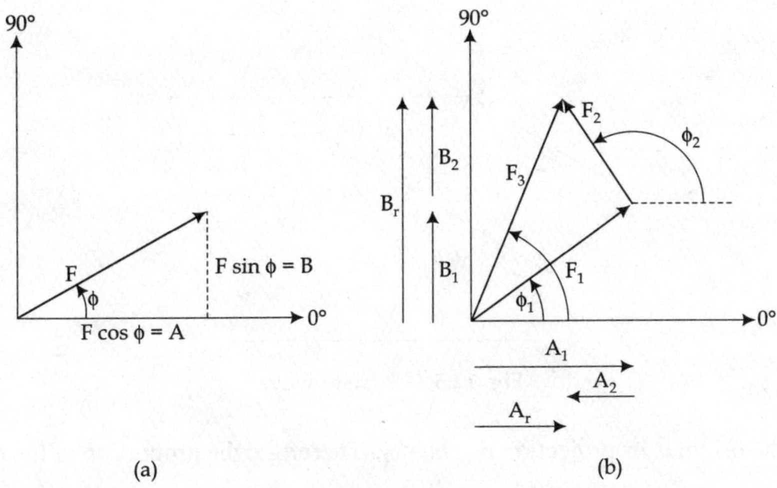

Fig. 12.4 Amplitudes and phases in vector representation

It is to be noted that multiplying twice by i is equivalent to multiplication by -1. In turn multiplying by -1 corresponds to a change of direction on a straight line (i.e. equivalent to a rotation of 180°) which can turn a vector F into $-F$. Hence, multiplying once by i may be taken as equivalent to a rotation of 90°. Thus in an orthogonal system of axes, a real number can be taken as a point say on the horizontal axis and an imaginary number as a point say on the vertical axis. Accordingly, the complex number $A + iB$ is represented as projection of its real and imaginary parts on respective axes as shown in Fig. 12.4a. Similarly, Fig. 12.4b illustrates the projections of real and imaginary parts of the wave vector on x and y axes. Based on such similarities, the wave vector can be represented as a complex number.

$$F = A + iB$$
$$= |F| \cos \phi + i|F| \sin \phi$$
$$= |F| (\cos \phi + i \sin \phi)$$

or
$$F = |F| . \exp (i\phi) \tag{13}$$

where the amplitude $|F|$ and the phase ϕ have been represented by one symbol F, a complex number.

12.4 PHASE OF A WAVE IN THREE DIMENSIONS (UNIT CELL)

In the preceding sections, we discussed the wave motion and related parameters for one dimensional system. Making use of the basic concepts, let us obtain the phase of a wave scattered by a point in a crystal which can be assumed to be made up of a large number of equidistant parallel (hkl) planes as shown in Fig. 12.5. Here, the first order diffraction from a plane (hkl) with interplanar spacing d_{hkl} will correspond to a phase difference of 2π. Further, the phase scattered by any irrational plane lying between the origin and the first rational plane (hkl) is directly proportional to the distance of the irrational plane from the origin. Let a point p_{xyz} lies on irrational plane whose

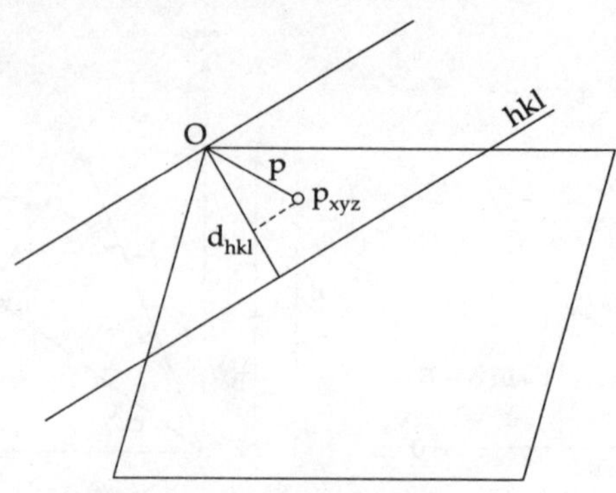

Fig. 12.5 Phase of wave

distance from the origin is its projection p_{xyz} on d_{hkl}. Therefore, the proportion of the two phases can be written as

$$\frac{\phi}{2\pi} = \frac{p_{xyz}}{d_{hkl}} \tag{14}$$

where $1/(d_{hkl}) = \sigma_{hkl}$ is the length of the reciprocal lattice vector, eq. 14 can be written as

$$\frac{\phi}{2\pi} = \sigma_{hkl} \cdot p_{xyz}$$

$$= (ha^* + kb^* + lc^*) \cdot (xa + yb + zc)$$

$$= (hx + ky + lz)$$

or $$\phi = 2\pi(hx + ky + lz) \tag{15}$$

This is the phase of the wave scattered by a point atom on (hkl) plane in the unit cell. In a similar manner, the phase of the wave scattered by j^{th} atom on (hkl) plane in the unit cell is given by

$$\phi_j = 2\pi(hx_j + ky_j + lz_j) \tag{16}$$

12.5 THE STRUCTURE FACTOR OF A CRYSTAL

Generallizing eq. 13 for X-ray diffraction from a unit cell of a crystal, the complex member in terms of three indices hkl for each reflection can be expressed as

$$F(hkl) = |F(hkl)| . \exp[i\phi(hkl)] \tag{17}$$

where $|F(hkl)|$ is called the amplitude of the structure factor of reflection with the indices h, k and l and ϕ is the phase angle. Substituting the value of ϕ from eq. 16 into 17, we obtain

$$F(hkl) = |F(hkl)| . \exp 2\pi i \, (hx_j + ky_j + lz_j)$$

$$= A(hkl) + iB(hkl) \tag{18}$$

where

$$A(hkl) = \sum_{j=1}^{N} f_j \cos 2\pi(hx_j + ky_j + lz_j) \tag{19}$$

$$B(hkl) = \sum_{j=1}^{N} f_j \sin 2\pi(hx_j + ky_j + lz_j) \tag{20}$$

and the phase angle $\phi(hkl)$ associated with each structure factor is given by

$$\tan \phi(hkl) = \frac{B(hkl)}{A(hkl)} \tag{21}$$

In an ordinary optical or electron microscope, the radiation scattered by the object (to be studied) can be recombined by the lens system to give a magnified image of the scatterer. In such a system, the flow of scattered radiation continues beyond the lens through the specimen and therefore the relationships between phases of the waves are maintained when they are recombined by the lens.

However, in X-ray (or neutron) diffraction, the recombination of the scattered waves is not possible due to absence of such a lens system. Further, in practice the scattered radiation is intercepted by either a photographic plate/camera or a detecting system. Therefore, the relationships between the phases of the scattered waves are lost. Only the intensities (not the phases) of the diffracted waves can be measured. However, it is possible to simulate the recombination of scattered X-rays with the help of Fourier synthesis.

12.6 THE FOURIER SYNTHESIS

It is well known that a periodic function (within reasonable limitations) can be represented by an appropriate sum of cosine and sine terms known as Fourier series. Since crystals are three dimensionally periodic and contains periodic distribution of electrons (scattering material particles), a three dimensional Fourier summation ("synthesis") can be made to obtain the image of the electron distribution analogous to a lens in optical system.

Let us begin with the structure factor expression in terms of fractional coordinates (x_j, y_j, z_j) of jth atoms on the (hkl) plane of a unit cell, i.e.

$$F(hkl) = \sum_{j=1}^{N} f_j \exp[2\pi i(hx + ky + lz] \tag{22}$$

where f_j is the atomic scattering factor related to electron density function ρdV. From eq. 22, it is clear that $F(hkl)$ is expressed in term of Fourier series, convensionally known as forward Fourier transform (FT). It also provides us the knowledge of arrangement of (N) atoms in the unit cell and resulting amplitude and phase of scattered waves.

Now as far as the X-ray diffraction is concerned, a crystal is just a periodic distribution of electrons in three dimensions. Consequently, the electron density $\rho(xyz)$ at any point in the crystal can be represented by three dimensional Fourier summation (synthesis), known as reverse Fourier transform (FT^{-1}),

$$\rho(xyz) = \frac{1}{V} \sum_{h,k,l} F(hkl) \exp[-2\pi i(hx + ky + lz)] \tag{23}$$

where $F(hkl)$ is a complex number containing both amplitude and phase as seen before. The term $1/V$ is necessary in order to give the correct units (structure factors like atomic scattering factors, have units of electrons, but electron density is electrons per $Å^3$). The summation is performed over all values of h, k and l. Combining eqs. 17 with 23 we get,

$$\rho(xyz) = \frac{1}{V} \sum_{h,k,l} |F(hkl). \exp[i\phi(hkl)]. \exp[-2\pi i(hx + ky + lz)] \tag{24}$$

Eq. 24 can be understood as the image of the electron density which is responsible for the diffraction pattern, is obtained by adding together all diffracted beams (for all hkl), with their correct amplitudes and phases.

From eq. 23 (or 24), we can see that if we know the amplitude and phases of the scattered waves (in reciprocal space), we can calculate the electron density $\rho(xyz)$ at any point xyz in the unit cell (in real space). If this is done at sufficiently large number of points in the unit cell, an electron density map can be constructed and by identifying the highest density regions, atoms can be located.

A comparison of eq. 22 with eq. 23 (or 24) suggests that they differ in sign of their exponent. It is a mathematical consequence of the fact that they are Fourier transform of each other. They also differ in the sense that the electron density can be sampled at any point xyz in real space (unit cell), while the Fourier transform of the unit cell content can be sampled only in reciprocal space, namely the reciprocal lattice points. In other words, the Fourier transform of the cell content is only directly observable if Bragg's law is satisfied.

In section 5.9, we have seen that the X-ray diffraction introduces a center of symmetry even if the crystal is non-centrosymmetric. Accordingly, let us consider the contributions to the summation on the right side of eq. 24 by expending it in terms sine and cosine for the indices (hkl) and $(\bar{h}\,\bar{k}\,\bar{l})$ and taking into account the Friedel's law. This will be

$$F_{hkl} \exp[-2\pi i(hx + ky + lz)] + F_{\bar{h}\,\bar{k}\,\bar{l}} \exp[-2\pi i(hx + ky + lz)]$$
$$\underset{I}{} \qquad\qquad\qquad\qquad \underset{II}{}$$
$$= |F(hkl) \exp[-2\pi i(hx + hy + lz) + i\phi(hkl)] + |F(hkl)| \exp[2\pi i(hx + ky + lz) - i\phi(hkl)] + II$$
$$= 2|F(hkl) \cos[2\pi(hx + ky + lz) - \phi(hkl)]$$
$$= |F(hkl)| \cos[2\pi(hx + ky + lz) - \phi(hkl) + |F_{\bar{h}\,\bar{k}\,\bar{l}}| \cos[2\pi(hx + ky + lz) - \phi(\bar{h}\,\bar{k}\,\bar{l})] \tag{25}$$

Therefore, eq. 23 (or 24) can be written as

$$\rho(xyz) = \frac{1}{V} \sum_{h,k,l} |F(hkl)| \cos[2\pi(hx + ky + lz) - \phi(hkl)] \tag{26}$$

This form of three dimensional Fourier series is sometimes advantageous because the phase angle $2\pi\phi(hkl)$ for each coefficient appears explicitly. However, in general we can obtain only the structure factor amplitudes $|F|$ and not the phase angle directly from the experimental measurements. We must derive $\phi(hkl)$ either from values of $A(hkl)$ and $B(hkl)$ or by purely analytical methods.

The calculation of electron density from the Fourier summation of observed structure amplitude, $|F_o|$ can be used as a systematic way of improving the values of those atomic coordinates not fixed by the space group considerations. However, the calculation of three-dimensional electron density map is a difficult task, and is possible only with very fast computer. In their absence, the calculation is limited to two dimensions or even to one dimension (only for simple structures). The corresponding projected electron densities (obtained from eq. 26) per unit area and per unit length, respectively are:

$$\rho_p(x, y) = \frac{1}{A} \sum_{h,k} |F(hko)| \cos\left[2\pi(hx + ky) - \phi(hko)\right] \tag{27}$$

and

$$\rho_l(x) = \frac{1}{d_{100}} \sum_{h,k} |F(hoo)| \cos\left[2\pi hx - \phi(h)\right] \tag{28}$$

One-dimensional summations are not very useful for complicated structures (in this case the peaks representing different atoms overlap). However, two dimensional summations are a compromise between the amount of computation one can undertake and the amount of information necessary to solve a structure.

12.7 SPECIAL STRUCTURE FACTORS DUE TO SYMMETRY

In principle, structure factors for a given crystal structure can always be calculated by using the general expression given by eq. 22. However, in practice there exist many cases when the general expression can be modified to simpler forms by taking into account the symmetries that are present. The modified forms of the structure factor based on symmetry considerations are known as special structure factors. In order to do this, let us change the unit cell content into an asymmetric unit content. An 'asymmetric unit' may be defined as the unique and independent part of the structure (usually only a fraction of the unit cell) which depends on the amount of the symmetry present.

Now, dividing j atoms into m asymmetric units where each asymmetric unit contains n unique atoms, then eq. 22 can be written as

$$F(hkl) = \sum_{n} f_n \left[\sum_{n} \exp\left\{2\pi i\left(hx_{m,n} + ky_{m,n} + lz_{m,n}\right)\right\} \right] \tag{29}$$

In short, this can be written as

$$F(hkl) = \sum_{n} f_n (T_{hkl})_n \tag{30}$$

where

$$(T_{hkl})_n = \sum_{n} \exp\left[2\pi i(hx_{m,n} + ky_{m,n} + lz_{m,n})\right] \tag{31}$$

Since eq. 31 holds for an atom placed anywhere in the unit cell, the subscript n can be dropped. Further, consider T_{hkl} as a function of xyz for the space group described by the symmetry elements relating m equivalent positions. Let us derive special structure factors for some simple cases.

Case I: Center of Symmetry

In this case, $m = 2$ and the equivalent positions are x, y, z and $\bar{x}, \bar{y}, \bar{z}$. T_{hkl} is

$$T_{hkl} = \cos 2\pi(hx + ky + lz) + i \sin 2\pi (hx + ky + lz)$$
$$+ \cos 2\pi(-hx - ky - lz) + i \sin 2\pi(-hx - ky - lz)$$
$$= \cos 2\pi(hx + ky + lz) + i \sin 2\pi (hx + ky + lz)$$
$$+ \cos 2\pi(hx + ky + lz) - i \sin 2\pi(hx + ky + lz)$$

Since $\cos(-x) = \cos x$ and $\sin(-x) = -\sin x$, therefore

$$T_{hkl} = 2 \cos 2\pi(hx + ky + lz) \tag{32}$$

and
$$F_{hkl} = 2 \sum_n f_n \cos 2\pi(hx_n + ky_n + lz_n) \tag{33}$$

Case II: 2-fold Rotation (z-axis)

In this case, $m = 2$ and the equivalent positions are x, y, z and $\bar{x}, \bar{y}, \bar{z}$. T_{hkl} is

$$T_{hkl} = \cos 2\pi(hx + ky + lz)\, i \sin 2\pi (hx + ky + lz)$$
$$+ \cos 2\pi(-hx - ky + lz) + i \sin 2\pi(-hx - ky + lz)$$

Using the trigonometric relationships

$$\cos(A \pm B) = \cos A \cos B \pm \sin A \sin B$$
$$\sin(A \pm B) = \sin A \cos B \pm \cos A \sin B$$

and simplifying, we get

$$T_{hkl} = 2 \cos 2\pi(hx + ky) . \exp(2\pi ilz) \tag{34}$$

and
$$F_{hkl} = 2 \sum_n f_n \cos 2(hx_n + ky_n) . \exp(2\pi ilz) \tag{35}$$

Case III: $m \perp$ z-axis

In this case, the two equivalent positions are x, y, z and $x, y, \bar{z}$. T_{hkl} is

$$T_{hkl} = \cos 2\pi(hx + ky + lz) + i \sin 2\pi (hx + ky + lz)$$
$$+ \cos 2\pi(hx + ky - lz) + i \sin 2\pi(hx + ky - lz)$$

Using the above trigonometric relationships and simplifying, we get

$$T_{hkl} = 2 \exp[2\pi(hx + ky)] \cos 2\pi lz \tag{36}$$

and
$$F_{hkl} = 2 \sum_n f_n \exp[2\pi(hx_n + ky_n)] \cos 2\pi lz_n \tag{37}$$

Case IV: 2-fold Screw Axis Along [001], 2_1

In this case, $m = 2$ and the equivalent positions are: x, y, z and $\bar{x}, \bar{y}, z + \dfrac{1}{2}$. T_{hkl} is

$$T_{hkl} = \cos 2\pi(hx + ky + lz) + i \sin 2\pi(hx + ky + lz)$$

$$+ \cos 2\pi\left(hx + ky + lz + \frac{l}{2}\right) + i \sin 2\pi\left(-hx - ky + lz + \frac{l}{2}\right)$$

Substitute $hx + ky = \delta$ and then use the above trigonometric relationships and simplify it. Also taking into account the fact that l is always an integer so $2\pi l/2$ is always an integral multiple of π. This implies that $\sin 2\pi l/2$ will be zero, while $\cos 2\pi l/2 = (-1)^l$.

In other words

$$\cos 2\pi/2 = +1 \text{ if } l \text{ is even}$$

$$= -1 \text{ if } l \text{ is odd.}$$

After simplification, T_{hkl} is

$$T_{hkl} = \cos 2\pi(\delta + lz) + i \sin 2\pi(\delta + lz)$$

$$+ \cos 2\pi(-\delta + lz) \cos 2\pi\left(\frac{l}{2}\right) + i \sin 2\pi(-\delta + lz) \cos 2\pi\left(\frac{l}{2}\right) \qquad (38)$$

In general, the terms do not cancel each other and $T_{hkl} \neq 0$. However, for the special case, $h = k = 0$, T_{hkl} becomes

$$T_{hkl} = \cos 2\pi lz + i \sin 2\pi lz$$

$$+ (\cos 2\pi lz + i \sin 2\pi lz) \cos 2\pi\left(\frac{l}{2}\right)$$

$$= 2\,(\cos 2\pi lz + 2 \sin 2\pi lz) \quad \text{for } l \text{ is even}$$

$$= 0 \qquad\qquad\qquad\qquad \text{for } l \text{ is odd} \qquad (39)$$

Hence,

$$F_{hkl} = 2 \sum_n f_n \exp(2\pi ilz) \qquad \text{for } l \text{ is even}$$

$$= 0 \qquad\qquad\qquad\qquad \text{for } l \text{ is odd} \qquad (40)$$

Case V: *a*-glide Oriented Along [010]

In this case $m = 2$, and the equivalent positions are: x, y, z and $x + \dfrac{1}{2}, \bar{y}, z$. T_{hkl} is

$$T_{hkl} = \cos 2\pi(hx + ky + lz) + i \sin 2\pi(hx + ky + lz)$$

$$+ \cos 2\pi\left[h\left(x + \frac{1}{2}\right) - ky + lz\right] + i \sin 2\pi\left[h\left(x + \frac{1}{2}\right) - ky + lz\right]$$

Substituting $hx + lz = \delta$ and then using trigonometric relationships for simplification, we get

$$T_{hkl} = \cos 2\pi(\delta + ky) + i \sin 2\pi(\delta + ky)$$

$$+ \cos 2\pi(\delta - ky) \cos 2\pi \frac{h}{2} - \sin 2\pi(\delta - ky) \sin 2\pi \frac{h}{2}$$

$$+ i \sin 2\pi(\delta - ky) \cos 2\pi \frac{h}{2} + i \cos 2\pi(\delta - ky) \sin 2\pi \frac{h}{2}$$

$$= \cos 2\pi(\delta + ky) + i \sin 2\pi(\delta + ky)$$

$$+ \cos 2\pi(\delta - ky) \cos 2\pi \frac{h}{2} + i \sin 2\pi(\delta - ky) \cos 2\pi \frac{h}{2} \qquad (41)$$

In general, terms do not cancel each other and $T_{hkl} \neq 0$. However, for the special case, $k = 0$, T_{hkl} is

$$T_{hkl} = (\cos 2\pi\delta + i \sin 2\pi\delta) + (\cos 2\pi\delta + i \sin 2\pi\delta) \cos 2\pi \frac{h}{2}$$

$$= 2\,(\cos 2\pi\delta + i \sin 2\pi\delta) \qquad \text{for } h \text{ is even}$$

$$= 0 \qquad \text{for } h \text{ is odd} \qquad (42)$$

Hence,

$$F_{hkl} = 2 \sum_{n} f_n \exp\,(2\pi i\delta) \qquad \text{for } l \text{ is even}$$

$$= 0 \qquad \text{for } l \text{ is odd} \qquad (43)$$

In a similar manner, special structure factors can be obtained for all symmetry elements related to point groups and space groups. For details one can refer International Tables for *X*-ray crystallography.

12.8   SPECIAL STRUCTURE FACTORS DUE TO LATTICE CENTERING

Another way of modifying the general structure factor expression into special form is by taking into account the lattice centering. Non primitive lattices contain symmetries like screw axes and glide plane and certain class of reflections are absent in their diffraction patterns. Such systematic absences are known as space group extinctions or simply extinctions. Let us consider some simple cases of centerings and derive the corresponding structure factors.

Case I: Body Centering in a Cubic Lattice

In this case, the pair of related points have the coordinates as: (x, y, z) and $\left(x + \dfrac{1}{2}, y + \dfrac{1}{2}, z + \dfrac{1}{2}\right)$. T_{hkl} is

$$T_{hkl} = \cos 2\pi(hx + ky + lz) + i \sin 2\pi(hx + ky + lz)$$

$$+ \cos 2\pi\left[h\left(x + \frac{1}{2}\right) + k\left(y + \frac{1}{2}\right) + l\left(z + \frac{1}{2}\right)\right]$$

$$+ i \sin 2\pi\left[h\left(x + \frac{1}{2}\right) + k\left(y + \frac{1}{2}\right) + l\left(z + \frac{1}{2}\right)\right] \tag{44}$$

Substituting $hx + ky + lz = \delta$ and then using trigonometric relationships for simplification, we get

$$T_{hkl} = \exp(2\pi i\delta) + \exp(2\pi i\delta) \exp \pi i(h + k + l)$$

$$= \exp(2\pi i\delta)\left[1 + \exp \pi i(h + k + l)\right]$$

$$= 2 \exp(2\pi i\delta) \qquad \text{for } h + k + l \text{ is even}$$

$$= 0 \qquad \text{for } h + k + l \text{ is odd} \tag{45}$$

$$\text{Hence,} \quad F_{hkl} = 2\sum_{n} f_n (2\pi i\delta) \qquad \text{for } h + k + l \text{ is even}$$

$$= 0 \qquad \text{for } h + k + l \text{ is odd} \tag{46}$$

Case II: Face Centering in a Cubic Lattice

In this case, the four related points have the coordinates as: (x, y, z) $\left(x + \frac{1}{2}, y + \frac{1}{2}, z\right)$, $\left(x + \frac{1}{2}, y, z + \frac{1}{2}\right)$ and $\left(x, y + \frac{1}{2}, z + \frac{1}{2}\right)$. T_{hkl} is

$$T_{hkl} = \cos 2\pi(hx + ky + lz) + i \sin 2\pi(hx + ky + lz)$$

$$+ \cos 2\pi\left[h\left(x + \frac{1}{2}\right) + k\left(y + \frac{1}{2}\right) + lz\right] + i \sin 2\pi\left[h\left(x + \frac{1}{2}\right) + k\left(y + \frac{1}{2}\right) + lz\right]$$

$$+ \cos 2\pi\left[h\left(x + \frac{1}{2}\right) + ky + l\left(z + \frac{1}{2}\right)\right] + i \sin 2\pi\left[h\left(x + \frac{1}{2}\right) + ky + l\left(z + \frac{1}{2}\right)\right]$$

$$+ \cos 2\pi\left[hx + k\left(y + \frac{1}{2}\right) + l\left(z + \frac{1}{2}\right)\right] + i \sin 2\pi\left[hx + k\left(y + \frac{1}{2}\right) + l\left(z + \frac{1}{2}\right)\right]$$

Substituting $hx + ky + lz = \delta$ and then using trigonometrical relationships for simplifications, we get

$$T_{hkl} = \exp(2\pi i\delta) + \exp(2\pi i\delta).\exp \pi i(h + k) + \exp(2i\delta).\exp \pi i(h + l)$$

$$+ \exp(2\pi i\delta).\exp \pi i(k + l)$$

$$= \exp(2\pi i\delta)[1 + \exp \pi i(h + k) + \exp \pi i(h + l) + \exp \pi i(k + l) \tag{47}$$

Since we know that

$$\exp(\pi i m) = (-1)^m$$

$$= \text{negative, for m is odd}$$

$$= \text{positive, for m is even}$$

Hence,

$$T_{hkl} = 4 \exp(2\pi i\delta), \qquad \text{for } h, k \text{ and } l \text{ are all odd or all even}$$

$$= 0 \qquad \text{for mixed } hkl \tag{48}$$

and

$$F_{hkl} = 4 \sum f_n \exp(2\pi i\delta), \quad \text{for } h, k \text{ and } l \text{ all odd or all even}$$

$$= 0 \qquad \text{for mixed hkl} \tag{49}$$

12.9 ANOMALOUS SCATTERING

So far in our discussion of scattering factor or structure factor calculation we assumed that the phase of the waves depends only on the position of the scattering atom, that is the scattering process at the atom does not introduce any additional relative phase change. In general, it is true that such changes are negligible and that is the reason the Friedel's law normally holds. However, there are certain important exceptions: that is the atoms that have an absorption edge (Fig. 10.2) close to the wavelength (or frequency) of the incident X-rays (radiation) do introduce an additional phase change leading to anomalous scattering (different from the scattering due to free elections.) The scattering factor for those atoms can be represented as

$$f = f_o + f' + if'' \tag{50}$$

where f_o is the normal scattering factor, f' is a real correction term (usually negative) and f'' is the imaginary correction term both appearing due to anomalous scattering.

The details of the quantitative effect produced by anomalous scattering on the structure factor depend on whether the crystal is centrosymmetric or not, and on whether the X-ray wavelength used in shorter or longer than that of the absorption edge.

For centrosymmetric structures, the effect of anomalous scattering on the pairs of opposite reflections h, k, l and $-h, -k, -l$ are equal, so they have the same intensity and the Friedel's law is obeyed. However, for non-centrosymmetric structures, the effect of anomalous scattering on the opposite pair of reflections are not equal, so they have different intensities and are known as Friedel pairs or Friedel opposites. The effect of anomalous scattering is illustrated in Fig. 12.6 with the help of a vector diagram. This shows the amplitudes and phases of waves scattered by three atoms A, B and C of a non centrosymmetric structure. In Fig. 12.6a it is observed that although the resultant F_{hkl} and $F_{\bar{h}\,\bar{k}\,\bar{l}}$ differ in phase, there magnitude are same, the observed intensities are equal and the Friedel's law is obeyed. The real and imaginary parts of the scattering factors (eq. 50) is represented in Fig. 12.6b (where i introduces a phase difference of $\pi/2$). Finally, the effect of anomalous scattering introduced in Fig. 12.6a is shown in Fig. 12.6c.

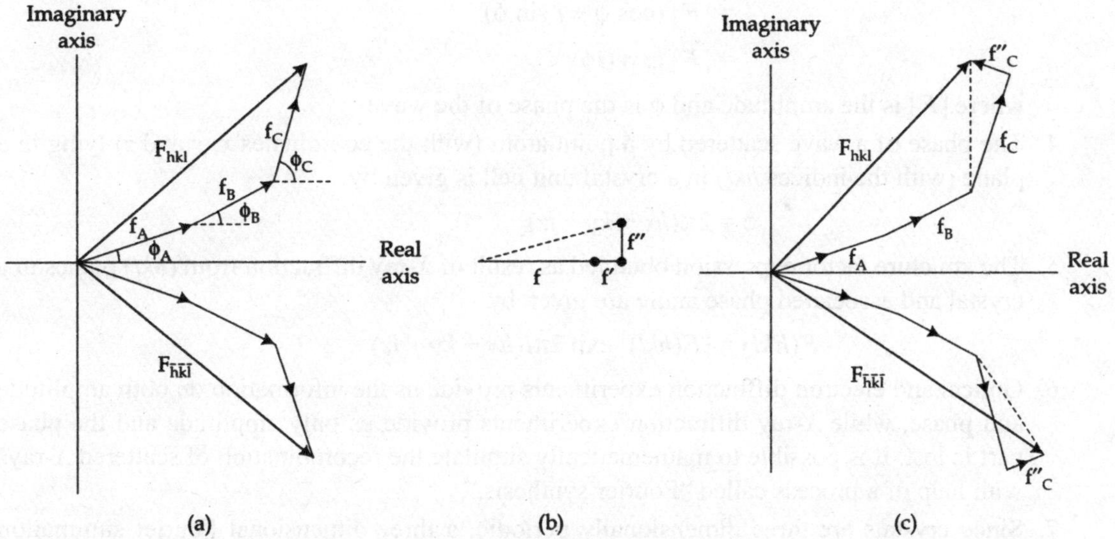

Fig. 12.6 The effect of anomalous scattering. (a) The structure factors F_{hkl} and $F_{\bar{h}\,\bar{k}\,\bar{l}}$ for a non-centrosymmmetric structure shown on a vector diagram. (b) The real and imaginary parts of the scattering factor of an anomalously scattering atom, showing how a phase change is introduced. (c) The effect on F_{hkl} and $F_{\bar{h}\,\bar{k}\,\bar{l}}$ of introducing an anomalous scatterer into the structure shown in (a).

An important consequence of the existence of this effect is that the absolute configurations of non centrosymmetric crystal structures (e.g. the optically active compounds) can be determined.

12.10 SUMMARY

1. Depending upon the instant of counting time for the particle executing S.H.M., the displacement of the wave can be algebraically represented as

$$x_r = |F| \cos \omega t$$

or

$$x_r = |F| \cos (\omega t + \phi)$$

or

$$x_r = |F| \cos (\omega t - \phi)$$

2. According to the principle of superposition of waves the amplitude of the resultant wave in any direction is the algebric sum of individual waves scattered in that direction. It holds for any number of waves regardless of their amplitude, phase or frequency.

3. Wave can be represented in different ways such as graphical, algebric, vectorial and exponential, out of which the exponential representation is easiest to handle. The wave vector can also be represented as a complex number

$$F = A + iB$$

$$= |F| (\cos \phi + i \sin \phi)$$

$$= |F| \exp (i\phi)$$

where $|F|$ is the amplitude and ϕ is the phase of the wave.

4. The phase of a wave scattered by a point atom (with the coordinates x, y and z) lying in a plane (with the indices hkl) in a crystal unit cell is given by

$$\phi = 2\pi(hx + ky + lz)$$

5. The structure factor expression obtained as result of X-ray diffraction from (hkl) planes in a crystal and associated phase angle are given by

$$F(hkl) = |F(hkl)| \exp 2\pi i(hx + ky + lz)$$

6. Optical and electron diffraction experiments provide us the information on both amplitude and phase, while X-ray diffraction experiments provide us only amplitude and the phase part is lost. It is possible to mathematically simulate the recombination of scattered X-rays with help of a process called "Fourier synthesis."

7. Since crystals are three dimensionally periodic, a three dimensional Fourier summation (synthesis) is made to obtain the image of electron distribution analogous to a lens in optical system.

8. Determination of three dimensional Fourier synthesis is a difficult task, and is possible only with very fast computers. In their absence, the calculation is limited to two dimensions or even to one dimension (only for simple structures) which is a good compromise.

9. The structure factor expressions can be simplified by taking into account either the symmetry or the lattice centering before using for final calculation. They are termed as special structure factors.

10. The atoms that have an absorption edge close to the wavelength of the incident X-ray beam introduces an additional phase change leading to anomalous scattering (which is different from the scattering due to free electron).

12.11   DEFINITIONS

Amplitude: The maximum numerical value of a periodic function measured from its mean or based value. It is half the peak-to-trough value.

Anomalous scattering: A change of phase takes place on scattering of radiation (X-ray) that is strongly absorbed by one or more kinds of atoms present in the crystal.

Asymmetric unit: The smallest part of the crystal structure (usually only a fraction of the unit cell) from which the complete crystal structure can be obtained by applying the space group symmetry operations.

Complex number: An expression of the form $A + iB$ where A and B are real numbers and $i = \sqrt{-1}$.

Electron density: The number of electrons per unit volume, usually represented as per Å^3.

Fourier series: Any periodic function $f(t) = f(t + T)$ can be represented by an infinite series of the form.

$$f(t) = a_0/2 + a_1 \cos 2\pi\,(t/T) + a_2 \cos 2\pi\,(2t/T) + \cdots + b_1 \sin 2\pi\,(t/T) + b_2 \sin 2\pi\,(2t/T) + \cdots$$

$$= a_0/2 + \sum_{n=1}^{\infty} a_n \cos 2\pi\,(nt/T) + \sum_{n=1}^{\infty} b_n \sin 2\pi\,(nt/T)$$

$$= \sum_{n=-\infty}^{\infty} c_n \exp\,[2\pi i(nt/T)]$$

is known as Fourier series.

Fourier Synthesis: The summation of sine and cosine waves to give a periodic function. For example, the computation of an electron density map from the waves of known amplitude, phase and frequency.

Fourier transform: In the pair of equations

$$f(x) = \int_{-\infty}^{+\infty} \exp\,(2\pi ixy)\, g(y)\, dy$$

and

$$g(y) = \int_{-\infty}^{+\infty} \exp\,(-2\pi ixy)\, f(x)\, dx$$

$g(y)$ is the Fourier transform of $f(x)$ and $f(y)$ is the Fourier transform of $g(x)$. In X-ray diffraction, the structure factor $F(hkl)$ and electron density $\rho(xyz)$ are Fourier transform of each other.

Phase: In general, it is the difference in position of the crests of the two waves of the same wavelength travelling in the same direction. Also it is the advancement of crest of a given wave with respect to the starting point.

Phase Problem: It is the problem of determining the phase angle which should inherently be associated with each structure factor, so that an electron density map can be simulated with the help of Fourier synthesis. This problem arises due to the fact that X-ray diffraction experiments provide only intensities (square of the amplitudes) and the phases are lost due to interception of diffracted beam by photographic film/plate or counter.

Superposition of waves: When two or more waves travel through a medium, their combined effect is found by the principle of superposition. According to this principle the resultant displacement at any point is the algebric (or vector) sum of the displacements by individual waves.

REVIEW QUESTIONS AND PROBLEMS

1. Define the term 'Structure factor'. Derive an expression for generalized structure factor.
2. Define structure factor and show that it is a complex number. Derive exponential form of the structure factor.
3. What do you understand by the term "phase of a wave"? Determine the generalized form of the phase in three dimensions (unit cell).

4. What do you understand by phase problem in crystallography? Discuss some of the phase determination procedures.

5. What do you understand by the terms "Fourier transform" and "Fourier synthesis"? Explain forward Fourier transform (FT) and reverse Fourier transform (FT^{-1}). Obtain general form of electron density equation.

6. What do you understand by Fourier transform? Obtain electron density equation for a crystal and give its physical interpretation.

7. What do you understand by the term "Fourier synthesis"? Obtain electron density equation for a centrosymmetric crystal.

8. Obtain structure factor expression for crystals having a 2-fold rotation (z-axis) and a mirror plane $\perp$ to z-axis.

9. What do you understand by the term "Special structure factor"? Derive an expression for a centrosymmetric crystal.

10. Obtain structure factor expressions for crystals have a 2-fold screw axis along z-axis and a glide oriented along y-axis.

CHAPTER 13

CRYSTAL STRUCTURE ANALYSIS

13.1 INTRODUCTION

The main objective of the crystal structure analysis is to locate the atomic positions in the unit cell with the help of Fourier synthesis and thus completely determine the structure. However, it is observed that there is no any single method which can be used to determine the structure of all chemical compounds. Instead, there exist a number of different methods, each of which is suitable for a particular class/type of compounds.

In this chapter, we shall discuss some of the important and widely used methods but not in great details. For the sake of making a continuity of account, we shall begin our discussion with trial and error method which is useful only to simple type of crystal structures.

13.2 TRIAL AND ERROR METHOD

Historically, this is the first and perhaps the simplest method used to tackle only the simpler crystal structures in the early days of the subject when it was in the developing stage. In general, this method is based on some useful guesses which could be made by observing some special features (such as symmetry etc.) in the diffraction pattern. For example, Bragg solved the structure of alkali halides by observing that the crystals were of face centered type. Based on this, he assumed that the alkali atoms occupied the corner positions and the halide atoms the face centers. Using similar intuitive approach, crystal structures of many *fcc* metals like copper, aluminum, silver etc. and *bcc* metals like iron, tungsten, molybdenum etc. were solved in the earlier days.

Although, hardly any body may be using this method now in its original form but many of its concepts are still an important part of more sophisticated methods. The main object of this method is to propose a or a few "trail structure" based on the available *X*-ray diffraction data and some valuable bold guesses. The term "trial structure" here means that the proposed structure is only an approximation to the correct or 'true structure'. Each trial structure is then checked by comparing the calculated structure factors with the observed values and the value of R-factor is calculated by using eq. 18 of Chapter 11. The process is repeated until a lowest acceptable value of R is obtained when the particular trial structure is treated as representing the true structure.

Prima-facie, the trial-and-error method seems to be an easier method. However, practically it is fairly teduis even for a single unknown, and becoming more so as the number of unknowns increases.

13.3 THE PATTERSON FUNCTION

Form section 12.6, we know that the electron density $\rho(xyz)$ at any point in a crystal is represented by three-dimensional Fourier synthesis, i.e.

$$\rho(xyz) = \frac{1}{V} \sum_h \sum_k \sum_l F(hkl) \exp[-2\pi i(hx + ky + lz)] \tag{1}$$

where $F(hkl)$ is a complex number and contains both amplitude and phase. From section 11.6, we also know that the intensity of diffracted beam is proportional to the square of the amplitude, i.e.

$$I(hkl) \propto |F(hkl)|^2$$

Keeping in view these factors and to avoid the phase problem, Patterson in 1935 proposed a modified Fourier synthesis by using the square of the amplitudes as the coefficients of the Fourier series. Since all the coefficients $[|F(hkl)|^2 = F(hkl)\cdot F(\overline{h}\,\overline{k}\,\overline{l})]$ in the series are positive, there is no phase problem and therefore there is no any doubt about the structure. Such a Fourier series is known as Patterson synthesis or Patterson map or more popularly as Patterson function and is given by

$$P(uvw) = \frac{1}{V} \sum_h \sum_k \sum_l |F_o(hkl)^2\cdot\exp[-2\pi i(hu + kv + lw)] \tag{2}$$

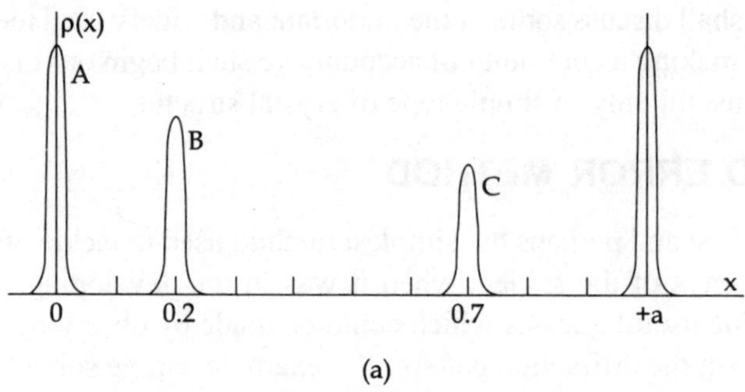

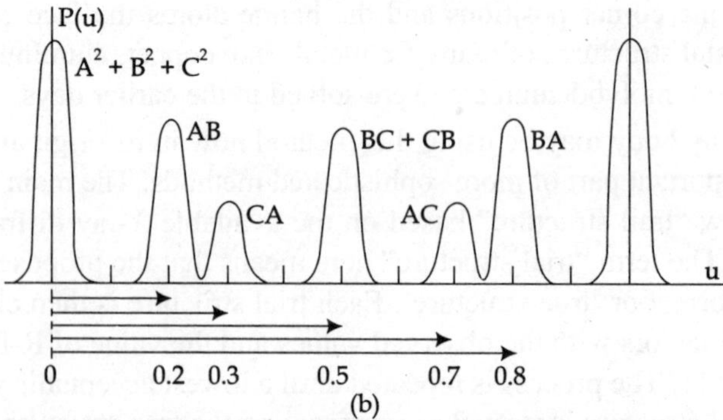

Fig. 13.1 Illustration of the Patterson function in terms of a one-dimensional crystal (a) The electron density function $\rho(x)$ (b) The corresponding Patterson function

Comparing the above two equations, we observe that eq. 1 has electron density peaks at atomic sites while eq. 2 has peaks of positive density not at atomic positions but at different positions such that they lie between the pair of atoms in the structure. A peak at a point uvw in the Patterson map implies that there are two atoms one each at $x_i y_i z_i$ and $x_j y_j z_j$ in the structure such that

$$u = x_j - x_i$$
$$v = y_j - y_i$$
$$w = z_j - z_i \tag{3}$$

The height of the Patterson peaks is proportional to the product of the numbers of electrons in each of the two atoms involved. Consequently, the Patterson peaks between the heavy atoms will be very prominent. As a result of this, the Patterson method provides a powerful means of locating heavy atoms in, say metal oxides, coordination compounds or organometallics. The knowledge of heavy atom positions can be used to locate the lighter atoms in the unit cell. However, the Patterson peaks show where the atoms lie relative to each other, but not where they lie relative to the unit cell origin, which is what we are interested to know.

Now according to Friedel's law we have $|F(hkl)|^2 = |F(\bar{h}\,\bar{k}\,\bar{l})|^2$ which is directly obtainable from measured integrated intensities, and hence the Patterson function is real and centrosymmetric. This may be written as

$$P(uvw) = \frac{1}{V} \sum_h \sum_k \sum_l |F_o(hkl)|^2 \cos 2\pi(hu + kv + lw) \tag{4}$$

Two dimensional (or one dimensional) form can be easily obtained. Further, the Patterson function is maximum corresponding to a null vector, i.e.

$$P(0, 0, 0) = \frac{1}{V} \sum_h \sum_k \sum_l |F_o(hkl)|^2 \tag{5}$$

The Patterson function at any point in a crystal can also be obtained in terms of the product of electron densities and their summation through a process known as 'convolution'. For example, the Patterson function at a point u, v, w may be thought of as a convolution of the electron density at all points x, y, z in the unit cell with the electron density at points $x + u, y + v, z + w$, i.e.

$$P(u, v, w) = V \int_0^1 \int_0^1 \int_0^1 (x, y, z)\, \rho(x + u, y + v, z + w)\, dx\, dy\, dz \tag{6}$$

Substituting the values of ρ and simplifying after integration of eq. 6, we obtain eq. 2 and hence eq. 4. Two dimensional (or one dimensional) form can be easily obtained. In order to see the use of Patterson function to locate some of the atoms in crystal structure determination, we note down some characteristic features and properties of the Patterson function. They are:

1. Each atom forms a pair (hence a vector) with every other atoms including with itself. So a unit cell containing N atoms (or peaks) given N^2 vectors (or Patterson peaks). Of these, N self vectors or null vectors (each atom of itself) have zero length and hence concentrate at the origin as the largest peak in any Patterson map. There are $N(N - 1)$ other peaks

distributed throughout the unit cell. Thus if there is a set of atoms in positions $r_j (j = 1$ to $N)$ they will give rise to a set of Patterson peaks at positions $r_i - r_j$ for $i = 1$ to N and $j = 1$ to N.

2. The vectors between atom A and atom B is exactly equal and opposite to the vector between atom B and atom A. This implies that a Patterson map is always centrosymmetric even if the crystal itself is not. In the event that the structure is centrosymmetric there is a systematic overlap of some peaks to give double weight peaks. Thus, for atoms with coordinates r_j, r_i, r_i and $-r_j$, the Patterson single peaks occur at $\pm 2r_i$, $\pm 2r_j$ and double peaks $\pm(r_i - r_j)$ and $\pm(r_i + r_j)$ apart from the peaks at null vectors. There are a total of $N(N-1)/2$ independent peaks in the unit cell.

3. The Patterson peaks have a similar shape as the electron density peaks, but about twice as wider.

4. As a consequence of 1 and 3, usually there is a considerable overlap of peaks and not all of them could be resolved as separate maxima.

To understand it better, let us obtain the Patterson function of a one-dimensional crystal by taking a specific case. For this purpose, let us consider three atoms in the unit cell lying at $x = 0.0$, 0.2 and 0.7 and the corresponding electron densities $\rho(x)$ are shown at A, B, and C in Fig. 13.1a. From eq. 6 one-dimensional form of Patterson function can be defined as

$$P(u) = \int_0^1 \rho(x)\, \rho(x + u)\, dx \tag{7}$$

Making use of the above properties, we can obtain various Patterson peaks as shown in Fig. 13.1b. Here the first strong peak at $u = 0$ represents the fact that all the three atoms are at zero distance from themselves and superimpose at the origin, the peak at $u = 0.2$ represents that the atom B is displaced from A by 0.2 units and the peak at $u = 0.3$ represents that the atom A is displaced from C by 0.3 units and so on. The figure also show its centrosymmetric nature.

13.4   THE HEAVY ATOM METHOD

When a crystal contains a large number of light atoms and a few heavy atoms then its structure can be conveniently solved by a method called the heavy atom method. The use of the Patterson function in determining such structures depends on its ability to resolve the interatomic vectors (peaks) between heavy atoms from all other atoms.

Since the height of the Patterson peaks is proportional to the product of the electrons of the concerned pair of atoms (or the atomic numbers), peaks related to heavy atoms stand out strongly against the background of heavy-light and light-light atom peaks. In such cases, it is usually possible to find a self-consistent set of atomic positions for the heavy atoms which explain the large Patterson peaks. For example, the vectors (peaks) between symmetry related heavy atoms often lie

in special positions, with some coordinates equal to 0 or $\dfrac{1}{2}$ can be recognised.

One has to be very careful in the selection of the heavy atoms because the observations show that on one hand the heavier atoms are easier to locate by means of Patterson map and also produce better phase and intensities of reflections, but on the other hand, the dominance of a very heavy atom (compared to other atoms) becomes excessively large so as to make the comparison of $|F_o|$ and $|F_c|$ relatively insensitive to the positions of light atoms. As a compromise, a good criterion is

$$\sum_{\text{heavy atoms}} f^2 = \sum_{\text{light atoms}} f^2 \tag{8}$$

In such cases, the phase angle for most reflections turn out to be nearly equal to that for the heavy atom(s).

We know that crystals are either centrosymmetric or non-centrosymmetric. For centrosymmetric crystal structures all that is necessary is that the heavy atom contribution give the correct sign of a sufficient number of structure factors. On the other hand, for non- centrosymmetric structures the phase angle is determined as if the heavy atom(s) alone controls it. Using these phase angles (usually 0 or π), a trial crystal structure is generated using eq.1. The trial atomic positions generated in this way are then used to determine a more precise set of phase angles, and the crystal structure is re-determined. This process may be repeated until no change in the crystal structure is found by further iteration.

13.5 ISOMORPHOUS REPLACEMENT

In general, if two or more chemical substances have the same crystal structure (space group) but containing different heavy atoms are said to be isomorphous. They have almost the same atomic arrangement, with only slight differences in their unit cell dimensions. Salts of alkaloids, derivatives of pencillin and porphyrins with different metal ions (or non at all) in the chelated positions are some of the well known examples of isomorphous compounds. This method is particularly useful in solving complex structures.

Let us consider two isomorphous compounds A and B in which the scattering effect of light atoms (say L) in both the compounds are same, but the scattering effect of replaceable (e.g. heavy) atoms (say R) are different. Since the atomic positions in the two cases are the same, they contribute identical phase but the magnitude of their structure factors are different. They are:

$$F(hkl)_A = F(hkl)_L + F(hkl)_{AR} \tag{9}$$

$$F(hkl)_B = F(hkl)_L + F(hkl)_{BR} \tag{10}$$

and
$$\Delta F(hkl) = F(hkl)_A - F(hkl)_B = F(hkl)_{AR} - F(hkl)_{BR} \tag{11}$$

Now suppose the crystal is centrosymmetric, for which we have seen in the last section that the phase angles of the $F(hkl)$ can be either 0 or π and the corresponding values of the $F(hkl)$ is either $+|F(hkl)|$ or $-|F(hkl)|$. The correct sign of the structure factor can be determined as follows:

1. $F(hkl)_{AR}$ and $F(hkl)_{BR}$ can be computed from the known heavy atom positions, the right side of eq. 11 is known, i.e.

$$F(hkl)_{AR} - F(hkl)_{BR} = 2(f_{AR} - f_{BR}) \sum_j \cos 2\pi (hx_j + ky_j + lz_j) \tag{12}$$

where x_j, y_j, z_j are non- centrosymmetrically related positions of the heavy atoms in A (or B) compound.

2. $F(hkl)_A$ and $F(hkl)_B$ can be determined from diffraction intensities.

3. The left side of eq. 11 is one of the four possible sign combinations which can be computed. They are:

$$\{\pm|F(hkl)_A|\} - \{\pm|F(hkl)_B|\} \tag{13}$$

The sign combination which gives best agreement with eq. 11 is accepted.

4. Once the correct signs of $F(hkl)$ are found, the crystal structure is solved using eq. 1

The procedure is considerably more difficult for non- centrosymmetric structures.

13.6   SUPERPOSITION METHOD

In section 13.3, the overlap of Patterson peaks were found to depend on the number of atoms in the unit cell. When the overlapping peaks are resolved, the crystal structure is solved.

However, when the peaks remain unresolved (as the case may be in most of the structure analysis), the recovery of the atomic arrangement from the Patterson map is not automatic.

M.J. Burger proposed superposition method to accomplish this recovery. This method involves the calculation of the Patterson function at a large number of points and superimposing two maps such that the origin of one coincides with a suitable peak of the other, and preparing a new map where the value of each point is lower than the two superimposed values. Determination of a minimum function is an essential part of this procedure. The minimum function (new map) contains much lesser peaks than the Patterson function, since in the new map only peaks appearing simultaneously on both Patterson maps are retained and hence the minimum function may closely resemble the actual structure. The same process can be applied to the superposition of several Patterson maps at a time to get better results.

In practice, the problem is not so simple as it appear to be. A number of difficulties do appear when an attempt is made to apply the theoretically simple superposition processes. However, most of these are related to the overlap of Patterson peaks. The success of this method depends on the accuracy with which the superposition peaks coincide with the Patterson peaks and the use of multiple superpositions to get a single minimum function from the overlapped peaks. This method has been extremely successful on some rather complicated structures. For example, the structure of celloboise ($C_{12}O_{11}H_{22}$) containing 23 independent atoms other than hydrogen, has been successfully solved by this method.

13.7   DIRECT METHODS

The Patterson method described above (sec. 13.3) is very powerful for solving structures in which the unit cell contains a small proportion of heavy atoms and a large proportion of small atoms. However, the Patterson map can become very congested if all the atoms have approximately the same atomic number (i.e. atoms of approximately same size) as for examples in most organic compounds. Under these circumstances, the so called 'direct methods' approach is preferred. This is a general name given to methods which attempt to determine the approximate phases from the measured intensities. These methods make use of the fact that the intensities of reflections contain structural information and that the electron density in a real crystal can be negative.

In the beginning, the use of direct methods were limited to centrosymmetric crystals where the phasing is done simply by assigning a + or − sign to each observed amplitude. Here, we shall discuss only two frequently used direct methods. They are: (i) Inequality relationship and (ii) Statistical method.

Inequality Relationship

In 1948, D. Harker and J.S. Kasper showed that there exist relationships between the structure factors (F's or $|F^2|$'s) for symmetrical crystals which could be expressed in the form of inequalities. These inequality relationships occasionally lead to definite information about the phases and sings of structure factor. In order to derive these relationships, let us first define the unitary structure factor as

$$U(hkl) = \frac{F(hkl)}{\sum\limits_{j=1}^{N} f_j} \tag{14}$$

where f_j are the actual scattering factors for atoms and $F(hkl)$ is the structure factor given by

$$F(hkl) = \sum_{j=1}^{N} f_j \exp 2\pi i(hx_j + ky_j + lz_j) \tag{15}$$

From the scattering factors, let us define a term n_j as

$$n_j = \frac{f_i}{\sum\limits_{j=1}^{N} f_j} \tag{16}$$

so that the unitary structure factor (eq. 14) reduces to

$$U(hkl) = \sum_{j=1}^{N} n_j \exp 2\pi i(hx_j + ky_j + lz_j) \tag{17}$$

Now in order to derive the required inequality relationship, let us make use of the Cauchy inequality, which is given by

$$\left| \sum_{j=1}^{N} a_j b_j \right|^2 \leq \left(\sum_{j=1}^{N} |a_j|^2 \right) \left(\sum_{j=1}^{N} |b_j|^2 \right) \tag{18}$$

Substituting $a_j = (n_j)^{1/2}$ and $b_j = (n_j)^{1/2} \exp 2\pi i(hx_j + ky_j + lz_j)$ and apply the Cauchy inequality to the unitary structure factor, we obtain

$$\left| \sum_{j=1}^{N} n_j \exp 2\pi(hx_j + ky_j + lz_j) \right|^2 \leq \sum_{j=1}^{N} n_j |\exp 2\pi i(hx_j + ky_j + lz_j)|^2$$

or
$$|U(hkl)|^2 \leq \sum_{j=1}^{N} n_j \sum_{j=1}^{N} n_j \,|\exp 2\pi i(hx_j + ky_j + lz_j)|^2 \tag{19}$$

Since we observe from eqs. 16 and 17 that

$$\sum_{j=1}^{N} n_j = 1, \text{ and}$$

$$|\exp 2\pi i(hx_j + ky_j + lz_j)| = |\cos 2\pi i(hx_j + ky_j + lz_j) + i \sin 2\pi i(hx_j + ky_j + lz_j)|$$
$$= 1$$

Therefore, eq. 19 reduces to

$$|U(hkl)|^2 \leq 1 \tag{20}$$

This is a fairly obvious inequality relationship which is not actually very useful one because in whatever way the unitary structure factor $U(hkl)$ is defined, it must be true. However, more useful inequality relationships can be derived by taking into account the presence of symmetry elements in the crystal. Let us derive a few of them for some simple cases.

Case I. Center of Symmetry

In this case, the atoms appear in pairs with coordinates (x, y, z) and $(-x, -y, -z)$, corresponding to which the unitary structure factor equation becomes

$$U(hkl) = \sum_{j=1}^{N} n_j \cos 2\pi(hx_j + ky_j + lz_j) \tag{21}$$

Substituting $a_j = (n_j)^{1/2}$ and $b_j = (n_j)^{1/2} \cos 2\pi(hx_j + ky_j + lz_j)$ in eq. 18, we obtain

$$\left| \sum_{j=1}^{N} n_j \cos 2\pi(hx_j + ky_j + lz_j) \right|^2 \leq \sum_{j=1}^{N} n_j \cos^2 2\pi(hx_j + ky_j + lz_j) \tag{22}$$

Again using $\sum n_j = 1$ and the trigonometric identity

$$\cos^2 A = \frac{1 + \cos 2A}{2}$$

eq. 22 reduces to

$$|U(hkl)|^2 \leq \frac{1}{2} \sum n_j[1 + \cos\{2\pi \times 2(hxj + kyj + lzj)\}]$$

$$\leq \frac{1}{2}\left[1 + \sum n_j \cos 2\pi(2hx_j + 2ky_j + 2lz_j)\right]$$

$$\leq \frac{1}{2}[1 + U(2h, 2k, 2l)] \tag{23}$$

This inequality can be used to get the sign of $U(2h, 2k, 2l)$.

Case II. 2-fold Axis (|| to z)

In this case, the atoms appear in pairs with coordinates (x, y, z) and $(-x, -y, z)$, corresponding to which the unitary structure factor equation becomes

$$U(hkl) = \sum n_j \exp\,(2\pi i\,lz_j) \cos 2\pi(hx_j + ky_j) \tag{24}$$

Substituting $a_j = (n_j)^{1/2} \exp\,(2\pi i\,lz_j)$ and $b_j = (n_j)^{1/2} \cos 2\pi(hx_j + ky_j)$ in eq. 18, we obtain

$$|U(hkl)|^2 \leq \frac{1}{2}\,[1 + U(2h, 2k, 0)] \tag{25}$$

In a similar manner, the unitary structure factor equation is written down for a given symmetry element (space group), the summation terms are partitioned in all possible ways and then for each partition an inequality relationship can be obtained. With the help of this, the sign of the structure factor can be determined.

Statistical Method

In order to predict phases, the statistical method ideally requires that:

   (i) All atoms have the same X-ray scattering power.

   (ii) The distribution of atoms within the unit cell is quasi-random.

The first step in the procedure is to place all the structure factors on an absolute scale, normalized for the effects of angular dependence of f_x and the variation in f_x from one compound to another. Normalized structure factors are calculated according to

$$|E(hkl)|^2 = \frac{F(hkl)^2}{\epsilon \sum\limits_{n} f_n^2} \tag{26}$$

where the summation is over all the n atoms in the unit cell; ϵ is an integer which is generally unity but may assume other values for special sets of reflections in certain space groups. The resultant $|E|$ values typically fall in the range 0-5 and for Fourier calculations $E(hkl)$ has the same phases as $F(hkl)$.

The prediction of phases is largely based upon the Sayre probability relationship:

$$S(hkl) \sim S(h'k'l') \cdot S(h - h', k - k', l - l') \tag{27}$$

where S means the 'sign of' and $\sim$ means 'is probably equal to', e.g.

h, k, l	442
h', k', l'	110
$h - h', k - k', l - l'$	332

Thus, if $E(442)$ and $E(110)$ are both negative phases, $E(332)$ is likely to be positive. The probability P that this relationship holds depend upon the $|E(hkl)|$:

Table 13.1   Probabilities calculated according to eqn. 28

$|E_{av}|$	$|E_{av}|^3$	P		
		$N = 20$	$N = 60$	$N = 100$
3.0	27.0	1.0	1.0	0.99
2.6	17.0	1.0	0.99	0.97
2.3	12.2	1.0	0.96	0.92
1.8	5.8	0.93	0.81	0.76
1.5	3.4	0.82	0.71	0.66

$$P = \frac{1}{2} + \frac{1}{2}\tan h\left\{\frac{1}{N}|E(hkl).E(h'k'l').E(h-h',k-k',l-l')|\right\} \tag{28}$$

where N is the number of atoms in the cell and

$$\tan hx = \frac{e^x - e^{-x}}{e^x + e^x} \tag{29}$$

Table 13.1 shows the probability P as a function of N and the average $|E|$ value in the triple product calculated by using eq. 28. It is observed that for a given value of $|E_{av}|^3$, the predication becomes less precise as the number of atom in the unit cell increases.

We will now describe the step to be followed for prediction of phases in centrosymmetric structures.

1. Prepare a list of E values in the triple product (as in Table 13.1) in decreasing order of magnitudes.

2. Analyse the statistical distribution of E values to determine if the structure is centrosymmetric or non-centrosymmetric. This is necessary only if there is an ambiguity in the space group determined from systematic absences.

3. Select the best starting reflections from larger $|E|$'s (greater than 3.0) so that the initial prediction will be reliable. A computer program can be used for this purpose.

4. Select three origin fixing reflections. They are arbitrarily assigned + phases in order to define the position of the cell origin along the three crystallographic axes, but others may be + or -1. The phase prediction will be performed for each of the total possibilities, one of which is correct.

5. The correct set of phases is used to calculate an E-map.

$$F(xyz) = \frac{1}{V}\sum_h\sum_k\sum_l E(hkl)\cos 2\pi(hx + ky + lz) \tag{30}$$

This is essentially equivalent to electron density map for centrosymmetric structuress and should reveal the atomic positions. With some more difficulty, this method can also be applied to non centrosymmetric structures.

13.8 SUMMARY

1. A number of methods have been proposed to solve crystal structures, each of which is suitable for a particular class/types of compounds.
2. Very simple crystal structures can be solved by trial and error method. Patterson function may be interpreted directly, if only a few atoms are involved. It is a powerful method.
3. If there is one or a few (heavy) atoms in the unit cell, a method called the heavy atom method is used to solve the structure.
4. Complicated structures may require the chemical labour of preparing isomorphous derivatives to solve the structures belonging to the same space group.
5. Superposition method is the one in which two or more Patterson maps are suitably superimposed to get the minimum function and hence the crystal structure.
6. For equal atom structures, direct methods are more commonly used. These methods make used of the facts that the intensities of reflections contain structural information and that the electron density in a real crystal cannot be negative.

13.9 DEFINITIONS

Direct Phase Determination: A method of deriving relative phases of diffracted beams of stronger reflections and the relationships among their indices and among their structure factor amplitudes obtained on the basis that structure is composed of atoms and that the electron density must be positive or zero everywhere. Only certain values of the phases are consistent with these conditions.

Electron density map: A representation of electron density in the form of contours around the atomic sites in a crystal structure obtained from Fourier synthesis.

E-map: A Fourier map equivalent to an electron density map with phases derived by 'direct methods' using normalized structure factor amplitudes $E(hkl)$ in place of $F(hkl)$ in Fourier summation. Resulting peaks are sharper than the peaks obtained with F(hkl) values.

Harker-Kasper inequalities: Symmetry (space group) dependent inequality relationships among unitary structure factors that help to determine the phases of certain intense reflections in a centrosymmetic structure under direct methods.

Heavy atom method: A method of determining phase angle in which the phases calculated from the position of heavy atom are used to compute the first approximate electron density map. If necessary, successive approximate electron density maps are computed to obtain the entire structure.

Isomorphism: It is the similarity of crystal shape, unit cell dimensions and crystal structure between substance of similar chemical composition.

Isomorphous replacement method: A method of obtaining phases, from measurements of intensities of reflections from two or more isomorphous crystals.

Patterson function: A map obtained from the summation of a Fourier series which has the squares of the structure factor amplitudes as coefficients. Ideally, the position of maxima in the Patterson map represents the end points of vectors between the atoms.

Probability relationships: Equations representing the probability that a phase has a certain value. They form the basis of phase determination in direct methods.

Trial-and error method: A method which involves postulating an approximately correct (or a trial) structure, calculating its structure factor and comparing the same with the observed structure factor. The number of trial structures increase with the increase of unknown parameters.

REVIEW QUESTIONS AND PROBLEMS

1. Enumetrate various methods of crystal structure analysis. Discuss in detail the trial and error method.
2. Describe Patterson method in detail. How this method is useful in locating heavy atoms?
3. The Patterson method is treated to be most powerful method in crystal structure analysis, justify. Enumerate the characteristic features and properties of the Patterson function.
4. Briefly explain various stages involved in crystal structure analysis. Describe heavy atom method in detail. How the position of light atoms are determined?
5. Discuss isomorphous replacement method of crystal structure analysis in detail.
6. Discuss the superposition method of crystal structure analysis in detail.
7. What do you understand by the term "direct methods" in crystallography? Discuss the case of inequality relationships, which are helpful in knowing the crystal structure. Explain the role of crystal symmetry.
8. What do you understand by the term "normalized structure factor? Making use of the Sayre probability relationships, explain the utility of this method in crystal structure analysis.

CHAPTER 14

CRYSTAL STRUCTURE REFINEMENTS

14.1 INTRODUCTION

In the last chapter, we discussed some of the more frequently used method of crystal structure determination. Some of them often provide us an approximate crystal structure in the sense that the atoms in the unit cell are only approximately localized. However, in some other cases (e.g. Patterson method), a partial structure is obtained where only some atomic positions are known relatively accurately but not all (including that of hydrogen atoms, which have little electron density and are not usually found until later, if at all). The approximate structure or a partial structure obtained from any of the crystal structure methods (with $R \sim 0.3$ to 04) serves as our initial model or trial structure which needs to be refined by using some suitable techniques.

Like several crystal structure methods, there are many methods of crystal structure refinements. However, here also we shall limit our discussion to only a few more frequently used methods, that too not in details. Advance texts may be referred for that purpose.

14.2 SUCCESSIVE FOURIER SYNTHESES

Successive Fourier Syntheses can be done by using either forward Fourier transform (FT) or reverse Fourier transform (FT^{-1}), these equations have been discussed in suction 12.5. Thus, let us start with the model or trial structure and calculate the corresponding structure facture factors and electron density function one by one.

Starting with the model structure and using eq. 12.21, we can calculate the improved structure factors F_c for each observed structure factor F_o through:

$$\text{model structure} \xrightarrow{\quad FT \quad} \text{set of } F_c \tag{1}$$

This calculation provides us values for the amplitudes and phases of F_c (i.e. $|F|$ and ϕ_c), where as we have only amplitudes F_o (i.e. $|F|$) and no ϕ_o.

On the other hand, to obtain the electron density function corresponding to the calculated structure factor amplitudes $|F_c|$ and phase ϕ of the model structure, a reverse Fourier transform (FT^{-1}) is carried out through:

$$|F_c| \text{ with } \phi_c \text{ (of model structure)} \xrightarrow{\;FT^{-1}\;} \rho \tag{2}$$

This is effectively no progress. However, the combination of experimentally observed amplitudes $|F_o|$ (which carry information about the true structure) with the calculated phases ϕ_c (which are taken to be $\sim \phi_o$ values) will certainly produce an improved electron density function through:

$$|F_o| \text{ with } \phi_c \, (\sim \phi_o) \xrightarrow{\;FT^{-1}\;} \rho \text{ for a new model structure} \tag{3}$$

In general, if the errors in the calculated phases ϕ_c are not too large, the electron density function in eq. 3 will show the atoms of the starting model structure together with the atoms not previously known. Therefore, this provides us an improved model structure. A forward Fourier transform (*FT*) of the new model structure (eq. 3) will provide us a new set of $|F_c|$ and ϕ_c (the previous set is discarded). The new model structure will have less discrepancies between $|F_c|$ and $|F_o|$ and hence a lower value of R is expected.

The entire process is repeated until no further change in the value of phases (or the value of R) results. This procedure is particularly useful when the phases have been determined by considering only some of strong scattering power atoms, because the resulting electron density map will help in locating the low scattering power atoms. Further, the procedure is quite straightforward and fairly rapid for centrosymmetric structures, because the phase angles are either $0°$ or $180°$. However, for non-centrosymmetric structure where the phases can have any values from $0°$ to $360°$, the convergence to finally correct synthesis is considerably slow.

14.3   DIFFERENCE FOURIER SYNTHESIS

We know that a true electron density synthesis requires the use of an infinite series. However, in the above discussed successive Fourier syntheses, series consisting of only finite number of terms (which is observed in practice) are used. As a result, series termination errors are introduced in the resulting maps which fail to locate the atomic peaks (positions) correctly. Fig. 14.1 is an illustration of this view point, where P_t (equivalent to P_o) and P_a (equivalent to P_c) represent the true (equivalent to observed) and assumed (equivalent to calculated) the positions of an atom, $\rho_t(\rho_o)$ $\rho_a(\rho_c)$ represent the corresponding electron density curves and ρ_{map} $(\rho_o - \rho_c)$ represents the electron density map obtained from the Fourier synthesis.

In order to overcome the above discrepancy, A.D. Booth proposed that a difference synthesis $(|F_o| - |F_c|)$ with ϕ_c rather than $|F_o|$ with ϕ_c be carried out. This virtually eliminates series termination errors with the assumption that the magnitudes of the unobserved terms in ρ_o are the same as those omitted form ρ_c. A one-dimensional difference synthesis can be expressed as

$$\Delta\rho = \rho_o - \rho_c = \frac{1}{2}\sum_h \left[F_o(h) - F_c(h) \right] \exp 2\pi i h x \tag{4}$$

In an ideal situation, if the assumed structure is correct (i.e. when the position of the atom P_a coincides with P_t), the difference Fourier map should be featureless (i.e. it should be zero everywhere). However, in practice the random errors in the observed data make this only approximately true even for fully refined structures.

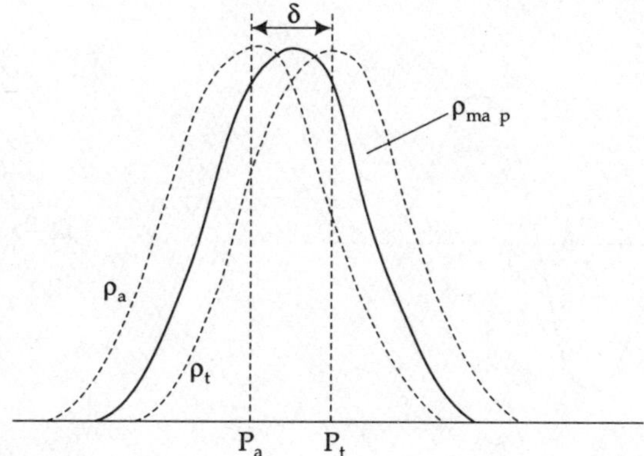

Fig. 14.1   Refinement by Fourier methods. The relationship between the assumed position, P_a, the true position P_t and the electron density peak due to a slightly misplaced atom.

Let us consider some of the possible ways due to which the assumed atom position may differ from the true position and the appearance of the difference map. For simplicity, Let us confine to a two-dimensional situation.

(i) Position Error

When the assumed position of the atom is slightly misplaced with respect to the true position as shown in Fig. 14.2, the difference synthesis (eq. 4) shows a characteristic structure, the assumed atom position will lie on a steep gradient on the difference map and the necessary correction is to be applied in the direction of the gradient (i.e. towards a more positive region).

The correction in the x-coordinate can be applied with sufficient accuracy by using the equation

$$\Delta x = -\frac{\text{Slope}}{\text{Curvature}} = \frac{\partial \Delta \rho / \partial x}{\partial^2 \rho / \partial x^2} \tag{5}$$

In a similar manner, Δy and Δz can be calculated to get the value of Δr. However, in practice it is usually satisfactory to shift the atom by $k.d\rho/d\gamma$, where the constant k (for three-dimensional work) is of the order of 0.02-0.01 for light atoms, and less for heavy atoms, and is adjusted empirically in the course of refinement.

In general, the difference Fourier map showing the electron density peaks (positive peaks) or the holes (negative peaks) at the positions of the model structure indicates the incorrect assignment of too little or too much assumed density to the atom.

(ii) Missing Atoms

We observed earlier that the missing atoms in difference map should appear as positive maxima. However, their heights are usually smaller than that of the corresponding atomic number of the missing atoms because of their approximate phases.

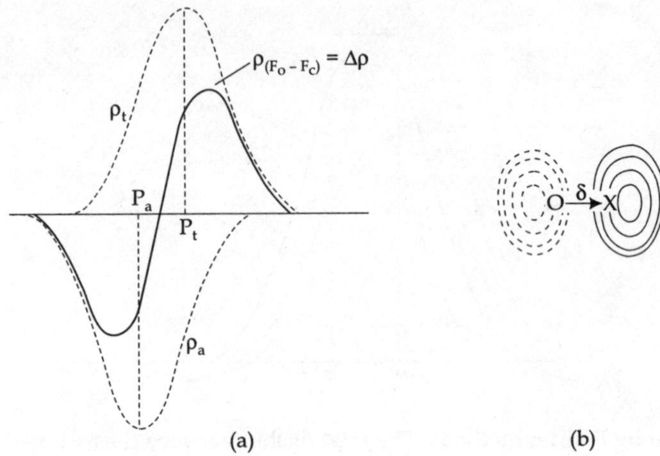

Fig. 14.2   (a) Position error of a model atom (b) corresponding difference synthesis

The lack of truncation errors in the difference Fourier map (when the model structure is corrected for other important errors) allows the correct localization of light atoms (including the hydrogen atoms) even in the presence of heavier atoms.

(iii) Errors in the Thermal Parameters

From section 11.5 we know that because of thermal motion the electron density function around each atomic nucleus becomes wider. Fig. 14.3a illustrates the effects of errors solely in the thermal parameters under two different situations. That is the case of overestimated situation when the root mean-square amplitude of vibration is supposed to be too large and the case of underestimated situation when the root mean-square amplitude of vibration is too small (or the thermal motion is neglected). In the first case, a small positive maximum surrounded by a negative ring will appear.

There is yet another situation where the thermal motion is supposed to be isotropic in nature, the assumed electron density ($\equiv \rho_c$) distribution will be spherical in shape. However, we know that in reality thermal motion is anisotropic whose electron density ($\equiv \rho_o$) distribution will be non-spherical (i.e. normally elliptical in shape) as shown in Fig. 14.3b. The difference map shows a quadrupole-type distribution characteristic of anisotropy.

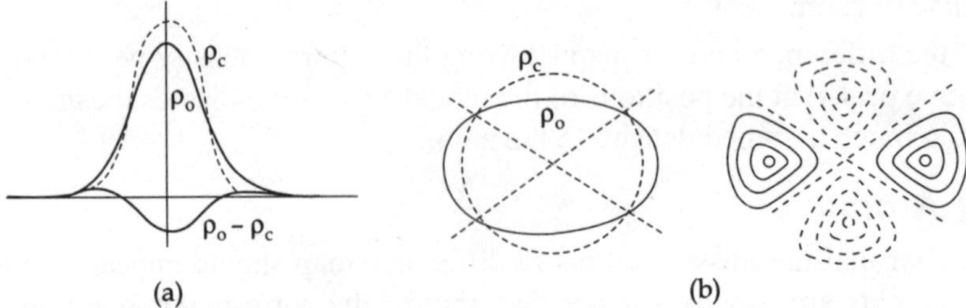

Fig. 14.3   Thermal parameter errors and their effect on the difference syntheses; for the atom of the model it is assumed: (a) too small an isotropic motion (b) an isotropic motion when an anisotropic model should be assumed

14.4   LEAST-SQUARES REFINEMENT

In principle, the least squares refinement of crystal structures is similar to finding a 'best-fit' straight line through a set of point on a graph. However, this is more complicated because (i) there are many variable parameters instead of just two (the gradient and intercept) for a straight line graph, and (ii) the equation relation to experimental data to the parameters to be refined (i.e. the Fourier transform) is far from linear.

The mathematical basis for this procedure is about two hundred years old and consists of the proposition that the best fit between a large number of experimental quantities (independent parameters) and a finite set of functional equations (properly related to these quantities) is obtained when the sum of the squares of the discrepancies between the observed and calculated values is a minimum, i.e.

$$D = \sum_{hkl} w_{hkl}(|F_o| - |F_c|)^2$$

or

$$D = \sum_{hkl} w_{hkl}(F_o^2 - F_c^2)^2 \qquad (6)$$

where w_{hkl} is the appropriate weight assigned to each reflection according to the expected reliability of its measurement. Based on the value of $|F_o|$ several weighting schemes have been proposed. First such scheme suggested by Hughes (1941) was

$$w = \text{constant} = 1/16\, F_{\min}^2 \qquad \text{for } |F_o| \le 4F_{\min}$$

$$= 16\, F_{\min}^2 / |F_o|^2 \quad \text{for } |F_o| > 4F_{\min} \qquad (7)$$

where $F_{\min}$ is close to the minimum observed amplitude. Later, Cruickshank (1970) proposed another scheme, i. e.

$$w = \frac{1}{(a + |F_o| + c\,|F_o|^2)} \qquad (8)$$

where $a = 2F_{\min}$ and $c = 2/F_{\max}$ as initial guesses and $F_{\max}$ is close to the maximum observed amplitude.

The first of the two in eq. 6 (refinement on F) has historically been most commonly used, but the popularity of the second (refinement on F^2) is now increasing and is in many ways superior.

As far as the mathematical formulation is concerned, we simply confine to mention that the standard process of finding the least-squares solutions of a set of equations is done by using matrix algebra as developed in sec. 5.4.

The refinement of larger molecules having complex structures (where thousands of parameters are to refined) is normally done by using two approximations. They are: (i) diagonal least-squares approximations, and (ii) block-diagonal least-squares approximation. The first one is based on the fact that the diagonal elements of the matrix are the sums of the squares, so they are positive and progressively large. On the other hand, the non-diagonal elements are the sums of the products, they

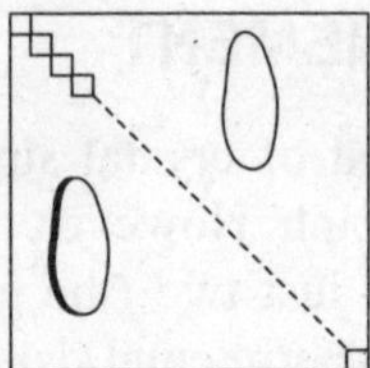

Fig. 14.4 Diagonal matrix

may either be positive or negative and expected to be smaller. In extreme cases, the non-diagonal elements are considered negligible as compared to the diagonal elements (Fig. 14.4).

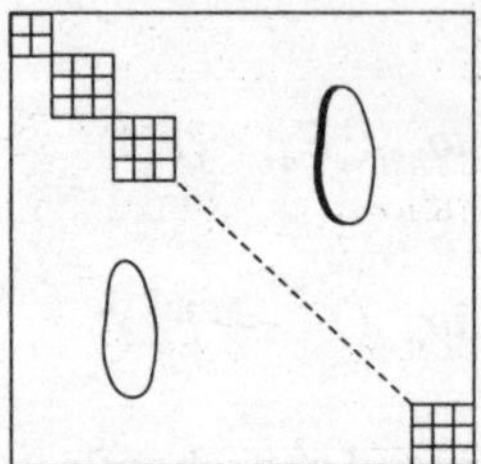

Fig. 14.5 Block-diagonal matrix using a 2 × 2 scale and thermal parameters, and 3 × 3 block for parameters of each atom

The block-diagonal least-squares approximation is based on the fact that this retains the non-diagonal elements involving correlation among the positional and among the thermal parameters of the same atom but neglects those involving correlation between positional and thermal parameters (Fig.14.5).

These approximations help reduce the computer time and strorage need. A very larger saving of computer time has been achieved by the application of fast Fourier transform (*FFT*) algorithm. Actually, the *FFT* based computer programs allow full matrix refinement of protein structures at reasonable cost.

In practice, the crystallographers need not get deeply involved in the mathematical complexities of refinement by the Least-squares method because there are some excellent standard computer programs available, viz. SHELX, XTAL, NRCVAX, CRYSTALS and many other commercial programs supplied with the most single crystal diffractometers. They make the refinement process quite automatic once the data are provided. A fairly complete account of the least-squares refinement methods can be found in Computing Methods in Crystallography, edited by Rollet (1961).

The crystallographers, however, need to keep a close check on the progress of the refinement, i. e. the temperature factor must be closely watched and bond length calculated from time to time to ensure that the structure continues to make crystal chemical sense. It is always better to check the final results of least-squares refinements with a difference Fourier map. The refined structure should therefore satisfy the following criteria:

   (i) It should be chemically sensible.

  (ii) It should give reasonable agreement between the observed and calculated structure factors. The value of R should be fairly low.

(iii) The difference Fourier map should be essentially featureless.

14.5 CONSTRAINED LEAST-SQUARES REFINEMENT

The refinement of protein structures (with very few exceptions) cannot be carried out accurately using the classical least-squares method, not due to computational problem but due to limited number of available X-ray data. The accuracy of the derived parameters is found to affect under any of the following conditions.

1. When the ratio between the number of observations and the number of parameters to be refined is low. The ratio is particularly unfavourable for biological macromolecules.

2. When heavy and light atoms coexist in the unit cell, even a modest error on the heavy atom parameters will cause large error on the light atom parameters.

3. Presence of high thermal motion, structural disorder, etc.

However, the presence of some rigid molecular groups like tetrahedral SiO_4 group, a planar five or six (e.g. benzene) membered ring, amino acid subunits in polypeptides etc. the number of parameters to be refined is considerably reduced. This helps increase the observations to parameters ratio. Thus, the process in which the observation to parameters ratio is enhanced by decreasing the number of variables is termed as the constrained least-squares. On the other hand, if the enhancement of the ratio is achieved by artificially increasing the number of observations, the process is called the restrained least-squares.

For each above mentioned rigid group of say n atoms, the number of independent positional parameters reduces from $3n$ to 6, three needed to locate the origin (x_0, y_0, z_0) of the coordinate system and three more (ω, ψ, ϕ called torsion angle) to specify its orientation with complete specification of a phenyls ring needs 18 positional variables in classical sense actually reduces to six (three transnational and three rotational) if the rigid group is considered. Thermal parameters are usually considered to be isotropic, and to keep the number of parameters small, a single overall thermal parameter is assumed for the whole group.

The constrained (or rigid body) refinement is a well-known and widely used method when the geometry of a group of atoms is accurately known, and the entire group is treated as a rigid entity. The success of this method depends on the accuracy of the assumed geometry. The constrained/restrained least-squares method has been discussed in detail in Fundamentals of Crystallography (1992).

14.6 AUTOMATION OF STRUCTURE ANALYSIS

The bulk of experimental and theoretical methods available today permits the solution of practically any crystal structure. Therefore, one can speak of an automation of the crystal structure analysis (with the exception of some complex molecular or biological structures) both from experimental and computational points of view.

With the advent of fast computers it has become possible to compute the structure factors, phases, electron density functions and perform structural refinements using different form of Fourier transform /syntheses. The availability of a number of standard programs are the added advantage towards automation of structural analysis.

Now, it is possible to summarize the discussions in the last four chapters related to crystal structure analysis in the form of a flow chart as shown in Fig. 14.6. The main operations performed are placed in boxes, results obtained are shown on the right and the approximate time required to perform an operation is indicated on left.

Time taken	Operation	Information obtained
Minutes	1. Select a suitable crystal and mount it for X-ray study	
Minutes/hours	2. Obtain unit cell geometry and preliminary symmetry information	$a\ b\ c\ \alpha\ \beta\ \gamma$ crystal system, space group? information on molecular symmetry?
Hours/days	3. Mesure intensity data	List of: $h\ k\ l\ I\ \sigma(I)$
Minutes	4. Data reduction (various corrections applied to data)	List of: $h\ k\ l\ F\ \sigma(F)$ or $h\ k\ l\ F^2\ \sigma(F^2)$
Minutes/hours	5. Solve the structure (a) Patterson methods (b) Direct methods (c) Other methods	Some or all non-H atom positions
Minutes/hours	6. Complete the structure—find all the atoms: Fourier and difference Fourier syntheses	All atom positions (approximate)
Minutes-days	7. Refine the structure model	Atom positions and displacement parameters
?	8. Interpret the result	Molecular geometry, packing arrangement, etc....

Fig. 14.6   A flowchart for the steps involved in a crystal structure determination

14.7 SUMMARY

1. Like crystal structure determination, there are many methods of crystal structure refinement.

2. Successive Fourier syntheses can be performed by using either forward Fourier transform (FT) or reverse Fourier transform (FT^{-1}).

3. Forward Fourier transform (FT) on the model structure gives rise to amplitudes and phases of F_c. On the other hand, a reverse Fourier transform (FT^{-1}) on the model structure provides the knowledge of electron density function.

4. Difference Fourier synthesis provides better results. A peak in the difference map represents too little electrons and a trough (negative peak) represents too much electrons associated with the atoms of a model structure.

5. In an ideal situation, if the model structure is correct, the difference Fourier map should be featureless.

6. In any crystal structure analysis, there are many more observations than the parameters to be determined. The best parameters corresponding to the model structure are found to by minimizing the sum of the squares of the discrepancies between the observed and calculated values, i.e.

$$D = \sum_{hkl} w_{hkl}(|F_o| - |F_c|)^2$$

or

$$D = \sum_{hkl} w_{hkl}(F_o^2 - F_c^2)^2$$

where w_{hkl} is appropriate weight assigned to each reflection.

7. In large organic molecules, invariably the presence of rigid molecular groups like tetrachedral SiO_4 group, a planar five or six (e.g. benzene) membered ring, amino acid subunits in polypeptides, etc. are observed. Their presence help reduce the number of parameters to be refined considerably and hence increase the observations to parameters ratio. The use of such properties in structural refinements is known as constrained least-squares.

8. With the advent of fast computers and the availability of a number of standard computer programs, the automation of crystal structure analysis in general is possible.

14.8 DEFINITIONS

Least-squares calculation: A statistical method of obtaining the best fit of a large number of observations to a given equation. This is done by minimizing the sum of the squares of discrepancies between the observed and calculated values.

Refinement: A process of improving the parameters of an approximate (or trial) structure until the best fit of calculated structure factor amplitudes to those observed is obtained. The process usually requires many stages.

Weight of a measurement: A number assigned to express the relative precession of each measurement. In least-squares refinement, the weight should be proportional to the reciprocal of the estimated standard deviation of the measurement.

REVIEW QUESTIONS AND PROBLEMS

1. Enumerate various crystal structure refinements. Discuss in detail the successive Fourier syntheses.

2. Explain the importance of structural refinements in crystal structure analysis. Discuss the difference Fourier synthesis in detail.

3. Discuss least-squares refinement in detail. Explain the utility of various weighting schemes that are used.

4. What do you understand by the term "Constrained" least squares refinement? Discuss its need and utility. Elaborate the role of R-factor in crystal structure refinement.

5. Discuss the basis of automation of crystal structure analysis. Prepare a flow chart for complete crystal structure determination.

BIBLIOGRAPHY

A.H. Cotrell, Theory of Dislocations, Gordan and Breach Science Publishers (1964).

A.K. Cheetham, Diffraction Methods (Solid State Chemistry Techniques) ed. by A.K. Cheetham and P. Day, Oxford Scientific Publication, Oxford Univ. Press (1987).

A.O.E.Animalu, Intermediate Quantum Theory of Quantum Solids, Prentice Hall, Inc. (1977).

A.R.Verma and O.N. Srivastava, Crystallography for Solid State Physics, Wiley Eastern Limited (1987).

A.R. Verma and P.Krishana, Polymorphin and Polytypism in crystals, John Wiley and Sons (1966).

B. Henderson, Defects in crystalline Solids, Crane, Russale & Company (1972).

B.E. Warren, X-ray Diffraction, Addison- Wesley, Reading, Mass. (1969).

Charles Kittel, Introduction to Solid State Physics, Wiley Eastern Limited, 5th ed. (1976).

C. Giacovazzo etal., Fundamentals of Crystallography, edited by C. Giacovazzo, IUCr., Oxford University Press (1992).

C.S. Barret and T.B. Massalski, Structure of Metals, Mc Graw-Hill Book Company (1996).

C. Hammond, The Basics of Crystallography and Diffraction, 2nd ed. IUCr.Oxford University Press (2001).

D.E. Sands, Introduction to Crystallography, W.A. Benjamin, Inc. (1969).

D. Harker and J.S Kasper, Acta Cryst., 1 (1948) 70.

D.Hull, Introduction to Dislocations, Pergamon Press (1965).

D.M.Adams, Inorganic Solids: An Introduction to concepts in Solid State Structural Chemistry, John Wiley & Sons Limited (1974).

D. Sayre, Acta Cryst., 5 (1952) 60.

F.C. Philips, An Introduction to Crystallography, John Wiley (1963).

G. Burns and A.M. Glazer, Space Groups for Solid State Scientists, Academic Press, New York (1978).

H.W. Streitwolf, Group Theory in Solid State Physics, Mac-Donald, London (1967).

International Tables for X-ray Crystallography, Vol.I, Published by IUCr., The Keynoch Press, Birmingham (1965).

J.A.K. Tareen and T.R.N. Kutly, A Basic Course in Crystallography, University Press (India). Limited (2001)

J.M. Schultz, Diffraction for Materials Scientists, Prentice-Hall Inc. New Jersey (1982).

J.P. Gulsher, K.N. Trueblood, Crystal Structure Analysis: A Primer, Oxford University Press (1985).

J.P. Herth and J.Lothe, Theory of Dislocations, Mc Graw- Hill Book Company (1968).

L.V. Azaroff, Elements of X-ray Crystallography, Mc Graw – Hill Book Company (1968).

M.J. Buerger, X-ray Crystallography, John Wiley & Sons (1966).

M.J. Buerger, Crystal Structure Analysis, Wiley , New York (1960).

M.J. Buerger, Elementary Crystallography: An introduction to the Fundamental Geometrical Features of crystal, Wiley, New York (1963).

M.M. Julian, Foundation of Crystallography with Computer Applications, CRC Press (2008).

M.M. Woolfson, An Introduction to X-ray Crystallography, Vikas Publishing House Pvt. Ltd. (1978).

M.T. Dove, Structure and Dynamics: An Atomic View of Materials, Oxford University Press (2008).

N.F.M. Henry, H. Lipson and W.A. Wooster, The Interpretation of X-ray Diffraction Photographs, 2nd ed. Macmillan, New York (1961).

S.K. Chatterjee, X-ray Diffraction: Its Theory and Applications, Prentice Hall of India Ltd. (1999).

W.H. Zachariasen, Acta Cryst, 23 (1967) 588.

SUBJECT INDEX